Nanotechnology in Modern Animal Biotechnology

Sanjay Singh • Pawan Kumar Maurya
Editors

Nanotechnology in Modern Animal Biotechnology

Recent Trends and Future Perspectives

Springer

Editors
Sanjay Singh
Division of Biological and Life Sciences
Ahmedabad University
Ahmedabad, Gujarat, India

Pawan Kumar Maurya
Biochemistry
Central University of Haryana
Mahendegarh, Haryana, India

ISBN 978-981-13-6003-9 ISBN 978-981-13-6004-6 (eBook)
https://doi.org/10.1007/978-981-13-6004-6

Library of Congress Control Number: 2019936162

This Springer imprint is published by the registered company Springer Nature Singapore Pte Ltd.
The registered company address is: 152 Beach Road, #21-01/04 Gateway East, Singapore 189721, Singapore

Preface

Nanotechnology is advancing at a fast pace with its ramifications felt in almost every field, ranging from materials science to food, forensic, agriculture, and life sciences, including biotechnology and medicine. Nanotechnology is already being harnessed to address many of the key problems in animal biotechnology. It is expected that next decade nanotechnology would have tremendous growth in the applications and product developments in the area of animal biotechnology (e.g. animal nutrition, health, disease diagnosis, and drug delivery). Nanotechnology provides the ability to manipulate the materials at atomic and molecular levels and also arrange atom-by-atom on a scale of ~1–100 nm to create, new materials and devices with fundamentally new functions and properties arising due to their small scale. These properties are significantly different than their bulk counterparts. Nanobiotechnology, on the other hand, is a recent field that arose due to the concourse of nanotechnology and biology where nanotechnology provides the tools and techniques to work from nanoscale principles to investigate, understand, and transform the biological systems.

This book, *Nanotechnology in Modern Animal Biotechnology: Recent Trends and Future Perspectives*, has been compiled to comprehensively present an overview of the recent progress at the interface of nanotechnology and animal biotechnology. This book contains seven chapters, which comprehensively cover various aspects of applications of nanotechnology closely related to human health and animal biotechnology. These chapters are primarily focused on the use of nanoparticles for development of biosensors, drug/gene delivery vehicles, antimicrobials, anticancers, and anti-leishmaniasis.

Chapter 1 introduces the readers about the basics of engineered nanomaterials and outlines the merits of applying nanotechnology in the development of nanocarriers for drug delivery, nanofilms for wound healing, and nanocomposite systems for synergistic therapeutics and diagnostics. Exploiting metallic anisotropic nanomaterials, exhibiting near-infrared (NIR) absorbance pattern, is one of the well-explored areas; therefore, we have included a chapter on the design and applicability of gold-based nanostructures of different morphologies for efficient photothermal therapy. The NIR light-stimulated heat can be generated from the targeted gold nanostructures, which can potentially and selectively kill the cancer cells. Considering the significance of applications of nanotechnology in animal biotechnology, few chapters are included on the topics covering the recent perspectives and challenges of nanomedicines. In view of pathogenic microbes developing resistance towards the antibiotics, novel antimicrobial agents are required; therefore, we have included two chapters on nanomaterials-based broad-spectrum antibiotics. The safety concerns of nanomaterials are one of the biggest hurdles that need to be surmounted to realize the full clinical potential of nanotechnology. These issues are also covered in a few of the chapters with special emphasis given to the strategies of synthesis of nanoparticles using "safe-by-design" approach. The emergence of neglected tropical disease caused by the protozoan parasite has been prevailing in the developing countries for several decades, about over a billion people (one-sixth of the world's population) are infected with one or more such diseases. In this context, we have included a separate chapter providing comprehensive coverage on the recent developments in the therapy of leishmaniasis. Additionally, Chap. 7 is intended to provide meticulous information to readers about the incorporation of nanoparticles in consumer goods of various sectors like medical industry, 3D printing, water purification, electronics, coating, and surface treatment processes. Further, the effect of the release of nanoparticles towards the flora and fauna of the environment has also been discussed.

This book aims to provide a basic understanding of nanoscience and nanotechnology to the reader as well as deliver the information about the current developments and future prospects for the improvements of animal biotechnology.

Ahmedabad, Gujarat, India Sanjay Singh
Mahendegarh, Haryana, India Pawan Kumar Maurya

Acknowledgements

The editors wish to acknowledge with thanks the generous support and contributions of all the authors of the chapters included in this book. S Singh acknowledges the generous funding from Gujarat Institute of Chemical Technology, India, for CENTRA (Centre of Nanotechnology and Research Application) and Department of Science and Technology, Science and Engineering Research Board (SERB), India, for the project "Multi target antioxidant nanoconstruct for prostate cancer treatment" (Grant number – SB/YS/LS-159/2014). The research fellowships provided to S Singh under Endeavour Research Fellowship by the Government of Australia and YY Memorial Research Fellowship by the Union of International Cancer Control, Japan, are also acknowledged.

Contents

About the Editors

Dr. Sanjay Singh is currently an Associate Professor at the Division of Biological and Life Sciences, Ahmedabad University, Ahmedabad, Gujarat, India. Dr. Singh obtained his Ph.D. from CSIR-National Chemical Laboratory, Pune, India, and continued his research work as postdoctoral fellow at the University of Central Florida and Pennsylvania State University, USA. His research work includes the synthesis of novel anti-cancerous drug formulation and nanomaterials with biological enzyme-like characteristics. Utilizing these materials, he has established several strategies for cancer treatment and detection of biomolecules to protect mammalian cells and developed non-antibiotic antibacterial agents. He has been bestowed with several national and international awards such as Endeavour Research Fellowship (2015), YY Memorial Award (2016), International Association of Advanced Materials Scientist Medal (2017), and EMBO Fellowship (2018). He has published 1 edited book, 5 book chapters, and over 70 research articles in various reputed international journals. He has also worked as guest editor of few annual meeting of the Society for Redox Biology and Medicine (SfRBM) and currently serving as Associate Editor in Chief for the journals *3Biotech* (Springer Nature) and *OncoTargets and Therapy* (Dove Press).

Dr. Pawan Kumar Maurya is currently employed as Associate Professor and Head at the Department of Biochemistry, Central University of Haryana (Gov. of India), India. Dr. Maurya is also a Deputy Director at Deen Dayal Upadhyay Kaushal Kendra (DDUKK), Central University of Haryana, India. He did his B.Sc. in Biology (1998–2001), M.Sc. in Biochemistry (2001–2003) (First Rank), and Ph.D. in Biochemistry from Allahabad University (2003–2008). He did his postdoctoral training from Taipei Medical University (2011–2012), Taiwan, and Federal University of Sao Paulo (2014–2016), Brazil. Before joining Central University of Haryana, Dr. Maurya worked as a Lecturer at SBS University, Dehradun, and as an Assistant Professor at Amity Institute of Biotechnology, Amity University Uttar Pradesh, Noida, India. Dr. Maurya's research contributions are highly interdisciplinary. Dr. Maurya is working on accelerated aging in neuropsychiatric disorders, nanomedicine, and clinical biochemistry. He has published over 54 research articles in reputed journals and one book (Springer, Singapore). Dr. Maurya is a recipient of prestigious fellowship: Science without Borders (Government of Brazil). He is member of various academic bodies of Central University of Haryana.

Novel Therapeutics and Diagnostics Strategies Based on Engineered Nanobiomaterials

Srijeeb Karmakar, Varun Saxena, Pranjal Chandra, and Lalit M. Pandey

Abstract The emergence of nanotechnology has opened up new avenues of research focusing on diagnostic and therapeutic advancement. In light of that, many of the previous problems associated with treatment failure and progress of diseases are being addressed through nanotechnology. For instance, the application of a spectrum of nanomaterials has shown promising possibilities in slow and controlled drug release, targeted delivery, biocompatibility and synergistic delivery of multiple drugs. Engineered nanomaterials in this direction have further attracted researchers to exploit/tune the features required for a given application. This book chapter, therefore, is aimed at outlining the merits of applying nanotechnology in the development of nanocarriers for drug delivery, nanofilms for wound healing, nanocomposite systems for synergistic therapeutics and diagnostics.

Keywords Nanocomposite · Nanoparticles · Nanoemulsion · Drug delivery · Therapeutics · Biocompatibility

1 Introduction

Richard Feynman (Nobel Prize in physics 1965) in his lecture "there's plenty of room at the bottom" endorsed the notion that in future, manipulating individual atoms would appear as a more powerful form of synthetic chemistry. The idea of "swallowing the doctor" which conceptualized that miniature surgical robots could be injected into the body for diagnosis and therapy, was also presented as a great possibility (Feynman 1959). Down the lane, these ideas have laid the foundation of nanotechnology particularly driven towards improving human health conditions. The advent of nanotechnology has overlaid possible solutions to many of the problems related to treatment failure such as the non-specific distribution of therapeutic agents, non-specific targeting, limited therapeutic window,

S. Karmakar · V. Saxena · P. Chandra · L. M. Pandey (✉)
Department of Biosciences and Bioengineering, Indian Institute of Technology Guwahati, Guwahati, India
e-mail: lalitpandey@iitg.ac.in

S. Singh, P. K. Maurya (eds.), *Nanotechnology in Modern Animal Biotechnology*, https://doi.org/10.1007/978-981-13-6004-6_1

etc. (Hu et al. 2018a; Zamani et al. 2018; Zhang et al. 2018b). On the other hand, the essential features provided by nanomaterials that make them lucrative are site-specific drug delivery, easiness in modulating desired properties, co-delivery of synergistic agents, and controlled drug release (Koh et al. 2011; Sahoo et al. 2018). Besides these fundamental features, engineered nanomaterials offer a range of tunable properties desired for a given application (Tiwari et al. 2017).

Nanotechnology has also found a wide array application in the development of diagnostic and prognostic strategies for early detection and monitoring of various diseases through in vivo molecular imaging, optical coherence tomography, quantum dots (QDs) imaging, nanobiosensors, guided- surgeries etc. (Zhu et al. 2012a; Chandra et al. 2013; Chandra 2016; Chen et al. 2018a; Kumar 2018; Nazar 2018). Nanocomposites are being extensively investigated for their ability to display an integrated effect of diagnostic, prognostic and therapeutic function (Koh et al. 2011; Zhu et al. 2012b; Shim 2013; Prasad et al. 2016a; Hasan and Pandey 2017). In light of that, a nanocomposite of Fe_3O_4 and polydopamine was synthesized which displayed high efficiency in quenching fluorescence, near-infrared absorption and presented a dynamic surface for the functionalization of various molecules. The Nanocomposite was found to be an efficient theranostic agent for the detection of intracellular mRNA, and furthering its scope, also used in imaging-guided photothermal therapy (PTT) (Lin et al. 2014). Other nanomaterials under focus are QDs, which are extremely small semiconductors (size 2–10 nm) with tunable, size-dependent properties like photoluminescence, absorbance and other optical properties (Dabbousi et al. 1997; Jasieniak et al. 2009; Moreels et al. 2009). Due to these attractive features, QDs find application in the development of diagnostics. Employment of nanoparticles have also come up as an emerging strategy for the development of diagnostics, therapeutics and various other sensing systems (Chandra et al. 2009, 2010; Mahato et al. 2016; Prasad et al. 2016b; Tiwari et al. 2017). Polymeric nanoparticles are colloids of polymers which are derived from natural and synthetic sources. These nanoparticles have exhibited biocompatibility, controlled release of drug, non-thrombogenicity, blood integrity retention and carrying capacity of a wide range of molecules like peptides, peptidomimetics, proteins, nucleic acid and other organic therapeutic molecules (Neun and Dobrovolskaia 2011; Lim et al. 2018; Potter et al. 2018; Suktham et al. 2018).

This chapter aims at entailing the wide range of applications of engineered nanomaterials in medicine, diagnosis, and prognosis, with mentions about the feasibilities and adversities in the whole scenario. Also, the chapter would highlight the present state-of-art research, the lacuna involved which is limiting further development, and the future prospects of nanomaterials.

2 Salient Features of Nanomaterials

Nanomaterials possess specific surface as well as physical properties than that of their respective bulk materials. Majorly, antibacterial, optical and magnetic properties are the most explored properties for biomedical applications. The tunability of these properties makes them efficient for various biomedical applications. For example, the antibacterial activity of the nanomaterials can be tuned based on its particle size, morphology as well as the dopant concentrations (Saxena et al. 2018a). Plasmonic properties of nanomaterials can also be tuned for various optical applications such as molecular detection, spectroscopy etc. The optical properties of the nanomaterials vary with their sizes (Gomes et al. 2018). In a similar way, most of the NPs possess superparamagnetism. However, their saturation magnetization and coercivity values remained dependant on temperature, size, and shape (Aslani et al. 2018; Deka et al. 2018).

Recent progress in the fabrication of such nanomaterials is based upon the conglomeration of strategies reported earlier. The antimicrobial activity of oxidized form of Ag, i.e., Ag^{2+}, has been reported to permeate bacterial cell membranes inhibiting essential enzymes, blocking DNA replication and causing the death of various clinically relevant strains (Marsich et al. 2013). Nanocrystals of cellulose, hydroxyapatite (HAp) based ceramics and composite scaffolds made of synthetic polymers have been demonstrated to overcome many of the limitations associated with the desired mechanical strength required in bone tissue engineering (Saravanan et al. 2016; Sharma et al. 2016). The potentiality of a frequently used polymer to design nanocomposite matrix, carboxymethyl cellulose (CMC), to be functionalized/engineered with other materials has been accredited to its hydrophilicity and anionic nature. Especially, being structurally similar to Chitosan (CS), the two polymers can form strong ionic crosslinks leading to the formation of polyelectrolyte complexes which further improves hydrophilicity and swelling behavior of CS (Zhao et al. 2009; Baranwal et al. 2018; Bhatnagar et al. 2018). Recently fabrication of such polyelectrolyte complexing hydrogels based on CS/poly(glutamic acid)/alginate has been found to be efficient in colon-specific drug delivery (Chen et al. 2018b). CMC and gelatin (Gel), according to various reports, interact with each other via non-covalent bonds (H-bond) to form a polymeric network, a feature which has been exploited to develop HAp based three-dimensional nanocomposites. The features displayed by such engineered nanomaterials have been attributed to nanoscale surface chemistry which modulates aggregation, adsorption and other propensities of biomacromolecules (Pandey and Pattanayek 2011; Pandey 2012; Pandey et al. 2012, 2013; Pandey and Pattanayek 2013a, b; Hasan and Pandey 2017; Hasan et al. 2018a)

3 Biomedical Applications of Nanomaterials

In this section, various applications of engineered nanomaterials including bone tissue engineering, wound healing, drug delivery, stem cell therapy and cancer therapy are detailed.

3.1 Bone Tissue Engineering

Abshar et al. have presented a recent advancement in addressing the issues relating to bone transplants such as implant failure and microbial contamination leading to osteomyelitis. These issues mentioned above are ongoing bone tissue engineering problems causing severe damage to health and incur huge commercial losses. The aforementioned research group fabricated nanocomposite polymeric scaffolds of chitosan, carboxymethyl cellulose (CMC) and silver nanoparticles (AgNPs) which demonstrated increased mechanical strength equivalent to cancellous bones, tunable pore size (shown in Fig. 1) and a scaffold degradation rate optimum for angiogenesis and vascularization. Intriguingly, the involvement of AgNPs endowed the nanocomposite scaffold with the dual function of being biocompatible and antimicrobial at the same time. Furthermore, other salient features which were exhibited by the nanocomposite scaffold were improved biomineralization essential for bone growth (Hasan et al. 2018b).

CMC-Gelatin(Gel)-HAp nanocomposite system has shown its potential to be used as a regenerative bone graft in the load-bearing regions. Also, the particular hybrid composite promoted osteoblasts to carry out alkaline phosphatase activity alongside extracellular mineralization which are essential markers of cell

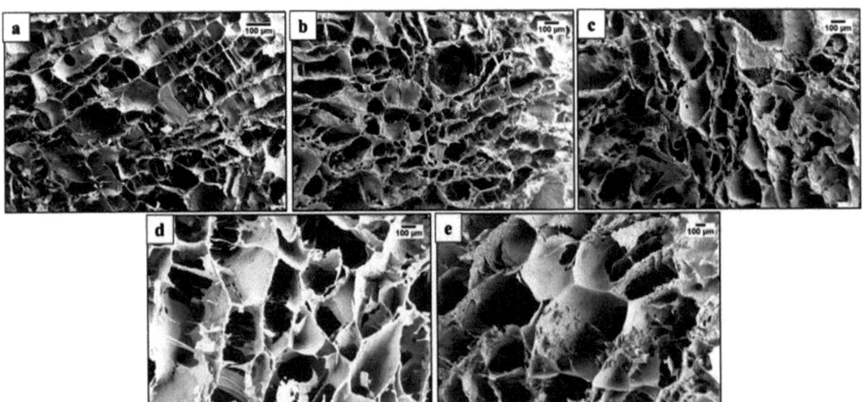

Fig. 1 **(a–e)** SEM images of the nanocomposite scaffolds displaying increasing pore size with increasing concentration of CNWs (Cellulose nanowhiskers). (Reproduced with permission from Hasan et al. 2018b)

proliferation and differentiation (Sarkar et al. 2018). Scaffolds have been synthesized composed of Gel, Hap and mesoporous silica, each known for favorable properties like osteoconductivity, biocompatibility, and drug delivery potential, respectively. The composite coupled with conductive polypyrrole (PPy) loaded with vancomycin as a model antibiotic has shown a promising application in regenerative medicine. In addition, the scaffolds showed enhanced mechanical strength, greater protein adsorption, a higher percentage of antibiotic release over a longer period along with greater osteoblast viability (Ezazi et al. 2018).

3.2 Stem Cell Therapy

Stem cell research is yet another field where nanomaterials have potential applications apart from their crucial possibilities in tissue engineering. For instance, the association of nanomaterials with stem cell therapy in tracking and treating neurological dysfunction is gaining increasing attention (Kim and De Vellis 2009; Dong et al. 2018; Zhang et al. 2018a). A recent study demonstrated higher osteogenesis of mesenchymal stem cells (MSCs) derived from bone marrow seeded on bioactive glass/gelatin scaffold (Kargozar et al. 2018). Another recent work displayed the design and fabrication of dual-functional persistent luminescence nanocomposite (LPLNP/PPT/TRAIL) which was found to induce apoptosis in glioblastoma cells (Wu et al. 2017). At yet another instance, adipose-derived stem cells were seeded on electrospun nanocomposite and were used as a chest wall graft in a murine model (Buschmann et al. 2017). Two-dimensional hexagonal layered double hydroxides (LDHs) nanocomposite was fabricated, and tonsil-derived MSCs adhered to it. Higher chondrogenic differentiation of MSCs was observed with improved cell-material interaction which has been presented as a potential injectable stem cell therapy (Lee et al. 2017b). Thus, the combination of nanocomposite and stem cell has shown improvement in diagnosis, prognosis and therapeutics in multiple central nervous system diseases (Deng et al. 2009; Baranes et al. 2015; Lv et al. 2017b). Such nanocomposites have proven their capacity in modulating stem cell proliferation and differentiation in vitro and in vivo as well (Shin et al. 2004; Liu et al. 2010; Shah et al. 2014). Such potential is exhibited due to various salient properties nanomaterials such as facile synthesis, controllable particle size, large surface-to-volume ratio, biocompatibility and tunable surface chemistry (Wei et al. 2017; Zhang et al. 2017). Furthermore, engineered nanomaterials have been reported to be an efficient gene and drug delivery vehicles (Roy et al. 1999; Panyam and Labhasetwar 2003). Nanomaterials are being used for a vast level of stem cell based therapies for biomedical applications. Recent advances in nanomaterial based stem cell therapy are listed in Table 1.

Table 1 Recent advances in nanomaterial based Stem cell therapy

S. N.	Stem cells	Applications	Materials used	Materials properties	References
1	Human mesenchymal stem cells (hMSC)	Differentiation into bone cells	Upconversion nano-transducer coated with mesoporous silica	Core-shell Diameter- 59 ± 5 nm	Kang et al. (2018)
2	Umbilical cord-derived mesenchymal stem cells (MSCs)	Tumor-targeted therapy	Doxorubicin loaded poly(lactic-co-glycolic acid) (PLGA)	Core-shell Diameter- 104.2 nm	Yang et al. (2018)
3	Adipose-derived mesenchymal stem cells (ADSCs)	Protection of stem cells in infarcted heart tissue	Poly(lactide-co-glycolide)-monomethoxy-poly-(polyethylene glycol) (PLGA-mPEG)	Spherical Hydrodynamic size-129.6 nm Zeta potential 3.83 mV	Ma et al. (2018)
4	hMSC	Gene delivery to hMSC	Iron oxide nanoparticles	Spherical Diameter- 15 nm	Xu et al. (2018)
5	hMSC	Magnetic resonance and photoacoustic imaging of brain tumor mediated by hMSC	Iron oxide coated with gold	Spherical Diameter ~82 nm	Qiao et al. (2018)
6	hMSC	Improvement in area and maturation level of hMSC for acute and functional cardiac integration	Silica nanoparticles	Spherical Diameter- 50 ± 2 nm Zeta potential −25 mV	Popara et al. (2018)
7	Canine bone marrow derived mesenchymal stem cells (cBM-MSCs).	Delivery of nerve growth factor	Chitosan nanoparticles	Spherical Diameter- 80–90 nm Zeta potential 38.7 ± 2.5 mV	Mili et al. (2017)

(continued)

Table 1 (continued)

S. N.	Stem cells	Applications	Materials used	Materials properties	References
8	Human embryonic stem cell-derived cardiomyocytes (hESC-CMs)	Photoacoustic imaging	Semiconducting polymer (PCPDTBT)	Spherical	Qin et al. (2018)
			Poly[2,6-(4,4-bis-(2-ethylhexyl)-4H-cyclopenta [2,1-b;3,4-b′] dithiophene)-alt-4,7 (2,1,3-benzothiadiazole)]	Diameter-48.6 ± 1.2 nm	
9	Human adipose stem cells (hASCs)	Differentiation into bone cells	Osteoinductive drug dex loaded nanodiamond-reinforced gelatin methacrylamide	Irregular shape	Pacelli et al. (2017)
				Diameter-5–10 nm)	
				Zeta potential	
				−9.4 to −12.3 mV	
10	Adult mesenchymal stem cells (MSCs)	Adult mesenchymal stem cells (MSCs)	Low oxygen content graphene nanoparticles	Layered	Elkhenany et al. (2017)
				1.0–1.2 nm thick (3–4 layer)	

3.3 *Potential Therapeutics for Neurodegeneration*

Various inorganic nanoparticles of Au, Ag, Fe, Cu, etc. functionalized with therapeutic molecules have been shown to be inhibitory and disaggregating towards recalcitrant amyloid fibrils found in neurodegenerative disease like Alzheimer's disease, Parkinson's disease, Huntington disease, etc. (Das et al. 2017; Singh et al. 2017, 2018a). The effect of particle size and tunable surface chemistry was demonstrated with glutathione capped gold nanoparticles (AuNPs) which were found to accelerate amyloid beta fibrillation at a larger size distribution and decelerate the process at smaller. Interestingly, gold nanoclusters (AuNCs) further exhibited complete inhibition of fibrillation of amyloid beta (Gao et al. 2017). While developing drugs for neurodegenerative and neurological diseases, one of the most difficult challenges is to develop drug and drug delivery systems which will be able to cross Blood Brain Barrier (BBB) (Abbott 2013). Small peptide inhibitors (for, eg. LPFFD and TGN) targeted towards Aβ and functionalized on Selenium nanoparticles (SeNPs) has shown dual property of inhibiting Aβ aggregation and crossing the BBB at the same time. In addition, the conjugate was also found to suppress the amyloid-beta mediated ROS (reactive oxygen species) generation and PC12 cell neurotoxicity (Yang et al. 2017). All these various features exhibited by nanocomposite-based scaffolds encourage the further development of nanobiomaterials to present solutions to human health problems. Antioxidant activity is one of the major activity being utilized for the theranostics and biological actions of the nanomaterials. Singh et al. explained that cerium oxide (CeO_2) are known to possess the excellent antioxidant activity for various biomedical applications. However, the physicochemical properties of CeO_2 are affected by the biological environment. Hence, their size and morphology should be tuned based on various applications to obtain specific antioxidant activities (Singh 2016).

3.4 Cancer Therapy

Polymeric nanoparticles offer a carrier matrix onto which the drug molecule can be encapsulated, entrapped or adhered with chemical or physical interaction (Reis et al. 2017). For example, PLGA-PEG (poly(lactic-co-glycolic acid)-block-polyethylene glycol) nanoparticles have been reported to encapsulate drug cargos and exhibit longer circulation time at tumor sites. However, they show non-specific off-target accumulation. Specificity and systemic toxicity reduction were achieved by linking a collagen IV derived peptide AXT050 which is both antiangiogenic and anti-tumorigenesis. The PLGA-PEG- AXT050 conjugates were found to enhance affinity towards triple-negative breast cancer cells through strong binding to integrin $\alpha_V\beta_3$ (a tumor-associated integrin). Such peptidomimetics capped on NP surfaces are thought to improve the efficacy of cancer nanomedicines (Arosio and Casagrande 2016; Bressler et al. 2018). PLGA, especially, has been approved by FDA for use in therapeutic drug delivery for protecting drug molecules from rapid elimination or degradation (Niwa et al. 1994; Bobo et al. 2016). PLGA NPs conjugated with docetaxel, an anti-breast-tumor drug, has been reported to enhance pharmacokinetics and pharmacodynamics (Rafiei and Haddadi 2017). PLGA nanospheres endowed with CS have exhibited an absorption-enhancing effect of opening intercellular tight junctions. The PLGA-CS nanospheres loaded with the peptide, elcatonin (a drug derived from calcitonin to remit pain caused by osteoporosis), showed a greater reduction of blood calcium level and enhanced elcatonin's pharmacological action up to 24 h. (Yamamoto et al. 2005). On similar lines, the oral delivery of elcatonin was improved by loading the drug on CS modified PLGA NPs with excellent muco-adhesion (Kawashima et al. 2000). A new method has been explored for the cancer treatment named as Hyperthermia, in which the temperature was hiked from the normal body temperature. Tumours are found to be highly vulnerable to elevated temperatures. Recently, Deka *et al.* designed biphasic α-Fe_2O_3-$GdFeO_3$ for the cancer treatment using its hyperthermic property. Authors reported a temperature generation upto 55.71 °C within 6 min of incubation of the synthesized nano-materials under an external magnetic field. Due to the in situ generation of high temperature cancer cells are found to be vulnerable, however, the minimal effect was observed over the normal osteosarcoma cells (Deka et al. 2018). Except to that of hyperthermia, and drug delivery, various other approaches such as nano-liposome containing PTEN plasmid and CeO_2 nanoparticles (for the restoration of the expression of lost PTEN) (Singh et al. 2018b), and non-viral siRNA delivery using nanoparticles (Singh 2013) have been explored for the cancer treatment. These methodologies work on the concept that the extremely smaller nanomaterials can easily be taken up by the cancer cells, but not by the healthy neighboring cells. Healthy cells carry selective import and export mechanism of the metal ions, whereby, cancerous cells do not possess such mechanisms due to their non-differentiative ability. Hence, smaller nanomaterials are extensively being explored for the delivery of anti-tumour biomolecules.

3.5 *Wound Healing Applications*

A recent work displayed an innovative approach in drug delivery towards wound healing where a nanocomposite film made of CS, polyvinylpyrrolidone (PVP), and cellulose nanowhiskers was fabricated showing sustained curcumin delivery. The sustenance in delivering drug was achieved through the increased swelling behavior of the polymeric matrix. Furthermore, the nanocomposite film was shown to be thermally stable with the excellent mechanical properties (Hasan et al. 2017). The curcumin release data fitted into a two-step release model (Shown in Fig. 2) demonstrated that curcumin desorbs from the outer surface of CS-PVP-CNW film at first, which is followed by its diffusion within the polymeric matrix and then, subsequent release. The efficiency of this innovative approach by combining CS, PVP, and CNWs was based upon the unique properties of each constituent. CS has been mentioned earlier to be biocompatible and antimicrobial whose biodegradability

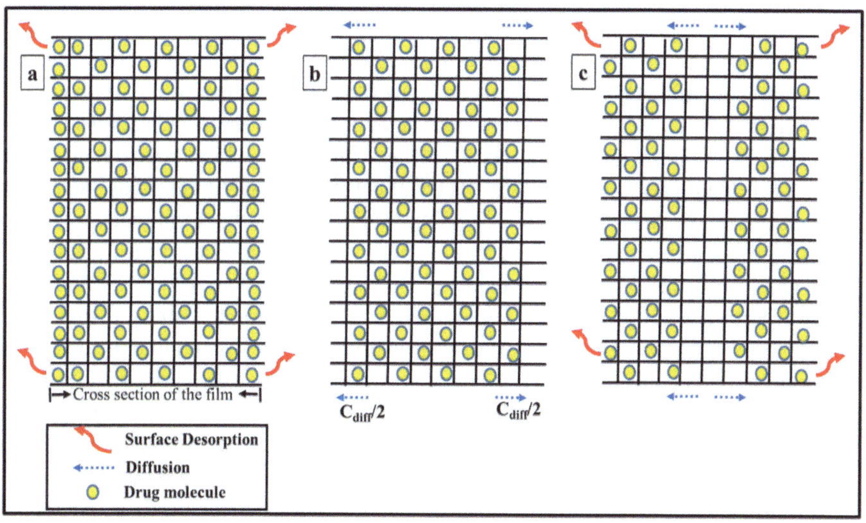

Fig. 2 Release model displaying different phases of drug release through the cross section of the engineered thin film. (**a**) Desorption of drug from the surface during first phase of drug release, (**b**) diffusion of therapeutic molecules from the inner polymeric matrix, and (**c**) simultaneous desorption and diffusion of therapeutic molecules. (Reproduced with permission from Hasan et al. 2017)

prevents posts-surgery implant removal (Bajpai et al. 2015). PVP, on the other hand, is a bio-inert, non-toxic and hydrophilic natural polymer which is capable of blending with CS. The carbonyl group of PVP establishes hydrogen bonds with the amino/hydroxyl groups of CS. The excellent blend of these polymers offers antimicrobial properties which are necessary to prevent treatment failure (Abou-Aiad et al. 2006). Sustained drug delivery was achieved by the use of CNWs which is found to regulate the swelling behavior of hydrophilic polymers (da Silva et al. 2012). The nanocomposite was loaded with curcumin as a model drug which is an established therapeutic compound for wound healing as it is antimicrobial, anti-inflammatory, antioxidant and contains anticancer activity as well (Aggarwal et al. 2003; Dai et al. 2018; Deck et al. 2018; Sanitá et al. 2018; Yen et al. 2018). Thus, nanocomposites offer a possibility of combining various properties of individual constituents and exhibiting their synergistic effect, thereby, opening avenues for more evolved therapies.

The major concern in the wound healing applications is the removal of Reactive Oxygen Species (ROS) at the wound site, which causes severe hindrance in wound healing. In this regard, Rather et al. designed CeO_2 nanoparticles functionalized with polycaprolactone gelatin nanofiber. Authors reported that the designed material maintained the fibrous morphology upto 14 days of incubation, and a sustained release of biological active Ce nanoparticles for 96 h, which helped in the scavenging of ROS for wound skin tissue regeneration (Rather et al. 2018). Nanomaterials also possess inherent self-cleaning properties. Various nanomaterials have been utilized for their broad spectrum of antibacterial activity. For example, Saxena et al. synthesized the aluminium doped ZnO (AZO) nanorods and explored their antibacterial activity. Authors explained that the antimicrobial properties of ZnO were enhanced after doping of Al. 15% doping of Al into the ZnO lattice provided the rod shaped morphology with an average size of 59 ± 16 nm, which consequently enhanced the antibacterial activity of ZnO. Authors also explained the antibacterial mechanism of the AZO nanorods. It was concluded that the AZO nanorods possessed 22.7 ± 0.3 mV zeta potential at 15% doping of Al, which contributed to the electrostatic interaction between the bacterial cell and the AZO nanorods. After this interactions, the released Zn^{2+} ions got internalized to the bacterial cell, which in turn killed bacterial cells (Saxena et al. 2018a). Similarly, Karim et al. designed CuO-nanorod-based NanoZyme catalyst with rod shaped morphology and sub-micron particle size (~10 μm). The synthesized nanoparticles showed enhanced antibacterial activity in the presence of visible light. Authors stated that the designed material had more affinity for H_2O_2, hence produced a large amount of ROS in the presence of visible light for the efficient killing of bacterial cells (Karim et al. 2018b). These studies suggested the particle size dependent antibacterial mechanism of metal ions based nanoantibiotics. Larger sized material showed ROS dependent antibacterial mechanism, whereas nanostructured material showed the antibacterial mechanism by the release of toxic ions, and did not require the presence of light.

4 Drug Delivery

In this section, we have summarized various nano-emulsion and nanoparticles based drug delivery systems.

4.1 Nano-emulsions

Therapeutic bioactive compounds which are insoluble in water require delivery systems for their administration to the biological site (Kwon 2003; Michal et al. 2015). In order to enhance bioaccessibility of the delivery system, several lipophilic and antimicrobial formulations have been developed over the years which display stability over a range of physicochemical parameters such as ionic concentration, pH, rotatory motion, temperature etc. (Kumar et al. 2008; Badoga et al. 2011; Ozturk et al. 2015; Sari et al. 2015). Recently, sesame oil derived natural proteins have been used to design biocompatible nanoemulsions enriched with ω-3 PUFA (polyunsaturated fatty acids) which are particularly focussed on increasing shelf-life of the drug (Shahidi and Ambigaipalan 2018). ω-3 PUFA includes α-linolenic acid, stearidonic acid, docosapentaenoic acid, docosahexaenoic acid (DHA) etc. which have been anticipated to alleviate several medical conditions like cancer, diabetes and Alzheimer's disease etc. (Molfino et al. 2016; Fiala et al. 2017; Yamagishi et al. 2017; Shahidi and Ambigaipalan 2018). The aforementioned study by Shahidi and Ambigaipalan incorporated protein concentrates derived from sesame oil as alternative natural surfactants for the preparation of nanoemulsion delivery systems with minimum cytotoxicity (Shahidi and Ambigaipalan 2018). In yet another recent study, anticancer properties of benzylisothiocynate were augmented by encapsulating the drug within decyl-β-D-glucopyranoside stabilized nanoemulsion with strong antimicrobial activity. Furthermore, the particular nanoemulsion was found to be non-toxic towards MDA-MB 231 breast cancer cell line under the concentration of 4 μg/ml (Kumar et al. 2018). The efficacy of Vitamin D (VD) absorption was reported to be enhanced using nanoscale delivery vehicle mimicking the gastrointestinal tract system. The nanoemulsion-based VD administration in vivo and in vitro resulted in an augmented performance in VD absorption, bioavailability, and also an expression of genes related to VD metabolism (Kadappan et al. 2018).

In light of the development of nanoemulsions for delivering drugs, a recent work requires special mention which presented a simple way of solving the delivery of water-insoluble molecules. Coconut oil, which is fairly available, inexpensive and endowed with antibiotic properties, was used to design nanoemulsion encapsulating the bioactive drug α-tocopherol or Vitamin E (VE) (Saxena et al. 2018b). The abovementioned nanoemulsion was scrutinized extensively with various biophysical and biochemical techniques like Fourier Transform Infrared Spectroscopy (FTIR), Dynamic Light Scattering (DLS), Phase contrast microscopy, etc. Furthermore, like any other nanomaterial which is intended to be administered to

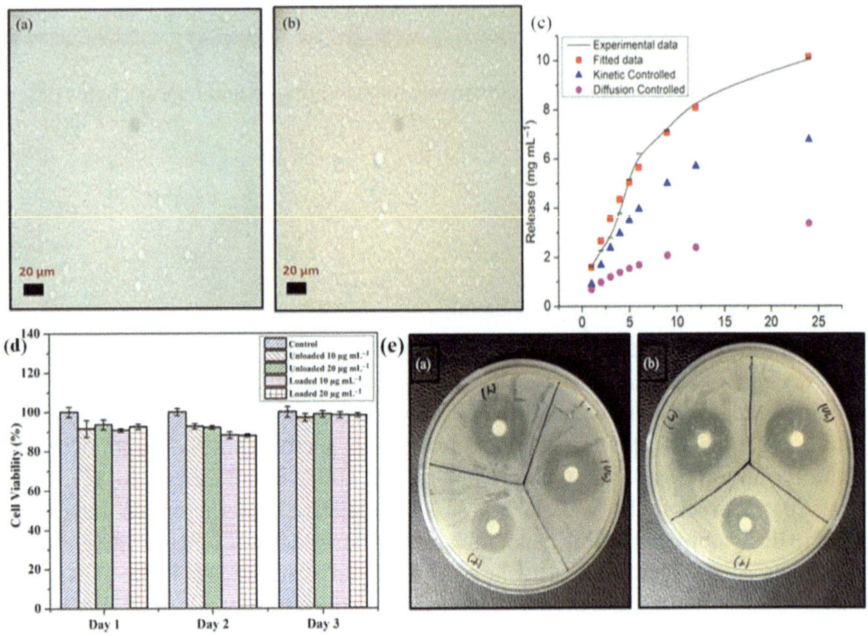

Fig. 3 (**a**) Microscopic image of the unloaded nanoemulsion carrier; and (**b**) the α-tocopherol-loaded nanoemulsion at 20X resolution (scale = 20 μm); (**c**) Rate profile of α-tocopherol release with time. A combination of diffusion controlled Higuchi model and first order kinetic model was used to fit release data; (**d**) Cytotoxicity studies (MTT assay) of both drug-loaded and unloaded nanoemulsions. (**e**) Antimicrobial potential of the engineered nanoemulsions; wherein (+) denotes positive control kanamycin, (L) denotes 9.5 mg mL^{-1} α-tocopherol-loaded nanoemulsion and (UL) denotes unloaded nanoemulsion **e**(a) zone of inhibition against *Escherichia coli*, **e**(b) zone of inhibition against E. hirae. These images are derived in part from an article published in International Journal of Polymeric Materials and Polymeric Biomaterials on 11 Aug 2017. (Saxena et al. 2018b)

the actual physiological system, the edible nanoemulsion was tested for its antimicrobial attributes, non-toxic properties, and bioaccessibility. The nano-antibiotic emulsion was found to be stable up to 230 °C, 8 h in basic and 6 h in acidic pH, 10,000 rpm for 60 min, and an ionic concentration as high as 500 mM. Figure 3a, b displays the microscopic image of the unloaded and α-tocopherol-loaded nanoemulsion. Figure 3c represents the release profile of α-tocopherol with time showing the release data fitted into two-models by combining a diffusion controlled Higuchi model and first order kinetic model. Figure 3d shows the cytotoxicity studies of unloaded and α-tocopherol-loaded nanoemulsion and the antimicrobial activity has displayed in 3e.

4.2 Inorganic Nanoparticles

Inorganic nanoparticles are being extensively researched on their therapeutic potentiality owing to their dynamic surface chemistry, permeability, site-specific targeting and flexibility in monitoring. Presently, a wide range of inorganic materials are being explored which includes iron oxide/magnetic nanoparticles, carbon nanotubes (CNTs), selenium/Gold/Silver/aluminum nanoparticles, quantum dots, etc. (Kango et al. 2013; Beach et al. 2017; Dimitriou et al. 2017; Nazıroğlu et al. 2017; Virani et al. 2017).

4.2.1 Gold Nanoparticles (AuNPs)

Gold nanoparticles (AuNPs) display dynamic absorption, optical scattering properties, and easiness in modulating shape, size, and capping composition. The salient properties of AuNPs make them promising candidates for developing nanomedicines aimed towards various treatment-resistant diseases like cancer and neurodegenerative diseases (Jain et al. 2006; Shim 2013; Ali et al. 2017a; Karmakar 2017). AuNP research has come a long way with recent progress presenting excellent promises to improve human health condition. For example, in a recent study, dopaminergic neurons which are severely dysregulated in Parkinson's disease were shown to be reprogrammed by AuNPs stimulated with electromagnetic field (EMF). Electromagnetized AuNPs mediated activation of histone acetyltransferase Brd2 resulting in histone H3K27 acetylation and robust activation of neuronal genes (Yoo et al. 2017). A recent study demonstrated that gold nanorods could be used in photothermal therapy to target cancer cell integrins to inhibit their migration by affecting cytoskeletal proteins (Ali et al. 2017b). On similar lines, synthesized AuNP composites have been recognized for their therapeutic properties towards Parkinson Disease (PD) models both in vitro and in vivo. AuNP was combined with chitosan (CTS), plasmid DNA (pDNA) and nerve growth factor (NGF) to form a target-specific multifaceted composite (CTS@GNP-pDNA-NGF). The AUNP composite was found to transfect into cells through endocytosis and inhibit the apoptosis of PC12 cells and dopaminergic neurons (Hu et al. 2018b).

4.2.2 Iron Oxide Nanoparticles (IONPs)

Iron oxide nanoparticles (IONPs), on the other hand, are gaining remarkable attention for their biocompatibility, magnetic/superparamagnetic properties, lesser cost of synthesis and antifungal activities (Cotin et al. 2018; Parveen et al. 2018). IONPs are efficient carriers of drugs owing to their dynamic surface chemistry and magnetic resonance. Their magnetic/supermagnetic properties can be exploited to achieve site-specific drug delivery in a number of medical conditions like cancer, arterial blood clots, hyperthermia, etc. to name a few (Peng et al. 2008; Yu et al.

2008; Laurent et al. 2011). Recent progress in investigating the use of IONPs in therapeutics has incited hope in the development of future medicines. For example, a recent study reported that IONPs loaded with lactonic sophorolipids (LSLs) and soluble TNF-a might open new avenues in the treatment of prostate cancer. The study focused on the synergistic effect of apoptosis initiation by TNF-a and the boasting capacity of LSLs towards immune response (Beach et al. 2017). In another study, self-assembled nanoparticles were synthesized for cancer theranostics. An et al. designed multifunctional Gallic acid, Fe, Bovine serum albumin and Paclitaxel based nanocomposite (GA-Fe@BSAPTX) to be utilized for Magnetic Resonance Imaging (MRI), PTT and tumor accumulation. It was reported that monodispersed nanomaterials with spherical morphology and ~115 nm size showed enhanced MRI applications, chemo-photothermal combined therapy and high biosafety. It was also reported that around 80% cells were viable at 200 μM concentration of the designed material with less haemotoxicity upto 24 h (An et al. 2018).

4.2.3 Silver Nanoparticles (AgNPs)

Silver nanoparticles (AgNPs) are endowed with antimicrobial and anticancer activity owing to their distinct crystallographic surface and large surface-to-volume ratios. Extensive research is being conducted to explore AgNP interaction with biological systems, potential toxicity, mechanism of action, and potential therapeutic effects (Franci et al. 2015; Wei et al. 2015; Zhang et al. 2016). A recent study conducted by Han et al. revealed that AgNPs are able to induce differentiation of teratocarcinoma stem cells pointing to the possibility of using AgNPs in differentiation therapy (Han et al. 2017). Another study evaluated the molecular and cellular mechanisms of spherical and stable AgNPs against *Candida albicans.* The study revealed that AgNPs not only can induce ROS generation in the pathogen but also can target which caused alteration of membrane fluidity and microenvironment, cellular morphology, ergosterol content, and altered ultrastructure (Radhakrishnan et al. 2018). Several strategies to fabricate AgNPs from several available plant sources are being developed lately (Kumar et al. 2017; Lekshmi et al. 2017; Yasir et al. 2018). A recent study demonstrated the bio-fabrication of AgNPs using the leaf extracts of *Taraxacum officinale* which mediated high cytotoxicity effect on the HepG2 cells (liver cancer cells). The AgNPs also showed strong antimicrobial activity against Pseudomonas syringae and Xanthomonas axonopodis (Saratale et al. 2018). Thus, AgNPs are gaining considerable impetus for further studies on their therapeutic effects and medical adoption.

4.2.4 Quantum Dots (QDs)

The optical and electronic properties of NPs are dependent on their size, dispersion, and morphology (Gan et al. 2005). QDs are extremely small semiconductor particles which differ in properties from their larger counterparts. QDs are extensively investigated to exploit their salient properties and to use in medical imaging and therapeutics (Zhao and Zeng 2015; Pohanka 2017). Graphene QDs functionalized with folic acid has been found to possess high loading capacity for the theranostic agent IR780 iodide which is otherwise insoluble in most pharmaceutically accepted solvents. The resultant conjugated species displayed a high amount of photothermia to kill cancer cells, showing great possibilities to improve photothermal therapy (PTT) (Li et al. 2017). In vitro studies on the therapeutic potential of WS_2 (Tungsten Sulfide) QDs have been studies lately which showed improved cell viability and removal of ROS against radiation-induced damages. In vivo studies on the same revealed effective protection of DNA and hematopoietic system from high energy radiation by removal of ROS (Bai et al. 2017). WS_2 QDs has been featured in another recent study as a multifunctional theranostic agent for dual-modal image-guided synergistic PTT (Yong et al. 2015). Magnetic mesoporous silica capped with graphene QDs and loaded with doxorubicin (DOX) (a chemotherapeutic drug) has been investigated on breast cancer 4T1 cells for magnetic hyperthermia capacity, drug release behavior, synergistic therapeutic and photothermal effect. The combined effect surpassed each therapy modes (chemotherapy, Radiation therapy, magnetic hyperthermia therapy) and established a strong anticancer activity (Yao et al. 2017).

5 Nanomaterials in Diagnostics

Nanotechnology has found a wide range of applications in the development of diagnostic and prognostic strategies for early detection and monitoring of various diseases through in-vivo molecular imaging, optical coherence topography, QD imaging, nanobiosensors, guided- surgeries etc. (Zhu et al. 2012a; Chandra et al. 2013; Chandra 2016; Chen et al. 2018a; Kumar 2018; Nazar 2018). Various nanomaterials such as nanocomposites, nanoparticles and QDs investigated in diagnostics are discussed in the following sections.

5.1 Nanocomposites

Nanocomposites have the ability to display the integrated effect of diagnostic, prognostic and therapeutic functions (Koh et al. 2011; Zhu et al. 2012b; Shim 2013; Prasad et al. 2016a; Hasan and Pandey 2017). In light of that, a nanocomposite of Fe_3O_4 and polydopamine was synthesized which displayed high efficiency in quenching fluorescence, near-infrared absorption and presented a dynamic surface for the functionalization of various molecules. The nanocomposite was used as an efficient theranostic agent for the detection of intracellular mRNA, and furthering its scope, it was also used in imaging-guided PTT (Lin et al. 2014). Recent progress with such nanocomposites demonstrated the synthesis of a Fe_3O_4 and polydopamine nanocomposite which was EGFR (epidermal growth factor receptor) antibody labeled and doxorubicin (DOX) loaded. The EGFR-specific DOX-loaded nanocomposite has shown multifunctionality in MRI and chemo-PTT (Mu et al. 2017). Optical imaging-guided PTT through mesoporous silica-based core shell nanostructure has shown efficient magnetic imaging and X-ray computed tomography. Furthermore, the synthesized nanocomposite displayed anticancer activity, biocompatibility and low toxicity (Lv et al. 2017a). Further progress focusing on eliminating the damage caused to normal cells by PTT has been achieved through the fabrication of carbon-coated core-shell upconversion nanocomposite. The particular nanocomposite adopted a strategy to kill the cancerous cells using microscopic temperature while keeping the temperature of the lesion low enough to protect normal cells (Zhu et al. 2016).

5.2 Nanoparticles

Employment of nanoparticles have come up as an emerging strategy for the development of diagnostics, therapeutics and various other sensing systems (Chandra et al. 2009, 2010; Mahato et al. 2016; Prasad et al. 2016b; Tiwari et al. 2017). Various biomolecules can be detected based on the red-shift in the absorbance, alteration in their conductance, and/or deflection in the cantilever (Savaliya et al. 2015). MRI presents a new frontier in the biomedical application of nanotechnology. MRI of the overexpression of tyrosine kinase Her-2/neu receptor, which plays an important role in staging breast cancer cells, has been achieved using IONPs. Commercially available streptavidin-functionalized IONP was used for this study, which yielded images displaying the expression of Her-2/neu receptor as a function of the contrast of the images (Artemov et al. 2003). Superparamagnetic IONPs coated with the bioactive molecule relaxin has shown the inhibition of hepatic stellate cell (HSC) contraction and activation and thereby, ameliorate liver fibrosis (Bansal et al. 2017). Vallabani et al. demonstrated the application of ATP-mediated peroxidase-like activity of Fe_3O_4 for the one step detection of glucose. Authors utilized the peroxidase-like activity of Fe_3O_4 for single step detection of glucose. The colorimetric detection limit was found to be 50 μM, and time span of <5 min at pH 7.4, removing the lag of required acidic pH of fundamental detection methods (Vallabani et al. 2017). Another study reported Ag nanoparticles (100 nm) embedded in a cotton fabric for detection of glucose in urine with a colorimetric detection limit of 0.08 mM (Karim et al. 2018a). Similarly, Au-CeO_2 (core-shell) nanoparticles showed the detection limit of 100 μM to 1 mM for glucose at a size of ~75 nm (Bhagat et al. 2018).

The detection of the unmetabolized anticancer drug, dounomycin, in urine samples was achieved through the development of a highly selective sensor using phosphatidylserine (PS) and aptamer. Aptamer and PS were immobilized on AuNPs which were functionalized with a conducting polymer, 2,2′:5′,2″-terthiophene-3′-(*p*-benzoic acid) (polyTTBA). The resultant sensor showed high selectivity towards dounomycin and exhibited increased stability and sensitivity (Chandra et al. 2011). A recent work demonstrated the development of an immunosensor to selectively detect wild type p53 and mutant $p53_{R175H}$ based on differential plasmonic resonance effect of AuNPs in association with various molecules (Bizzarri et al. 2018). On a similar line, immunosensing of Carcinoma Antigen 15-3 (CA 15-3), a breast cancer prognostic marker, has been developed using Au nanostructures coasted onto the organic substrate. It involved developing a multilayered film of Au nanostructure and thiolated graphene QDs to prepare a fully electrochemical transducer on Au surface, which provided a large surface to immobilize the antigen (Hasanzadeh et al. 2018).

5.3 Quantum Dots

Quantum dots are semiconductors which are extremely small nanocrystals (size 2–10 nm) with size dependent tunable properties like photoluminescence, absorbance and optical properties (Dabbousi et al. 1997; Jasieniak et al. 2009; Moreels et al. 2009). Due to these attractive properties, QDs find applications in the development of diagnostics. A recent study demonstrated the development of a pH responsive fluorescent graphene QDs, which underwent a sharp fluorescence shift at pH 6.8, the acidic microenvironment in solid tumors. This unique fluorescence switch was exploited to differentiate between cancer and normal cells allowing detection of solid tumors at an early developmental stage (Fan et al. 2017). Another recent work involved oligonucleotide-functionalized multicolor QDs for the detection of multiple fusion genes (TMPRSS2–ERG) expressed in prostate cancer (Lee et al. 2017a). DNA was functionalized on nitrogen doped Carbon QD to construct an ultrasensitive electrochemiluminescence biosensor for the detection of microRNAs by nicking enzyme-triggered cycling reaction (Liu et al. 2017).

6 Conclusions and Future Prospective

This chapter was an attempt to review recent advances in the field of nanomaterial-based therapeutic and drug delivery systems. The incentive behind employing nanomaterials in the field of medicine is the continuous growing demand for clinical solutions that would minimize the discomforting side effects of present therapeutics. Nevertheless, the research conducted is extensive, but there remains a lacuna of developing therapeutic and drug delivery systems that would be cost-effective, biocompatible and feasible for large-scale production at the same time. This requires ease of synthesis, low input of energy in synthesis procedure and least toxic effects to the environment. Various properties of nanomaterials have been focused based on specific applications. However, multifunctional nanomaterials, which can serve for both diagnostics and theranostics are yet to be explored in detail. Nanomaterials can be tuned for their specific biological properties for a vast degree of biomedical applications. These properties include bioactivity, biocompatibility and biodegradability. More importantly, a comprehensive framework about the biocompatibility and toxicity of the engineered nanomaterials and nanoparticles need to be constructed. Studies on the material structure relevant to its biological responses are to be studied for an insight into the nano-biotechnology. At the same time, the development of strategies for the removal of nanomaterials from the body needs to be considered. Regardless of the nascent form and limitations of the engineered therapeutic and drug delivery systems, engineered nanomaterials indeed have a great scope for future medicine.

References

Abbott, N. J. (2013). Blood–brain barrier structure and function and the challenges for CNS drug delivery. *Journal of Inherited Metabolic Disease, 36*(3), 437–449.

Abou-Aiad, T., Abd-El-Nour, K., Hakim, I., & Elsabee, M. (2006). Dielectric and interaction behavior of chitosan/polyvinyl alcohol and chitosan/polyvinyl pyrrolidone blends with some antimicrobial activities. *Polymer, 47*(1), 379–389.

Aggarwal, B. B., Kumar, A., & Bharti, A. C. (2003). Anticancer potential of curcumin: Preclinical and clinical studies. *Anticancer Research, 23*(1/A), 363–398.

Ali, M. R., Wu, Y., Ghosh, D., Do, B. H., Chen, K., Dawson, M. R., Fang, N., Sulchek, T. A., & El-Sayed, M. A. (2017a). Nuclear membrane-targeted gold nanoparticles inhibit cancer cell migration and invasion. *ACS Nano, 11*(4), 3716–3726.

Ali, M. R., Wu, Y., Tang, Y., Xiao, H., Chen, K., Han, T., Fang, N., Wu, R., & El-Sayed, M. A. (2017b). Targeting cancer cell integrins using gold nanorods in photothermal therapy inhibits migration through affecting cytoskeletal proteins. *Proceedings of the National Academy of Sciences, 114*(28), E5655–E5663.

An, L., Yan, C., Mu, X., Tao, C., Tian, Q., Lin, J., & Yang, S. (2018). Paclitaxel-induced ultrasmall Gallic acid-Fe@BSA self-assembly with enhanced MRI performance and tumor accumulation for cancer theranostics. *ACS Applied Materials & Interfaces, 10*(34), 28483–28493.

Arosio, D., & Casagrande, C. (2016). Advancement in integrin facilitated drug delivery. *Advanced Drug Delivery Reviews, 97*, 111–143.

Artemov, D., Mori, N., Okollie, B., & Bhujwalla, Z. M. (2003). MR molecular imaging of the Her-2/neu receptor in breast cancer cells using targeted iron oxide nanoparticles. *Magnetic Resonance in Medicine: An Official Journal of the International Society for Magnetic Resonance in Medicine, 49*(3), 403–408.

Aslani, A., Ghahremani, M., Zhang, M., Bennett, L. H., & Torre, E. D. (2018). Customizing magnetic and structural properties of nanomaterials. *IEEE Transactions on Magnetics*, 1–5.

Badoga, S., Pattanayek, S. K., Kumar, A., & Pandey, L. M. (2011). Effect of polymer–surfactant structure on its solution viscosity. *Asia-Pacific Journal of Chemical Engineering, 6*(1), 78–84.

Bai, X., Wang, J., Mu, X., Yang, J., Liu, H., Xu, F., Jing, Y., Liu, L., Xue, X., & Dai, H. (2017). Ultrasmall WS2 quantum dots with visible fluorescence for protection of cells and animal models from radiation-induced damages. *ACS Biomaterials Science & Engineering, 3*(3), 460–470.

Bajpai, S., Chand, N., Ahuja, S., & Roy, M. (2015). Curcumin/cellulose micro crystals/chitosan films: Water absorption behavior and in vitro cytotoxicity. *International Journal of Biological Macromolecules, 75*, 239–247.

Bansal, R., Nagorniewicz, B., Storm, G., & Prakash, J. (2017). Relaxin-coated superparamagnetic iron-oxide nanoparticles as a novel theranostic approach for the diagnosis and treatment of liver fibrosis. *Journal of Hepatology, 66*(1), S43.

Baranes, K., Shevach, M., Shefi, O., & Dvir, T. (2015). Gold nanoparticle-decorated scaffolds promote neuronal differentiation and maturation. *Nano Letters, 16*(5), 2916–2920.

Baranwal, A., Kumar, A., Priyadharshini, A., Oggu, G. S., Bhatnagar, I., Srivastava, A., & Chandra, P. (2018). Chitosan: An undisputed bio-fabrication material for tissue engineering and bio-sensing applications. *International Journal of Biological Macromolecules, 110*, 110–123.

Beach, J., Banerjee, T., Kallu, J., Higginbotham, R., & Gross, R. (2017). *Combination therapy of prostate cancer utilizing functionalized iron oxide nanoparticles carrying TNF-a and lactonic sophorolipids*.

Bhagat, S., Srikanth Vallabani, N. V., Shutthanandan, V., Bowden, M., Karakoti, A. S., & Singh, S. (2018). Gold core/ceria shell-based redox active nanozyme mimicking the biological multienzyme complex phenomenon. *Journal of Colloid and Interface Science, 513*, 831–842.

Bhatnagar, I., Mahato, K., Ealla, K. K. R., Asthana, A., & Chandra, P. (2018). Chitosan stabilized gold nanoparticle mediated self-assembled glip nanobiosensor for diagnosis of invasive aspergillosis. *International Journal of Biological Macromolecules, 110*, 449–456.

Bizzarri, A. R., Moscetti, I., & Cannistraro, S. (2018). Surface enhanced Raman spectroscopy based immunosensor for ultrasensitive and selective detection of wild type p53 and mutant p53R175H. *Analytica Chimica Acta, 1029*, 86–96.

Bobo, D., Robinson, K. J., Islam, J., Thurecht, K. J., & Corrie, S. R. (2016). Nanoparticle-based medicines: A review of FDA-approved materials and clinical trials to date. *Pharmaceutical Research, 33*(10), 2373–2387.

Bressler, E. M., Kim, J., Shmueli, R. B., Mirando, A. C., Bazzazi, H., Lee, E., Popel, A. S., Pandey, N. B., & Green, J. J. (2018). Biomimetic peptide display from a polymeric nanoparticle surface for targeting and antitumor activity to human triple-negative breast cancer cells. *Journal of Biomedical Materials Research Part A, 106*, 1753–1764.

Buschmann, J., Balli, E., Hess, S. C., Stark, W. J., Cinelli, P., Märsmann, S., Welti, M., Weder, W., & Jungraithmayr, W. (2017). Effects of seeding adipose-derived stem cells on electrospun nanocomposite used as chest wall graft in a murine model. *Injury, 48*(10), 2080–2088.

Chandra, P. (2016). *Nanobiosensors for personalized and onsite biomedical diagnosis*. The Institution of Engineering and Technology.

Chandra, P., Maurya, P. K., Kumar, P., Tripathi, P., & Srivastava, A. (2009). Diagnosis of rheumatic infections caused by group A Streptococcus pyogenes: future investigation by nanotechnology. *Digest Journal of Nanomaterials & Biostructures (DJNB), 4*(4), 645–650.

Chandra, P., Das, D., & Abdelwahab, A. A. (2010). Gold nanoparticles in molecular diagnostics and therapeutics. *Digest Journal of Nanomaterials & Biostructures (DJNB), 5*(2), 363–367.

Chandra, P., Noh, H.-B., Won, M.-S., & Shim, Y.-B. (2011). Detection of daunomycin using phosphatidylserine and aptamer co-immobilized on Au nanoparticles deposited conducting polymer. *Biosensors and Bioelectronics, 26*(11), 4442–4449.

Chandra, P., Son, N. X., Noh, H.-B., Goyal, R. N., & Shim, Y.-B. (2013). Investigation on the downregulation of dopamine by acetaminophen administration based on their simultaneous determination in urine. *Biosensors and Bioelectronics, 39*(1), 139–144.

Chen, X.-J., Zhang, X.-Q., Liu, Q., Zhang, J., & Zhou, G. (2018a). Nanotechnology: A promising method for oral cancer detection and diagnosis. *Journal of Nanobiotechnology, 16*(1), 52.

Chen, Y., Yan, X., Zhao, J., Feng, H., Li, P., Tong, Z., Yang, Z., Li, S., Yang, J., & Jin, S. (2018b). Preparation of the chitosan/poly (glutamic acid)/alginate polyelectrolyte complexing hydrogel and study on its drug releasing property. *Carbohydrate Polymers, 191*, 8–16.

Cotin, G., Piant, S., Mertz, D., Felder-Flesch, D., & Begin-Colin, S. (2018). *Iron oxide nanoparticles for biomedical applications*. In *Iron oxide nanoparticles for biomedical applications: Synthesis, functionalization, and application* (pp. 43–88). Cambridge, MA: Elsevier.

da Silva, J., Pereira, F. V., & Druzian, J. I. (2012). Cassava starch-based films plasticized with sucrose and inverted sugar and reinforced with cellulose nanocrystals. *Journal of Food Science, 77*(6), N14–N19.

Dabbousi, B. O., Rodriguez-Viejo, J., Mikulec, F. V., Heine, J. R., Mattoussi, H., Ober, R., Jensen, K. F., & Bawendi, M. G. (1997). (CdSe) ZnS core– shell quantum dots: Synthesis and char-

acterization of a size series of highly luminescent nanocrystallites. *The Journal of Physical Chemistry B, 101*(46), 9463–9475.
Dai, C., Ciccotosto, G. D., Cappai, R., Tang, S., Li, D., Xie, S., Xiao, X., & Velkov, T. (2018). Curcumin attenuates colistin-induced neurotoxicity in N2a cells via anti-inflammatory activity, suppression of oxidative stress, and apoptosis. *Molecular Neurobiology, 55*(1), 421–434.
Das, T., Kolli, V., Karmakar, S., & Sarkar, N. (2017). Functionalisation of polyvinylpyrrolidone on gold nanoparticles enhances its anti-amyloidogenic propensity towards hen egg white lysozyme. *Biomedicine, 5*(2), 19.
Deck, L. M., Hunsaker, L. A., Vander Jagt, T. A., Whalen, L. J., Royer, R. E., & Vander Jagt, D. L. (2018). Activation of anti-oxidant Nrf2 signaling by enone analogues of curcumin. *European Journal of Medicinal Chemistry, 143*, 854–865.
Deka, S., Saxena, V., Hasan, A., Chandra, P., & Pandey, L. M. (2018). Synthesis, characterization and in vitro analysis of α-Fe2O3-GdFeO3 biphasic materials as therapeutic agent for magnetic hyperthermia applications. *Materials Science and Engineering: C, 92*, 932–941.
Deng, X., Luan, Q., Chen, W., Wang, Y., Wu, M., Zhang, H., & Jiao, Z. (2009). Nanosized zinc oxide particles induce neural stem cell apoptosis. *Nanotechnology, 20*(11), 115101.
Dimitriou, N. M., Tsekenis, G., Balanikas, E. C., Pavlopoulou, A., Mitsiogianni, M., Mantso, T., Pashos, G., Boudouvis, A. G., Lykakis, I. N., & Tsigaridas, G. (2017). Gold nanoparticles, radiations and the immune system: Current insights into the physical mechanisms and the biological interactions of this new alliance towards cancer therapy. *Pharmacology & Therapeutics, 178*, 1–17.
Dong, F., Feng, E., Zheng, T., & Tian, Y. (2018). In situ synthesized silver nanoclusters for tracking the role of telomerase activity in the differentiation of mesenchymal stem cells to neural stem cells. *ACS Applied Materials & Interfaces, 10*(2), 2051–2057.
Elkhenany, H., Bourdo, S., Hecht, S., Donnell, R., Gerard, D., Abdelwahed, R., Lafont, A., Alghazali, K., Watanabe, F., Biris, A. S., Anderson, D., & Dhar, M. (2017). Graphene nanoparticles as osteoinductive and osteoconductive platform for stem cell and bone regeneration. *Nanomedicine: Nanotechnology, Biology and Medicine, 13*(7), 2117–2126.
Ezazi, N. Z., Shahbazi, M.-A., Shatalin, Y. V., Nadal, E., Mäkilä, E., Salonen, J., Kemell, M., Correia, A., Hirvonen, J., & Santos, H. A. (2018). Conductive vancomycin-loaded mesoporous silica polypyrrole-based scaffolds for bone regeneration. *International Journal of Pharmaceutics, 536*(1), 241–250.
Fan, Z., Zhou, S., Garcia, C., Fan, L., & Zhou, J. (2017). pH-responsive fluorescent graphene quantum dots for fluorescence-guided cancer surgery and diagnosis. *Nanoscale, 9*(15), 4928–4933.
Feynman, R. P. (1959). There's plenty of room at the bottom. *Miniaturization*, 282–296.
Fiala, M., Kooij, G., Wagner, K., Hammock, B., & Pellegrini, M. (2017). Modulation of innate immunity of patients with Alzheimer's disease by omega-3 fatty acids. *The FASEB Journal, 31*(8), 3229–3239.
Franci, G., Falanga, A., Galdiero, S., Palomba, L., Rai, M., Morelli, G., & Galdiero, M. (2015). Silver nanoparticles as potential antibacterial agents. *Molecules, 20*(5), 8856–8874.
Gan, Q., Wang, T., Cochrane, C., & McCarron, P. (2005). Modulation of surface charge, particle size and morphological properties of chitosan–TPP nanoparticles intended for gene delivery. *Colloids and Surfaces B: Biointerfaces, 44*(2–3), 65–73.
Gao, G., Zhang, M., Gong, D., Chen, R., Hu, X., & Sun, T. (2017). The size-effect of gold nanoparticles and nanoclusters in the inhibition of amyloid-β fibrillation. *Nanoscale, 9*(12), 4107–4113.
Gomes, M. A., Brandão-Silva, A. C., Avila, J. F. M., Alencar, M. A. R. C., Rodrigues, J. J., & Macedo, Z. S. (2018). Particle size effect on structural and optical properties of Y2O3:Nd3+ nanoparticles prepared by coconut water-assisted sol-gel route. *Journal of Luminescence, 200*, 43–49.
Han, J. W., Gurunathan, S., Choi, Y.-J., & Kim, J.-H. (2017). Dual functions of silver nanoparticles in F9 teratocarcinoma stem cells, a suitable model for evaluating cytotoxicity-and differentiation-mediated cancer therapy. *International Journal of Nanomedicine, 12*, 7529.

Hasan, A., & Pandey, L. (2017). Self-assembled monolayers in biomaterials. In *Nanobiomaterials* (pp. 137–178). Elsevier.

Hasan, A., Waibhaw, G., Tiwari, S., Dharmalingam, K., Shukla, I., & Pandey, L. M. (2017). Fabrication and characterization of chitosan, polyvinylpyrrolidone, and cellulose nanowhiskers nanocomposite films for wound healing drug delivery application. *Journal of Biomedical Materials Research Part A, 105*(9), 2391–2404.

Hasan, A., Saxena, V., & Pandey, L. M. (2018a). Surface functionalization of Ti6Al4V via self-assembled monolayers for improved protein adsorption and fibroblast adhesion. *Langmuir, 34*(11), 3494–3506.

Hasan, A., Waibhaw, G., Saxena, V., & Pandey, L. M. (2018b). Nano-biocomposite scaffolds of chitosan, carboxymethyl cellulose and silver nanoparticle modified cellulose nanowhiskers for bone tissue engineering applications. *International Journal of Biological Macromolecules, 111*, 923–934.

Hasanzadeh, M., Tagi, S., Solhi, E., Shadjou, N., Jouyban, A., & Mokhtarzadeh, A. (2018). Immunosensing of breast cancer prognostic marker in adenocarcinoma cell lysates and unprocessed human plasma samples using gold nanostructure coated on organic substrate. *International Journal of Biological Macromolecules, 118*(Pt A), 1082.

Hu, J., Youssefian, S., Obayemi, J., Malatesta, K., Rahbar, N., & Soboyejo, W. (2018a). Investigation of adhesive interactions in the specific targeting of Triptorelin-conjugated PEG-coated magnetite nanoparticles to breast cancer cells. *Acta Biomaterialia, 71*, 363–378.

Hu, K., Chen, X., Chen, W., Zhang, L., Li, J., Ye, J., Zhang, Y., Zhang, L., Li, C.-H., & Yin, L. (2018b). Neuroprotective effect of gold nanoparticles composites in Parkinson's disease model. *Nanomedicine: Nanotechnology, Biology and Medicine, 14*(4), 1123–1136.

Jain, P. K., Lee, K. S., El-Sayed, I. H., & El-Sayed, M. A. (2006). Calculated absorption and scattering properties of gold nanoparticles of different size, shape, and composition: Applications in biological imaging and biomedicine. *The Journal of Physical Chemistry B, 110*(14), 7238–7248.

Jasieniak, J., Smith, L., Van Embden, J., Mulvaney, P., & Califano, M. (2009). Re-examination of the size-dependent absorption properties of CdSe quantum dots. *The Journal of Physical Chemistry C, 113*(45), 19468–19474.

Kadappan, A. S., Guo, C., Gumus, C. E., Bessey, A., Wood, R. J., McClements, D. J., & Liu, Z. (2018). The efficacy of nanoemulsion-based delivery to improve vitamin D absorption: Comparison of in vitro and in vivo studies. *Molecular Nutrition & Food Research, 62*(4), 1700836.

Kang, H., Zhang, K., Pan, Q., Lin, S., Wong, D. S. H., Li, J., Lee, W. Y.-W., Yang, B., Han, F., Li, G., Li, B., & Bian, L. (2018). Remote control of intracellular calcium using upconversion nanotransducers regulates stem cell differentiation in vivo. *Advanced Functional Materials, 28*(41), 1802642.

Kango, S., Kalia, S., Celli, A., Njuguna, J., Habibi, Y., & Kumar, R. (2013). Surface modification of inorganic nanoparticles for development of organic–inorganic nanocomposites—A review. *Progress in Polymer Science, 38*(8), 1232–1261.

Kargozar, S., Mozafari, M., Hashemian, S. J., Brouki Milan, P., Hamzehlou, S., Soleimani, M., Joghataei, M. T., Gholipourmalekabadi, M., Korourian, A., & Mousavizadeh, K. (2018). Osteogenic potential of stem cells-seeded bioactive nanocomposite scaffolds: A comparative study between human mesenchymal stem cells derived from bone, umbilical cord Wharton's jelly, and adipose tissue. *Journal of Biomedical Materials Research Part B: Applied Biomaterials, 106*(1), 61–72.

Karim, M. N., Anderson, S. R., Singh, S., Ramanathan, R., & Bansal, V. (2018a). Nanostructured silver fabric as a free-standing NanoZyme for colorimetric detection of glucose in urine. *Biosensors and Bioelectronics, 110*, 8–15.

Karim, M. N., Singh, M., Weerathunge, P., Bian, P., Zheng, R., Dekiwadia, C., Ahmed, T., Walia, S., Della Gaspera, E., Singh, S., Ramanathan, R., & Bansal, V. (2018b). Visible-light-triggered reactive-oxygen-species-mediated antibacterial activity of peroxidase-mimic CuO nanorods. *ACS Applied Nano Materials, 1*(4), 1694–1704.

Karmakar, S. (2017). *Studies on effect of proline capped gold nanoparticles on Hen Egg White Lysozyme (HEWL) and Cytochrome C (Cyt C) amyloidogenesis.*

Kawashima, Y., Yamamoto, H., Takeuchi, H., & Kuno, Y. (2000). Mucoadhesive DL-lactide/glycolide copolymer nanospheres coated with chitosan to improve oral delivery of elcatonin. *Pharmaceutical Development and Technology, 5*(1), 77–85.

Kim, S. U., & De Vellis, J. (2009). Stem cell-based cell therapy in neurological diseases: A review. *Journal of Neuroscience Research, 87*(10), 2183–2200.

Koh, W. C. A., Chandra, P., Kim, D.-M., & Shim, Y.-B. (2011). Electropolymerized self-assembled layer on gold nanoparticles: Detection of inducible nitric oxide synthase in neuronal cell culture. *Analytical Chemistry, 83*(16), 6177–6183.

Kumar, C. S. (2018). *Nanotechnology characterization tools for biosensing and medical diagnosis*. Berlin: Springer.

Kumar, M., Misra, A., Babbar, A., Mishra, A., Mishra, P., & Pathak, K. (2008). Intranasal nanoemulsion based brain targeting drug delivery system of risperidone. *International Journal of Pharmaceutics, 358*(1–2), 285–291.

Kumar, B., Smita, K., Cumbal, L., & Debut, A. (2017). Green synthesis of silver nanoparticles using Andean blackberry fruit extract. *Saudi Journal of Biological Sciences, 24*(1), 45–50.

Kumar, R., Kaur, K., Pandey, S. K., Kumar, R., Uppal, S., & Mehta, S. (2018). Fabrication of benzylisothiocynate encapsulated nanoemulsion through ultrasonication: Augmentation of anticancer and antimicrobial attributes. *Journal of Molecular Liquids, 263*, 324–333.

Kwon, G. S. (2003). Polymeric micelles for delivery of poorly water-soluble compounds. *Critical Reviews™ in Therapeutic Drug Carrier Systems, 20*(5), 357–403.

Laurent, S., Dutz, S., Häfeli, U. O., & Mahmoudi, M. (2011). Magnetic fluid hyperthermia: Focus on superparamagnetic iron oxide nanoparticles. *Advances in Colloid and Interface Science, 166*(1–2), 8–23.

Lee, H., Kim, C., Lee, D., Park, J. H., Searson, P. C., & Lee, K. H. (2017a). Optical coding of fusion genes using multicolor quantum dots for prostate cancer diagnosis. *International Journal of Nanomedicine, 12*, 4397.

Lee, S. S., Choi, G. E., Lee, H. J., Kim, Y., Choy, J.-H., & Jeong, B. (2017b). Layered double hydroxide and polypeptide thermogel nanocomposite system for chondrogenic differentiation of stem cells. *ACS Applied Materials & Interfaces, 9*(49), 42668–42675.

Lekshmi, N. P., Sumi, S. B., Viveka, S., Jeeva, S., & Brindha, J. R. (2017). Antibacterial activity of nanoparticles from Allium sp. *Journal of Microbiology and Biotechnology Research, 2*(1), 115–119.

Li, S., Zhou, S., Li, Y., Li, X., Zhu, J., Fan, L., & Yang, S. (2017). Exceptionally high payload of the IR780 iodide on folic acid-functionalized graphene quantum dots for targeted Photothermal therapy. *ACS Applied Materials & Interfaces, 9*(27), 22332–22341.

Lim, E.-K., Chung, B. H., & Chung, S. J. (2018). Recent advances in pH-sensitive polymeric nanoparticles for smart drug delivery in cancer therapy. *Current Drug Targets, 19*(4), 300–317.

Lin, L.-S., Cong, Z.-X., Cao, J.-B., Ke, K.-M., Peng, Q.-L., Gao, J., Yang, H.-H., Liu, G., & Chen, X. (2014). Multifunctional Fe3O4@ polydopamine core–shell nanocomposites for intracellular mRNA detection and imaging-guided photothermal therapy. *ACS Nano, 8*(4), 3876–3883.

Liu, D., Yi, C., Zhang, D., Zhang, J., & Yang, M. (2010). Inhibition of proliferation and differentiation of mesenchymal stem cells by carboxylated carbon nanotubes. *ACS Nano, 4*(4), 2185–2195.

Liu, Q., Ma, C., Liu, X.-P., Wei, Y.-P., Mao, C.-J., & Zhu, J.-J. (2017). A novel electrochemiluminescence biosensor for the detection of microRNAs based on a DNA functionalized nitrogen doped carbon quantum dots as signal enhancers. *Biosensors and Bioelectronics, 92*, 273–279.

Lv, R., Yang, P., Chen, G., Gai, S., Xu, J., & Prasad, P. N. (2017a). Dopamine-mediated photothermal theranostics combined with up-conversion platform under near infrared light. *Scientific Reports, 7*(1), 13562.

Lv, Y., Cao, Y., Li, P., Liu, J., Chen, H., Hu, W., & Zhang, L. (2017b). Ultrasound-triggered destruction of folate-functionalized mesoporous silica nanoparticle-loaded microbubble for targeted tumor therapy. *Advanced Healthcare Materials, 6*(18), 1700354.

Ma, Q., Yang, J., Huang, X., Guo, W., Li, S., Zhou, H., Li, J., Cao, F., & Chen, Y. (2018). Poly(lactide-co-glycolide)-monomethoxy-poly-(polyethylene glycol) nanoparticles loaded with melatonin protect adipose-derived stem cells transplanted in infarcted heart tissue. *Stem Cells, 36*(4), 540–550.

Mahato, K., Prasad, A., Maurya, P., & Chandra, P. (2016). Nanobiosensors: Next generation point-of-care biomedical devices for personalized diagnosis. *Journal of Anaytical and Bioanalytical Techniques, 7*, e125.

Marsich, E., Bellomo, F., Turco, G., Travan, A., Donati, I., & Paoletti, S. (2013). Nano-composite scaffolds for bone tissue engineering containing silver nanoparticles: Preparation, characterization and biological properties. *Journal of Materials Science: Materials in Medicine, 24*(7), 1799–1807.

Michal, E. T., Lerner, D. J., & Pollman, M. J. (2015). *Local delivery of water-soluble or water-insoluble therapeutic agents to the surface of body lumens*. Google Patents.

Mili, B., Das, K., Kumar, A., Saxena, A. C., Singh, P., Ghosh, S., & Bag, S. (2017). Preparation of NGF encapsulated chitosan nanoparticles and its evaluation on neuronal differentiation potentiality of canine mesenchymal stem cells. *Journal of Materials Science: Materials in Medicine, 29*(1), 4.

Molfino, A., Amabile, M. I., Monti, M., Arcieri, S., Rossi Fanelli, F., & Muscaritoli, M. (2016). The role of docosahexaenoic acid (DHA) in the control of obesity and metabolic derangements in breast cancer. *International Journal of Molecular Sciences, 17*(4), 505.

Moreels, I., Lambert, K., Smeets, D., De Muynck, D., Nollet, T., Martins, J. C., Vanhaecke, F., Vantomme, A., Delerue, C., & Allan, G. (2009). Size-dependent optical properties of colloidal PbS quantum dots. *ACS Nano, 3*(10), 3023–3030.

Mu, X., Zhang, F., Kong, C., Zhang, H., Zhang, W., Ge, R., Liu, Y., & Jiang, J. (2017). EGFR-targeted delivery of DOX-loaded Fe3O4@ polydopamine multifunctional nanocomposites for MRI and antitumor chemo-photothermal therapy. *International Journal of Nanomedicine, 12*, 2899.

Nazar, H. (2018). The use of nanotechnology in disease diagnosis and molecular imaging. *Lung Cancer, 15*, 05.

Nazıroğlu, M., Muhamad, S., & Pecze, L. (2017). Nanoparticles as potential clinical therapeutic agents in Alzheimer's disease: Focus on selenium nanoparticles. *Expert Review of Clinical Pharmacology, 10*(7), 773–782.

Neun, B. W., & Dobrovolskaia, M. A. (2011). Method for in vitro analysis of nanoparticle thrombogenic properties. In *Characterization of nanoparticles intended for drug delivery* (pp. 225–235). New York: Springer.

Niwa, T., Takeuchi, H., Hino, T., Kunou, N., & Kawashima, Y. (1994). In vitro drug release behavior of D, L-lactide/glycolide copolymer (PLGA) nanospheres with nafarelin acetate prepared by a novel spontaneous emulsification solvent diffusion method. *Journal of Pharmaceutical Sciences, 83*(5), 727–732.

Ozturk, B., Argin, S., Ozilgen, M., & McClements, D. J. (2015). Formation and stabilization of nanoemulsion-based vitamin E delivery systems using natural biopolymers: Whey protein isolate and gum arabic. *Food Chemistry, 188*, 256–263.

Pacelli, S., Maloney, R., Chakravarti, A. R., Whitlow, J., Basu, S., Modaresi, S., Gehrke, S., & Paul, A. (2017). Controlling adult stem cell behavior using nanodiamond-reinforced hydrogel: Implication in bone regeneration therapy. *Scientific Reports, 7*(1), 6577.

Pandey, L. M. (2012). *Effect of solid surface with self assembled monolayers on adsorption of proteins*.

Pandey, L. M., & Pattanayek, S. K. (2011). Hybrid surface from self-assembled layer and its effect on protein adsorption. *Applied Surface Science, 257*(10), 4731–4737.

Pandey, L. M., & Pattanayek, S. K. (2013a). Properties of competitively adsorbed BSA and fibrinogen from their mixture on mixed and hybrid surfaces. *Applied Surface Science, 264*, 832–837.

Pandey, L. M., & Pattanayek, S. K. (2013b). Relation between the wetting effect and the adsorbed amount of water-soluble polymers or proteins at various interfaces. *Journal of Chemical & Engineering Data, 58*(12), 3440–3446.

Pandey, L. M., Le Denmat, S., Delabouglise, D., Bruckert, F., Pattanayek, S. K., & Weidenhaupt, M. (2012). Surface chemistry at the nanometer scale influences insulin aggregation. *Colloids and Surfaces B: Biointerfaces, 100*, 69–76.

Pandey, L. M., Pattanayek, S. K., & Delabouglise, D. (2013). Properties of adsorbed bovine serum albumin and fibrinogen on self-assembled monolayers. *The Journal of Physical Chemistry C, 117*(12), 6151–6160.

Panyam, J., & Labhasetwar, V. (2003). Biodegradable nanoparticles for drug and gene delivery to cells and tissue. *Advanced Drug Delivery Reviews, 55*(3), 329–347.

Parveen, S., Wani, A. H., Shah, M. A., Devi, H. S., Bhat, M. Y., & Koka, J. A. (2018). Preparation, characterization and antifungal activity of iron oxide nanoparticles. *Microbial Pathogenesis, 115*, 287–292.

Peng, X.-H., Qian, X., Mao, H., & Wang, A. Y. (2008). Targeted magnetic iron oxide nanoparticles for tumor imaging and therapy. *International Journal of Nanomedicine, 3*(3), 311.

Pohanka, M. (2017). Quantum dots in the therapy: Current trends and perspectives. *Mini Reviews in Medicinal Chemistry, 17*(8), 650–656.

Popara, J., Accomasso, L., Vitale, E., Gallina, C., Roggio, D., Iannuzzi, A., Raimondo, S., Rastaldo, R., Alberto, G., Catalano, F., Martra, G., Turinetto, V., Pagliaro, P., & Giachino, C. (2018). Silica nanoparticles actively engage with mesenchymal stem cells in improving acute functional cardiac integration. *Nanomedicine, 13*(10), 1121–1138.

Potter, T. M., Rodriguez, J. C., Neun, B. W., Ilinskaya, A. N., Cedrone, E., & Dobrovolskaia, M. A. (2018). In vitro assessment of nanoparticle effects on blood coagulation. In *Characterization of nanoparticles intended for drug delivery* (pp. 103–124). New York: Springer.

Prasad, A., Mahato, K., Chandra, P., Srivastava, A., Joshi, S. N., & Maurya, P. K. (2016a). Bioinspired composite materials: Applications in diagnostics and therapeutics. *Journal of Molecular and Engineering Materials, 4*(01), 1640004.

Prasad, A., Mahato, K., Maurya, P., & Chandra, P. (2016b). Biomaterials for biosensing applications. *Journal of Analytical and Bioanalytical Techniques, 7*, e124.

Qiao, Y., Gumin, J., MacLellan, C. J., Gao, F., Bouchard, R., Lang, F. F., Stafford, R. J., & Melancon, M. P. (2018). Magnetic resonance and photoacoustic imaging of brain tumor mediated by mesenchymal stem cell labeled with multifunctional nanoparticle introduced via carotid artery injection. *Nanotechnology, 29*(16), 165101.

Qin, X., Chen, H., Yang, H., Wu, H., Zhao, X., Wang, H., Chour, T., Neofytou, E., Ding, D., Daldrup-Link, H., Heilshorn, S. C., Li, K., & Wu, J. C. (2018). Photoacoustic imaging of embryonic stem cell-derived cardiomyocytes in living hearts with ultrasensitive semiconducting polymer nanoparticles. *Advanced Functional Materials, 28*(1), 1704939.

Radhakrishnan, V. S., Dwivedi, S. P., Siddiqui, M. H., & Prasad, T. (2018). In vitro studies on oxidative stress-independent, Ag nanoparticles-induced cell toxicity of *Candida albicans*, an opportunistic pathogen. *International Journal of Nanomedicine, 13*, 91.

Rafiei, P., & Haddadi, A. (2017). Pharmacokinetic consequences of PLGA nanoparticles in docetaxel drug delivery. *Pharmaceutical Nanotechnology, 5*(1), 3–23.

Rather, H. A., Thakore, R., Singh, R., Jhala, D., Singh, S., & Vasita, R. (2018). Antioxidative study of cerium oxide nanoparticle functionalised PCL-Gelatin electrospun fibers for wound healing application. *Bioactive Materials, 3*(2), 201–211.

Reis, C. P., Neufeld, R. J., & Veiga, F. (2017). Preparation of drug-loaded polymeric nanoparticles. In *Nanomedicine in cancer* (pp. 197–240). Singapore: Pan Stanford.

Roy, K., Mao, H.-Q., Huang, S.-K., & Leong, K. W. (1999). Oral gene delivery with chitosan–DNA nanoparticles generates immunologic protection in a murine model of peanut allergy. *Nature Medicine, 5*(4), 387.

Sahoo, S., Singh, D., Singh, P., & Minz, A. P. (2018). Advanced nanotherapeutic systems for drug delivery and imaging in cancer. In *Multifunctional nanocarriers for contemporary healthcare applications* (pp. 1–21). Hershey: IGI Global.

Sanitá, P. V., Pavarina, A. C., Dovigo, L. N., Ribeiro, A. P. D., Andrade, M. C., & de Oliveira Mima, E. G. (2018). Curcumin-mediated anti-microbial photodynamic therapy against Candida dubliniensis biofilms. *Lasers in Medical Science, 33*(4), 709–717.

Saratale, R. G., Benelli, G., Kumar, G., Kim, D. S., & Saratale, G. D. (2018). Bio-fabrication of silver nanoparticles using the leaf extract of an ancient herbal medicine, dandelion (*Taraxacum officinale*), evaluation of their antioxidant, anticancer potential, and antimicrobial activity against phytopathogens. *Environmental Science and Pollution Research, 25*(11), 10392–10406.

Saravanan, S., Leena, R., & Selvamurugan, N. (2016). Chitosan based biocomposite scaffolds for bone tissue engineering. *International Journal of Biological Macromolecules, 93*, 1354–1365.

Sari, T., Mann, B., Kumar, R., Singh, R., Sharma, R., Bhardwaj, M., & Athira, S. (2015). Preparation and characterization of nanoemulsion encapsulating curcumin. *Food Hydrocolloids, 43*, 540–546.

Sarkar, C., Kumari, P., Anuvrat, K., Sahu, S. K., Chakraborty, J., & Garai, S. (2018). Synthesis and characterization of mechanically strong carboxymethyl cellulose–gelatin–hydroxyapatite nanocomposite for load-bearing orthopedic application. *Journal of Materials Science, 53*(1), 230–246.

Savaliya, R., Shah, D., Singh, R., Kumar, A., Shanker, R., Dhawan, A., & Singh, S. (2015). Nanotechnology in disease diagnostic techniques. *Current Drug Metabolism, 16*(8), 645–661.

Saxena, V., Chandra, P., & Pandey, L. M. (2018a). Design and characterization of novel Al-doped ZnO nanoassembly as an effective nanoantibiotic. *Applied Nanoscience, 8*(8), 1925–1941.

Saxena, V., Hasan, A., Sharma, S., & Pandey, L. M. (2018b). Edible oil nanoemulsion: An organic nanoantibiotic as a potential biomolecule delivery vehicle. *International Journal of Polymeric Materials and Polymeric Biomaterials, 67*(7), 410–419.

Shah, S., Yin, P. T., Uehara, T. M., Chueng, S. T. D., Yang, L., & Lee, K. B. (2014). Guiding stem cell differentiation into oligodendrocytes using graphene-nanofiber hybrid scaffolds. *Advanced Materials, 26*(22), 3673–3680.

Shahidi, F., & Ambigaipalan, P. (2018). Omega-3 polyunsaturated fatty acids and their health benefits. *Annual Review of Food Science and Technology, 9*(1).

Sharma, C., Dinda, A. K., Potdar, P. D., Chou, C.-F., & Mishra, N. C. (2016). Fabrication and characterization of novel nano-biocomposite scaffold of chitosan–gelatin–alginate–hydroxyapatite for bone tissue engineering. *Materials Science and Engineering: C, 64*, 416–427.

Shim, Y. B. (2013). Gold nanoparticles and nanocomposites in clinical diagnostics using electrochemical methods. *Journal of Nanoparticles, 2013*, 1–12.

Shin, M., Yoshimoto, H., & Vacanti, J. P. (2004). In vivo bone tissue engineering using mesenchymal stem cells on a novel electrospun nanofibrous scaffold. *Tissue Engineering, 10*(1–2), 33–41.

Singh, S. (2013). Nanomaterials as non-viral siRNA delivery agents for cancer therapy. *BioImpacts: BI, 3*(2), 53.

Singh, S. (2016). Cerium oxide based nanozymes: Redox phenomenon at biointerfaces. *Biointerphases, 11*(4), 04B202.

Singh, A., Datta, P., & Pandey, L. M. (2017). Deciphering the mechanistic insight into the stoichiometric ratio dependent behavior of Cu (II) on BSA fibrillation. *International Journal of Biological Macromolecules, 97*, 662–670.

Singh, A., Hasan, A., Tiwari, S., & Pandey, L. (2018a). Therapeutic advancement in Alzheimer disease: New hopes on the horizon? *CNS & Neurological Disorders Drug Targets, 17*, 571–589.

Singh, S., Asal, R., & Bhagat, S. (2018b). Multifunctional antioxidant nanoliposome-mediated delivery of PTEN plasmids restore the expression of tumor suppressor protein and induce apoptosis in prostate cancer cells. *Journal of Biomedical Materials Research Part A, 106*, 3152–3164.

Suktham, K., Koobkokkruad, T., Wutikhun, T., & Surassmo, S. (2018). Efficiency of resveratrol-loaded sericin nanoparticles: Promising bionanocarriers for drug delivery. *International Journal of Pharmaceutics, 537*(1–2), 48–56.

Tiwari, S., Hasan, A., & Pandey, L. M. (2017). A novel bio-sorbent comprising encapsulated agrobacterium fabrum (SLAJ731) and iron oxide nanoparticles for removal of crude oil co-contaminant, lead Pb (II). *Journal of Environmental Chemical Engineering, 5*(1), 442–452.

Vallabani, N. V. S., Karakoti, A. S., & Singh, S. (2017). ATP-mediated intrinsic peroxidase-like activity of Fe3O4-based nanozyme: One step detection of blood glucose at physiological pH. *Colloids and Surfaces B: Biointerfaces, 153*, 52–60.

Virani, N. A., Davis, C., McKernan, P., Hauser, P., Hurst, R. E., Slaton, J., Silvy, R. P., Resasco, D. E., & Harrison, R. G. (2017). Phosphatidylserine targeted single-walled carbon nanotubes for photothermal ablation of bladder cancer. *Nanotechnology, 29*(3), 035101.

Wei, L., Lu, J., Xu, H., Patel, A., Chen, Z.-S., & Chen, G. (2015). Silver nanoparticles: Synthesis, properties, and therapeutic applications. *Drug Discovery Today, 20*(5), 595–601.

Wei, M., Li, S., & Le, W. (2017). Nanomaterials modulate stem cell differentiation: Biological interaction and underlying mechanisms. *Journal of Nanobiotechnology, 15*(1), 75.

Wu, S. Q., Yang, C. X., & Yan, X. P. (2017). A dual-functional persistently luminescent nanocomposite enables engineering of mesenchymal stem cells for homing and gene therapy of glioblastoma. *Advanced Functional Materials, 27*(11), 1604992.

Xu, Q., Zhang, T., Wang, Q., Jiang, X., Li, A., Li, Y., Huang, T., Li, F., Hu, Y., Ling, D., & Gao, J. (2018). Uniformly sized iron oxide nanoparticles for efficient gene delivery to mesenchymal stem cells. *International Journal of Pharmaceutics, 552*(1), 443–452.

Yamagishi, K., Ikeda, A., Chei, C.-L., Noda, H., Umesawa, M., Cui, R., Muraki, I., Ohira, T., Imano, H., & Sankai, T. (2017). Serum α-linolenic and other ω-3 fatty acids, and risk of disabling dementia: Community-based nested case–control study. *Clinical Nutrition, 36*(3), 793–797.

Yamamoto, H., Kuno, Y., Sugimoto, S., Takeuchi, H., & Kawashima, Y. (2005). Surface-modified PLGA nanosphere with chitosan improved pulmonary delivery of calcitonin by mucoadhesion and opening of the intercellular tight junctions. *Journal of Controlled Release, 102*(2), 373–381.

Yang, L., Sun, J., Xie, W., Liu, Y., & Liu, J. (2017). Dual-functional selenium nanoparticles bind to and inhibit amyloid β fiber formation in Alzheimer's disease. *Journal of Materials Chemistry B, 5*(30), 5954–5967.

Yang, N., Ding, Y., Zhang, Y., Wang, B., Zhao, X., Cheng, K., Huang, Y., Taleb, M., Zhao, J., Dong, W.-F., Zhang, L., & Nie, G. (2018). Surface functionalization of polymeric nanoparticles with umbilical cord-derived mesenchymal stem cell membrane for tumor-targeted therapy. *ACS Applied Materials & Interfaces, 10*(27), 22963–22973.

Yao, X., Niu, X., Ma, K., Huang, P., Grothe, J., Kaskel, S., & Zhu, Y. (2017). Graphene quantum dots-capped magnetic mesoporous silica nanoparticles as a multifunctional platform for controlled drug delivery, magnetic hyperthermia, and photothermal therapy. *Small, 13*(2), 1602225.

Yasir, M., Singh, J., Tripathi, M. K., Singh, P., & Shrivastava, R. (2018). Green synthesis of silver nanoparticles using leaf extract of common arrowhead houseplant and its anticandidal activity. *Pharmacognosy Magazine, 13*(Suppl 4), S840.

Yen, Y. H., Pu, C. M., Liu, C. W., Chen, Y. C., Chen, Y. C., Liang, C. J., Hsieh, J. H., Huang, H. F., & Chen, Y. L. (2018). Curcumin accelerates cutaneous wound healing via multiple biological actions: The involvement of TNF-α, MMP-9, α-SMA, and collagen. *International Wound Journal, 15*, 605–617.

Yong, Y., Cheng, X., Bao, T., Zu, M., Yan, L., Yin, W., Ge, C., Wang, D., Gu, Z., & Zhao, Y. (2015). Tungsten sulfide quantum dots as multifunctional nanotheranostics for in vivo dual-modal image-guided photothermal/radiotherapy synergistic therapy. *ACS Nano, 9*(12), 12451–12463.

Yoo, J., Lee, E., Kim, H. Y., Youn, D.-h., Jung, J., Kim, H., Chang, Y., Lee, W., Shin, J., & Baek, S. (2017). Electromagnetized gold nanoparticles mediate direct lineage reprogramming into induced dopamine neurons in vivo for Parkinson's disease therapy. *Nature Nanotechnology, 12*(10), 1006.

Yu, M. K., Jeong, Y. Y., Park, J., Park, S., Kim, J. W., Min, J. J., Kim, K., & Jon, S. (2008). Drug-loaded superparamagnetic iron oxide nanoparticles for combined cancer imaging and therapy in vivo. *Angewandte Chemie, 120*(29), 5442–5445.

Zamani, M., Rostami, M., Aghajanzadeh, M., Manjili, H. K., Rostamizadeh, K., & Danafar, H. (2018). Mesoporous titanium dioxide@ zinc oxide–graphene oxide nanocarriers for colon-specific drug delivery. *Journal of Materials Science, 53*(3), 1634–1645.

Zhang, X.-F., Liu, Z.-G., Shen, W., & Gurunathan, S. (2016). Silver nanoparticles: Synthesis, characterization, properties, applications, and therapeutic approaches. *International Journal of Molecular Sciences, 17*(9), 1534.

Zhang, G., Khan, A., Wu, H., Chen, L., Gu, Y., & Gu, N. (2017). The application of nanomaterials in stem cell therapy for some neurological diseases. *Current Drug Targets, 19*, 279–298.

Zhang, B., Yan, W., Zhu, Y., Yang, W., Le, W., Chen, B., Zhu, R., & Cheng, L. (2018a). Nanomaterials in neural-stem-cell-mediated regenerative medicine: Imaging and treatment of neurological diseases. *Advanced Materials, 30*(17), 1705694.

Zhang, F., Stephan, S. B., Ene, C. I., Smith, T. T., Holland, E. C., & Stephan, M. T. (2018b). Nanoparticles that reshape the tumor milieu create a therapeutic window for effective T cell therapy in solid malignancies. *Cancer Research, 78*, 3718–3730. canres. 0306.2018.

Zhao, M.-X., & Zeng, E.-Z. (2015). Application of functional quantum dot nanoparticles as fluorescence probes in cell labeling and tumor diagnostic imaging. *Nanoscale Research Letters, 10*(1), 171.

Zhao, Q., Qian, J., An, Q., Gao, C., Gui, Z., & Jin, H. (2009). Synthesis and characterization of soluble chitosan/sodium carboxymethyl cellulose polyelectrolyte complexes and the pervaporation dehydration of their homogeneous membranes. *Journal of Membrane Science, 333*(1–2), 68–78.

Zhu, Y., Chandra, P., & Shim, Y.-B. (2012a). Ultrasensitive and selective electrochemical diagnosis of breast cancer based on a hydrazine–Au nanoparticle–aptamer bioconjugate. *Analytical Chemistry, 85*(2), 1058–1064.

Zhu, Y., Chandra, P., Song, K.-M., Ban, C., & Shim, Y.-B. (2012b). Label-free detection of kanamycin based on the aptamer-functionalized conducting polymer/gold nanocomposite. *Biosensors and Bioelectronics, 36*(1), 29–34.

Zhu, X., Feng, W., Chang, J., Tan, Y.-W., Li, J., Chen, M., Sun, Y., & Li, F. (2016). Temperature-feedback upconversion nanocomposite for accurate photothermal therapy at facile temperature. *Nature Communications, 7*, 10437.

Gold Nanostructures for Photothermal Therapy

Prem Singh, Shounak Roy, Pallab Sanpui, Aditi Banerjee, and Amit Jaiswal

Abstract Gold nanostructures – due to their ease of synthesis and functionalization, unique tunable optical properties and stability – are widely being explored for their applicability in sensing, diagnostics, drug delivery and cancer therapy. Engineering different gold nanostructures with varying shape and size enable us to tune the localized surface plasmon resonance (LSPR) peak from visible to near infra-red (NIR) region of the electromagnetic spectrum, which can be exploited for biomedical applications. For example, gold nanorods show two peaks in their extinction spectra, corresponding to the transverse and longitudinal mode of surface electron oscillation on the influence of light. Similarly, other gold nanostructures with different morphologies like nanoshells, nanorattles, nanostars, nanopopcorns, nanoaggregates, etc. too have extinction band in the NIR region, which has a better tissue penetration depth. This strong optical absorbance of the gold nanostructures, especially in the NIR region and subsequent dissipation of energy in a nonradiative process can suitably be exploited for plasmonic photothermal therapy (PPTT). In this regard, NIR light stimulated heat can be generated from the targeted gold nanostructures and this can potentially be used to kill cancer cells. The present chapter discusses about the design and applicability of gold based nanostructures of different morphologies for efficient photothermal therapy.

Authors Prem Singh and Shounak Roy have equally contributed to this chapter.

P. Singh · S. Roy · A. Jaiswal (✉)
School of Basic Sciences, Indian Institute of Technology Mandi,
Kamand, Mandi, Himachal Pradesh, India
e-mail: j.amit@iitmandi.ac.in

P. Sanpui
Department of Biotechnology, BITS Pilani, Dubai Campus,
Dubai International Academic City, Dubai, UAE

A. Banerjee (✉)
Centre for Nanotechnology, Indian Institute of Technology Guwahati,
Guwahati, Assam, India
e-mail: aditi@iitg.ac.in

S. Singh, P. K. Maurya (eds.), *Nanotechnology in Modern Animal Biotechnology*, https://doi.org/10.1007/978-981-13-6004-6_2

Keywords Gold nanoparticles · Photothermal therapy · NIR · Cancer therapy · Plasmonic nanoparticles

1 Introduction

In the past few decades, thermal laser has been extensively used to kill cancerous cells. These lasers, emitting in the near infra-red (NIR) or visible region of the electromagnetic spectrum, induces moderate rise in the temperature (i.e. 42–49 °C) of cells or tissues (Brunetaud et al. 1995). Additionally, ultrasound waves, radio waves and microwaves can also be used to induce moderate temperature rise. This increase in temperature eventually causes destruction of concerned cells at the target site; and the phenomenon is termed medically as 'hyperthermia' (Huang and El-Sayed 2011). Compared to normal healthy tissues, tumor tissues are more sensitive to hyperthermia because of their un-ordered and unevenly distributed vascular structures making dissipation of heat difficult. Moreover, tumor tissues are more acidic, hypoxic and deficient in nutrients as compared to normal healthy tissues (Wust et al. 2002). By exploiting these imperfections of the tumor tissues, cancer cells can be killed selectively by hyperthermia at a temperature range between 40 and 44 °C. Hyperthermia causes DNA damage, protein denaturation and disruption of cell membrane resulting in the destruction of the tumor tissues (Hwang et al. 2014a, b). However, hyperthermia treatment for cancer cells requires high-power lasers (100–120 W) (Sultan 1990), which sometimes can lead to overexpression of heat shock proteins (Ciocca and Calderwood 2005) making the tumor cells more resistant to heat-based therapies (Calderwood and Ciocca 2008). Sometimes hyperthermia treatment lack specificity for tumor tissues. This may lead to exposure of normal tissues to hyperthermia that damages the normal healthy tissue adjacent to tumor tissues resulting in systemic toxicity (Hwang et al. 2014a, b).

Recently, photothermal therapy (PTT) has been widely opted to overcome the problems associated with these local heating or traditional hyperthermia treatment. Photothermal therapy is a non-invasive thermal treatment in which photon energy is converted to heat to achieve restricted and confined thermal damage at tumor sites. In this regard, noble metal particles (such as gold and silver) at nano-scale size have generated new interest in researchers because of their excellent property to generate local heating upon excitation of surface plasmon oscillations, which if targeted to tumour site can lead to cell death. Additionally, gold and silver nanoparticles have excellent light absorption and scattering efficiency. When the photon energy is converted to heat with the help of plasmonic nanostructures, it is termed plasmonic photothermal therapy (PPTT) (Chen et al. 2010a, b). For PPTT, absorption of NIR light by plasmonic nanoparticles is preferred because of its higher penetration depth arising from minimum absorption and scattering by other biomolecules and tissue components. This, in turn, can translate into better photothermal treatment efficiency as compared to visible light-mediated PTT (Ai et al. 2016).

Gold nanostructures have unique photophysical properties and are the best suited agents for PPTT because of flexibility in their optical and electronic properties.

Their optical and electronic properties can easily be tuned by changing their shape and size. Different shape and structure of gold nanostructures have its unique optical properties and have their own surface plasmon resonance (SPR) peak positions. For example, gold nanospheres have single SPR mode in the visible region while anisotropic nanoparticles like gold nanorods show multiple SPR modes in their extinction spectra (Kelly et al. 2003).

In this chapter, we mainly discuss on different gold nanostructures and their optical properties, impact on photothermal therapy by using these different morphologies of gold nanostructures e.g. nanorods, nanoshells, nanocages, nanorattles, nanostars, nanopopcorns and nanoaggregates. Along with this, we have also discussed the mechanism of cell death induced by PPTT.

2 Types of Gold Nanostructures and Their Optical Properties

Gold nanostructures show unique tunable optical properties and stability due to their ease of synthesis and functionalization. They have been widely explored in sensing, bioimaging, diagnostics, drug delivery, cancer therapy and other biomedical application (Nie et al. 2007; Hwang et al. 2014a, b; Abadeer and Murphy 2016). In 1857, Michael Faraday was the first scientist who made colloidal gold by reducing gold chloride with the help of phosphorous. He noticed that 'finely divided particles' formed upon reduction of gold salt solution with the help of reducing agent and the color of the solution became ruby red (Faraday 1857). Later in 1951, J. Turkevich and group reported a new approach for the synthesis of colloidal gold. They used sodium citrate as a reducing agent for their synthesis and proposed a nucleation and growth process in the synthesis of colloidal gold (Turkevich et al. 1951). They explained two basic steps involved in the synthesis of seed-mediated growth process: synthesis of gold seeds from the gold salt solution via chemical reduction method (nucleation) and then subsequently growth of the nanoparticles on the surface of the seeds in growth solution containing metal precursors, reducing agent and capping or shape directing reagents (growth). Since last few decades, researchers have been working on the synthesis of gold nanoparticles of different shape and sizes. They have also been engaged in understanding their optical property and to study their interaction with light. On exposure to light, the conduction band electrons at the surface of gold nanostructures start to oscillate due to absorption of energy. This oscillation is referred to as localized surface plasmon. This electron oscillation induces the charge separation around the metal particle surface which forms a dipole along the direction of the electric field of the radiation. When the frequency of localized surface plasmons matches with the frequency of the incident electromagnetic radiation, the amplitude of oscillation reaches its maxima and it results in resonance which is referred to as localized surface plasmon resonance (LSPR) (Henglein 1993; Papavassiliou 1979). Plasmonic nanoparticles have unique photophysical properties as compared to non-metal nanoparticles (Huang and El-Sayed 2011). The position and intensity of SPR band in the UV-Vis spectra depends on several factors which include nanoparticle shape, size, structure, type of metal and the nature of the surrounding medium (Mulvaney 1996). Gold nanospheres have SPR in their visible region while anisotropic nanoparticles like nanorods, nanoshells, nanocages, nanorattles, nanostars, nanopopcorns and nanoaggregates show a red shifted SPR or multiple SPR bands mainly in NIR region (~650 nm $\leq\lambda$ ~900 nm). This unique tunable SPR of gold nanoparticles makes them a promising agent for photothermal therapy. Figure 1 represents the TEM images and absorption spectra of various gold nanostructures.

Interaction of electromagnetic radiation behaves differently for different shape of metal nanoparticles. Different theories explained by different researchers exist which demonstrates the dependence of optical properties of metal nanoparticles on the size, shape and dielectric function of metal-host composite material. These theories include (i) The Mie theory: size-dependent optical properties, (ii) The Gans theory: shape-dependent optical properties and (iii) Maxwell-Garnett theory: effective medium theory (Khlebtsov and Dykman 2010; Link and El-Sayed 2000).

2.1 Gold Nanospheres

The optical properties of gold nanospheres mainly depend on their size. By changing the ratio of the seed to precursor, the size of the gold nanospheres can be varied. Gold nanospheres mainly show SPR peak in visible region of the electromagnetic spectrum. Spheres with size below 20 nm show SPR band around 520 nm (Khlebtsov and Dykman 2010). With increase in their size from 20 to 100 nm or more, the SPR band starts to shift towards longer wavelength known as the red shift. Along with this, broadening of the plasmonic peak is also observed with increase in size (Huang et al. 2007). Also, it is to be noted that the position of the SPR peak of the gold nanospheres is strongly influenced by the dielectric constant of the surrounding environment. So, the type of solvent and stabilizer/capping agent used during the synthesis can shift the position of SPR band of gold nanospheres. Apart from size dependence, a shift in SPR peak position of gold nanospheres have also been observed when electrostatic interactions between nanoparticles cause them to form nanoaggregates or assemblies in colloidal solution. Aggregation of plasmonic nanoparticles such as gold nanospheres results in plasmonic coupling of the surface plasmons of individual nanoparticles which leads to the shift in the SPR peak from visible to NIR region, thereby making such nanoaggregate systems efficient phototransducers suitable for PTT applications (Hu et al. 2006; Ghosh and Pal 2007). The SPR peak of Au nanospheres and aggregates are shown in Fig. 1i.

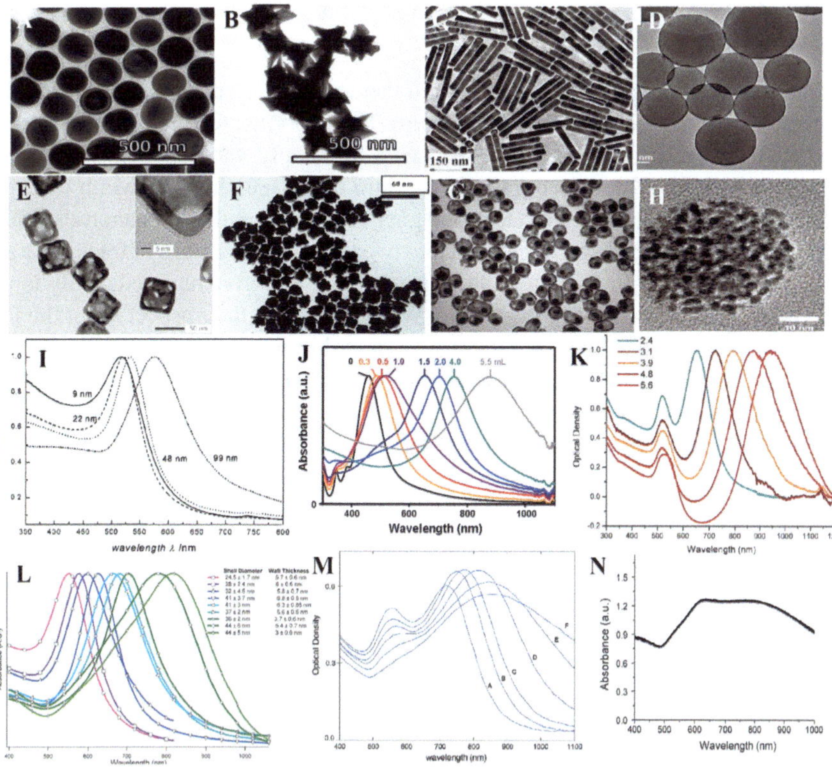

Fig. 1 TEM images of (**a**) Gold nanospheres, (**b**) Gold nanostars, (**c**) Gold nanorods, (**d**) Gold nanoshells, (**e**) Gold nanocages, (**f**) Gold nanopopcorns, (**g**) Gold nanorattles, (**h**) Gold nanoaggregates. Absorption spectra of (**i**) Gold nanospheres, (**j**) Gold nanocages, (**k**) Gold nanorods, (**l**) Gold nanoshells, (**m**) Gold nanostars, (**n**) IOC-Gold nanopopcorns (**a, b**) Reproduced with permission from Chu et al. 2014. Copyright 2014 Nature publishing group. (**c**) Reproduced with permission from Vigderman and Zubarev 2013. Copyright 2013 American Chemical Society. (**d**) Reproduced with permission from Gao et al. 2016. Copyright 2016 American Chemical Society. (**e**) Reproduced with permission from Yavuz et al. 2009. Copyright 2009 Nature publishing group. (**f**) Reproduced with permission from Lu et al. 2010. Copyright 2009 American Chemical Society. (**g**) Reproduced with permission from Mahmoud 2014. Copyright 2014 Royal society of chemistry. (**h**) Reproduced with permission from Jung et al. 2010. Copyright 2010 Royal society of chemistry. (**i**) Reproduced with permission from Link and El-Sayed 1999. Copyright 2009 American Chemical Society. (**j**) Reproduced with permission from Skrabalak et al. 2008. Copyright 2008 American Chemical Society. (**k**) Reproduced with permission from Huang et al. 2006. Copyright 2006 American Chemical Society. (**l**) Reproduced with permission from Schwartzberg et al. 2006. Copyright 2006 American Chemical Society. (**m**) Reproduced with permission from Khoury and Vo-Dinh 2008. Copyright 2008 American Chemical Society. (**n**) Reproduced with permission from Bhana et al. 2015. Copyright 2015 American Chemical Society

2.2 *Gold Nanorods*

Gold nanostructures used in photothermal therapy must have the property to absorb light in near infra-red regions (650–900 nm). The important feature of gold nanorods is that their optical properties can be tuned by simply varying their aspect ratio (Link et al. 1999). The aspect ratio is the ratio of the length to the width of a rod-shaped particle. Interaction of electromagnetic radiation with gold nanorods results in the splitting of the absorbance band into two SPR peaks because of transverse and longitudinal mode of oscillation of electrons. In case of gold nanorods, only longitudinal band is sensitive to aspect ratio. With increase in their aspect ratio, the longitudinal plasmon band ($\lambda \approx 650$–850 nm) starts shifting towards higher wavelength from the visible to NIR region ($\lambda \approx 650$–850 nm), whereas the transverse plasmon band undergoes slight blue shift (Hu et al. 2006; Lee and El-Sayed 2005). The SPR peak of Au nanorods are shown in Fig. 1k.

2.3 *Gold Nanoshells*

Gold nanoshells (GNSs) are basically core shell structures consisting of a dielectric or semiconductor core encapsulated in a thin nanosized metallic shell. In 1989, Neeves and Birnboim reported for the first time that these nanostructures can show absorbance over a wide range of the electromagnetic spectrum (Neeves and Birnboim 1989). Later in 1998, Oldenburg et al. experimentally demonstrated that the optical properties of the GNSs depend on the size of the inner core and thickness of the outer metallic shell (Oldenburg et al. 1998). By decreasing the ratio of shell thickness to the core radius, the researchers showed that the LSPR peak of GNSs shifted from visible to NIR region. A 1 nm decrease in shell thickness resulted in more than 100 nm shift in SPR peak towards NIR region. The SPR peak of GNSs are shown in Fig. 1l. Plasmonic coupling and energy difference between inner core and outer shell are responsible for the shift in SPR position form visible to NIR region in the electromagnetic spectrum.

2.4 *Gold Nanocages and Nanorattles*

Gold nanocages (GNCs) are a novel class of noble metal nanostructures synthesized through galvanic replacement of silver nanocubes by gold precursors, resulting in the formation of hollow structures with a thin porous wall. The morphological changes such as decrease in wall thickness, increase in porosity and decrease in pore size that are induced by the de-alloying process between silver nanocube and gold precursor solution, alters the optical properties of the resultant nanostructures. By increasing the amount of gold precursor during galvanic replacement, SPR peak of GNCs can be tuned from the visible region to NIR region Fig. 1j (Skrabalak et al. 2008).

Gold nanorattles (GNRTs) are another class of noble metal nanostructures that are synthesized by galvanic replacement reactions and consist of core-shell configuration in which the core may or may not move freely inside a porous or non-porous shell, thereby resembling a rattle (Mahmoud 2014; Jaiswal et al. 2014). Mahmoud A. M reported the synthesis of spherical gold nanorattles consisting of a gold nanosphere enclosed inside a non-porous hollow gold nanosphere and observed three LSPR peaks along with a shoulder in the visible and NIR region. Jaiswal et al. reported the synthesis of cubical gold nanorattles consisting of a gold octahedra enclosed inside a porous hollow gold nanocube and observed two LSPR peaks in the visible and NIR region (Jaiswal et al. 2014).

2.5 Gold Nanostars (GNSTs) and Gold Nanopopcorns (GNPs)

Gold nanostars (GNSTs) are multi-branched highly anisotropic nanoparticles having a small inner core covered with multiple sharp tips. Plasmons are located both at the core and also at the tips of nanostars. These plasmons undergo hybridization to produce enhanced electromagnetic field at the tips, that plays a very important role in shaping the optical properties of GNSTs (Barbosa et al. 2010). GNSTs with small core and multibranched sharp tips exhibit distinctive NIR-absorbing plasmon properties (Fig. 1m) and behave like an efficient PTT transducer agent. LSPR shift of GNSTs towards higher wavelength mainly depends on two factors: (i) number of sharp tips present on the core structure (ii) sharpness of the tips. (Rodríguez-Oliveros and Sánchez-Gil 2011)

Gold nanopopcorns (GNPs) are self-assembled anisotropic structures having multiple sharp tips on their outer surface grown by shape templating agent such as CTAB, Ascorbic acid etc. Having these sharp tips on the outer surface and anisotropic behaviour of gold nanopopcorns enable their LSPR peak in NIR region. Figure 1n shows the extinction spectrum of gold nanopopcorns with iron oxide cluster (IOC) core. Thus they have excellent NIR-absorbing plasmon properties and can convert light energy into thermal energy. Due to these properties, gold nanopopcorns are further exploited for PTT, PDT, drug delivery and imaging.

3 Mechanism of Heat Generation and Cell Death Induced by Plasmonic Photothermal Therapy (PTT)

3.1 Mechanism of Heat Generation

When a plasmonic nanoparticle having a significant SPR band in the NIR region (such as gold or silver) is irradiated by a NIR laser, the free electrons that are present on the surface of the plasmonic nanoparticle absorb energy and get excited creating a hot electron cloud. Through the process of electron-electron relaxation and electron-phonon coupling, these hot electrons having temperatures of several thousand-degree kelvin transfer their heat to the gold lattice, thereby increasing the lattice temperature to thousands of kelvin within picoseconds (Link and El-Sayed 2000). This hot lattice finally transfers the heat energy to its surrounding environment through the process of phonon-phonon interactions that results in increased temperature of the surrounding environment by tens of degrees (Link and El-Sayed 2000). Following heat dissipation, the nanoparticles return back to their normal surface temperature. When they are irradiated using a continuous wave laser, the plasmonic nanoparticles continue to dissipate heat into their surrounding environment, which is a requisite for PTT.

3.2 Mechanism of Cell Death

Photothermal therapy can induce cell death either by apoptosis or necrosis or by a combination of both (Abadeer and Murphy 2016; Huang and El-Sayed 2011). Photothermal therapy using plasmonic nanoparticles like gold and silver, induces hyperthermia that leads to a rise in temperature of the surrounding cellular environment. This increase in temperature of around 45–48 °C results in cellular damage including disruption of plasma membrane integrity, denaturation of proteins, damage of nucleic acids, ultimately leading to activation of cell death pathways.

3.2.1 Apoptosis

Apoptosis or programmed cell death as it is commonly known, is a caspase-mediated, controlled, regulated and active process of cell deletion that occurs in healthy adult tissues as well as during normal embryonic development to maintain normal cellular turnover by removing the worn out, decaying, damaged, infected and tumoral cells (Fink and Cookson 2005; Kono and Rock 2008). Apoptosis is characterized by cytoplasmic and nuclear condensation, chromatin and nucleic acid fragmentation, membrane blebbing followed by apoptotic body formation, mitochondrial dysfunction and translocation of specific signal molecules like phosphatidylserine (PS) from inner membrane to outer membrane surface for tagging the cells to be rapidly cleared by the circulating phagocytes (Fink and Cookson 2005; Kono and Rock 2008). However, the cells undergoing apoptosis do not lose their plasma membrane integrity, as a result of which the apoptotic cells or bodies do not release their intracellular contents (e.g. damage-associated molecular patterns/DAMPs) into their extracellular milieu, thereby preventing any unwanted inflammatory or immunogenic response.

3.2.2 Necrosis

On the contrary, necrosis is known to be a passive, accidental, injury or trauma-induced cell death process that inevitably leads to detrimental inflammatory and immunogenic responses (Fink and Cookson 2005; Kono and Rock 2008). Cells undergoing necrosis lose their plasma membrane integrity and rupture to release their intracellular danger signals (DAMPs) into the extracellular environment thereby activating the innate immune system to mount an unwanted immune response.

3.2.3 Secondary or Apoptotic Necrosis

In addition to this, there is another mode of cell death where the process is initiated by apoptosis but terminates by necrosis with unwanted side-effects. Such a process is referred to as secondary or apoptotic necrosis (Kono and Rock 2008) which occurs when the phagocytes fail to clear off the apoptotic bodies rapidly either due to their absence (as is the case during in vitro photothermal studies) or due to some pathological conditions. As a result, the apoptotic bodies start to lose their membrane integrity and release their intracellular DAMPs into the extracellular environment initiating necrosis and immunogenic responses (Fig. 2).

3.2.4 Factors Affecting Plasmonic Photothermal Therapy (PPTT)

In context of photothermal ablation of cancer cells using plasmonic nanostructures, the most common mode of in vitro cell death reported is necrosis (Ramos and Sanchez 2012; von Maltzahn et al. 2009; Lowery et al. 2006). However, there are also reports that suggest that the primary mechanism of cancer cell death by PPTT is apoptosis (Huang et al. 2010; Li and Gu 2010). The type of cell death pathway induced by plasmonic photothermal treatment depends on a number of factors such as nature of the plasmonic nanoparticle (size, shape, surface functionality, optical properties) being used as the photothermal transducer, the intensity or power of the laser, irradiation time and also on the location of the nanoparticle with respect to the target cells (i.e.; on the surface of the cells or inside the cells in cytoplasm or nucleus). Studies have shown that higher laser powers and pulsed lasers induced necrosis, whereas lower laser powers and continuous wave lasers triggered

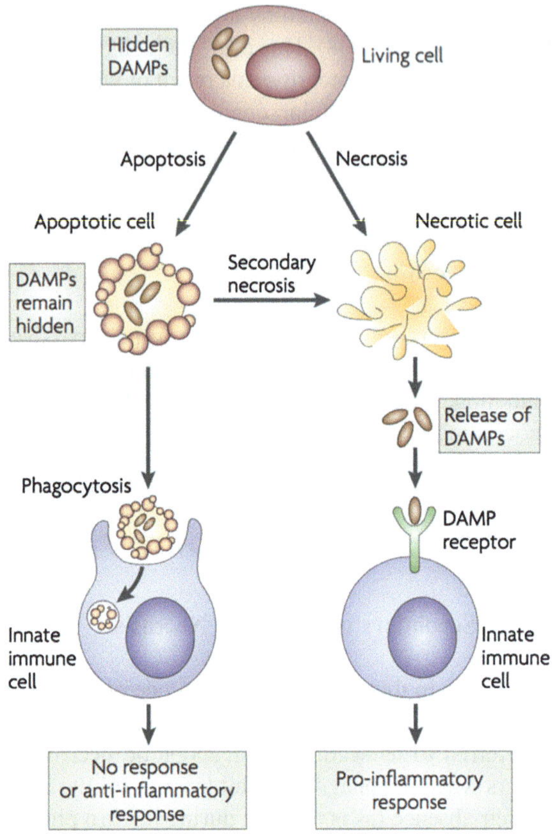

Fig. 2 Schematic showing the different pathways of cell death namely apoptosis, necrosis and secondary necrosis induced by plasmonic photothermal therapy. (Reproduced with permission from Kono and Rock 2008. Copyright 2008 Nature Publishing Group)

apoptosis (Huang et al. 2010). It was also observed that lower exposure time induced apoptosis, while higher exposure time induced necrosis (Huang et al. 2010). These observations have led the researchers to understand that by tuning the optical properties and surface functionalities of the plasmonic nanostructures along with the laser properties, one can carefully induce the target cancerous cells and tissues (both in vitro and in vivo) to specifically undergo apoptosis which is a more "cleaner" mode of cell death as compared to necrosis in terms of therapy, without causing any detrimental immunological side-effects.

3.2.5 Mechanism of Plasmonic Photothermal Therapy (PPTT)

To get a detailed picture about the mechanism of cell death induced by PPTT, de la Fuente and colleagues used gold nanoprisms (NPRs) as photothermal transducers to elucidate the intracellular signaling cascades of cell death that are stimulated in response to NIR laser irradiation in SV40 transformed murine embryonic fibroblast (MEF) cells (Pérez-Hernández et al. 2015). Using a series of carefully designed experiments and assays, the researchers showed apoptosis to be the major mediator of cell death by PTT under low energy irradiation, with necrosis being triggered at later stages once the cells have undergone apoptosis. Annexin V (AnnV) and 7-aminoactinomycin D (7AAD) were used as reporters to identify apoptotic (AnnV+7AAD-) and necrotic (AnnV+7AAD+) cells respectively. During apoptosis, phosphatidylserine (PS) is observed, instead of inner leaflet as in healthy cells, on the external leaflet of the plasma membrane. AnnV binds to PS on apoptotic cells specifically, whereas the fluorophore 7AAD is a DNA intercalating agent that indicates loss of membrane integrity, signifying direct necrosis or apoptotic necrosis. The researchers found that NIR irradiation at a high power of 30 W/cm^2 for 2 min directly triggered necrosis of the NPRs loaded MEF cells, whereas a lower power irradiation of 5 W/cm^2 for even 10 min induced primarily apoptosis which later switched to secondary necrosis with time as evidenced by the increase in the ratio of AnnV+7AAD+ to AnnV+7AAD- cells. FACS analysis of activation of the pro-apoptotic marker Caspase-3 in the NIR irradiated NPRs loaded MEF cells revealed high level of Caspase-3 expression in AnnV+7AAD- cells. These observations confirmed that at initial stages under low energy irradiation, PPTT induced apoptosis in cancer cells in vitro, with secondary necrosis getting triggered in the apoptotic bodies at later stages due to absence of phagocytes in in vitro experimental conditions.

Once apoptosis was established to be the primary mediator of PPTT induced cell death, the researchers next investigated whether the extrinsic or intrinsic pathway of apoptosis was involved in the process. The extrinsic pathway of apoptosis gets stimulated when specific death signals/ligands bind to their cognate "death" receptors on the cell membrane followed by the activation of a cascade of downstream signaling leading to cell death (Fink and Cookson 2005). However, the intrinsic pathway is activated by cell stress/injury such as DNA damage or heat shock which ultimately leads to the disruption of mitochondrial membrane potential followed by cell death (Fink and Cookson 2005). Since PTT mediates cell death by increasing the temperature in and around the target cells, the mitochondrial-mediated intrinsic pathway of apoptosis seems to be more appropriate for inducing cell death via PTT. To scientifically prove and validate this hypothesis, the researchers performed detailed studies using wild type MEFs (WT-MEFs) and mutant MEFs (KO-MEFs) cells that lacked expression of some of the key components of both intrinsic and extrinsic pathways like Bax/Bak, Caspase-3, Caspase-9 and Bid, and compared the results. The WT-MEFs upon irradiation with NIR laser showed AnnV+7AAD- staining pattern indicating apoptosis. On the contrary, all types of KO-MEFs strongly resisted apoptosis and cell death following irradiation as evidenced by lack of AnnV staining, thereby showing the requirement and involvement of Bax/Bak, Caspase-3, Caspase-9 and Bid in PTT induced cell death. Bax/Bak, Caspase-9 and Caspase-3 are all components of the intrinsic pathway of apoptosis that are activated sequentially to mediate cell death. The activated Bax/Bak undergoes oligomerization in the outer membrane of mitochondria which creates pores in the membrane thereby causing the release of cytochrome C into the cytoplasm. Cytochrome C then forms a complex with Apaf-1, dATP and pro-caspase 9 to form apoptosome. Formation of apoptosome leads to the activation of Caspase-9, which results in the cleavage and activation of Caspase-3 which executes the final stages of cell death (Fink and Cookson 2005). So, the resistance of the KO-MEFs to apoptosis upon NIR irradiation clearly demonstrated the involvement of the intrinsic pathway of apoptosis in PTT mediated cell death. However, Bid is a component of the extrinsic pathway of apoptosis which gets activated by Caspase-8 into tBid upon binding of ligands to the death receptors. The activated tBid then induces oligomerization of Bax/Bak in the mitochondrial outer membrane, thereby activating the intrinsic pathway (Fink and Cookson 2005). Hence, the resistance of Bid KO-MEFs to apoptosis upon NIR irradiation clearly showed that there was some crosstalk between the components of intrinsic and extrinsic pathway of apoptosis in PTT-mediated cell death. Although, the main cause of Bid activation in PTT-mediated apoptosis was not fully deciphered and was thought to be the result of lysosomal cathepsin mediated Bid cleavage that gets released from lysosomes after photothermal heating. Another key feature of intrinsic pathway of apoptosis is the loss of membrane potential ($\Delta\Psi_m$) of mitochondrial outer membrane. Flow cytometric analysis of $\Delta\Psi_m$ of mitochondrial outer membrane revealed that all KO-MEFs maintained intact $\Delta\Psi_m$, whereas 50% of the WT-MEFs lost their $\Delta\Psi_m$, again establishing the role of intrinsic pathway of apoptosis in PTT-mediated cell death.

Based on these observations, the researchers concluded that PTT using plasmonic nanostructures such as gold nanoprisms induced cell death in cancer cells in vitro primarily via the intrinsic pathway of apoptosis under low energy irradiation and less exposure time (Fig. 3). By increasing the laser power and irradiation time, necrosis could be selectively induced in place of apoptosis.

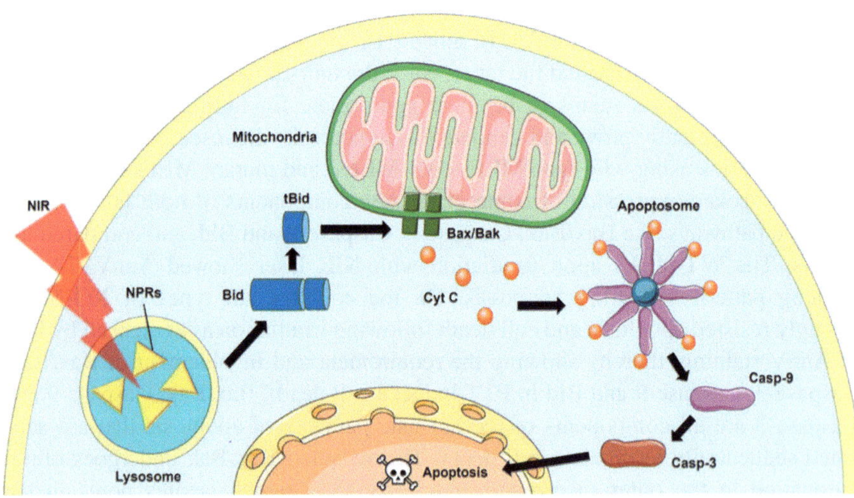

Fig. 3 Schematic showing the proposed mechanism of apoptosis induced by gold nanoprism mediated photothermal therapy. (Reproduced with permission from Melamed et al. 2015. Copyright 2015 American Chemical Society)

4 Plasmonic Photothermal Therapy by Different Gold Nanostructures

Gold nanoparticles of different morphologies have been widely exploited for the photothermal destruction of cancer cells and tissues both in-vitro and in-vivo. For efficient detection and successful killing of cancer cells, plasmonic PTT has been combined with several other anticancer therapies such as chemotherapy, photodynamic therapy and immunotherapy. The theranostic potential of these combined strategies can further be extended by incorporation of multimodal imaging capabilities. Such combinatorial anti-cancer therapeutic approaches have improved the efficacy and outcome of these strategies. Tables 1, 2, 3, and 4 summarizes examples of photothermal therapies, combined chemo-photothermal therapies, combined photothermal-photodynamic therapies (PTT/PDT) and immunotherapy, and multimodal imaging with PTT using different gold nanostructures respectively.

4.1 Gold Nanospheres

Gold nanospheres usually have their LSPR peak in the visible region of the electromagnetic spectrum, and hence are not suitable for NIR mediated PTT therapy. However, due to their strong absorbance in the visible region, gold nanospheres have been exploited for PTT using visible light emitting pulsed lasers and continuous lasers. (Pitsillides et al. 2003) were the first to report visible PPTT using anti-CD8 functionalized gold nanospheres targeted to T lymphocytes that caused thermal ablation of more than 90% of the cells upon irradiation with nanosecond visible pulsed laser (Pitsillides et al. 2003). Later, gold nanosphere mediated visible PPTT for targeting cancer cells in vitro were reported by Zharov et al. (2005) and El-Sayed group (2006), and Huang et al. (2006) by using nanosecond pulsed laser and continuous argon ion laser respectively. The application of visible PPTT is limited to only in vitro purpose because of very poor penetrability and high absorption of visible light by haemoglobin and water molecules present in tissues.

4.2 Gold Nanorods (GNRs)

For efficient photothermal action, it is mandatory that gold nanoparticles should have significant absorbance in the NIR region (650–900 nm) (Huang et al. 2006). As the shape of the particles changes from sphere to more elongated forms along one axis e.g. gold nanorods, the plasmon resonance band splits into two because of transverse and longitudinal oscillation of electrons. With increase in their aspect ratio, longitudinal plasmon band shift towards higher wavelength i.e. red shift (Link et al. 1999; Liz-Marzán 2004; Link and El-Sayed 2005). Gold nanorods with

Table 1 Photothermal therapy using different gold nanostructures

Gold nanostructure	Target cell line/ animal model	Nanoparticle design	Therapeutic outcome	References
Nanorods (GNR)	RAW 264,7	CD11b antibody functionalized GNR	Targeted in vitro photothermal destruction of murine macrophage cells	Pissuwan et al. (2007)
	DU145 cell	Aptamer functionalized GNR	In vitro photothermal ablation of prostate cancer stem cells	Wang et al. (2013)
	Malignant fibrous histiocytoma-like *KM-Luc/GFP cells*		Inhibition of lymph node metastasis	Okunoet al. (2013)
	Nu/nu mice bearing HSC-3 tumors		Dramatic reduction in tumour size following PPTT	Dickerson et al. (2008)
	Athymic nude mice bearing SCC7 tumors	Chitosan-conjugated, pluronic-based GNR	Complete resorption of NIR irradiated tumors	Choi et al. (2011)
Nanoshell (GNS)	KB cells	Anti-HER2 functionalized GNS	Targeted in vitro photothermal destruction of breast cancer cells	Sheikholeslami et al. (2011)
	BEL-7404 cells	A54 peptide functionalized GNS	Targeted in vitro photothermal destruction of liver cancer cells	Lu et al. (2010)
	PC-3, C4–2 cells	GNSs	Targeted in vitro photothermal destruction of prostate cancer cells	Stern et al. (2008)
	SKBR3 cells	Anti-HER2 functionalized GNS	Targeted in vitro photothermal destruction of breast cancer cells	Lowery et al. (2006)
	CB17-rkdcscid/J mice bearing canine TVT	GNSs	Irreversible tissue damage and tumor shrinkage using NIR irradiation of 35 W/cm^2 under magnetic resonance guidance	Hirsch et al. (2003)

(continued)

Table 1 (continued)

Gold nanostructure	Target cell line/ animal model	Nanoparticle design	Therapeutic outcome	References
Nanocages (GNC)	SKBR3 cells	Anti-HER2 functionalized GNC	Targeted in vitro photothermal destruction of breast cancer cells	Chen et al. (2007)
	LNCaP cells	A9RNA aptamer-conjugated GNC-carbon nanotube hybrids	Targeted in vitro photothermal destruction of 95% prostate cancer cells at 2 W/cm^2 NIR laser power	Khan et al. (2012)
	Athymic nu/nu mice bearing U87MG-wtEGFR tumors	Pegylated GNC	Irreversible photothermal destruction of tumors in vivo	Chen et al. (2010a, b)
Nanostars (GNST)	BT549 cells	TAT peptide functionalized GNST	Targeted in vitro photothermal destruction of breast cancer cells at ultralow NIR laser power intensity ($\geq$0.2 W/cm^2)	Yuan et al. (2012a, b)
	SKOV3 cells	Anti-HER2 functionalized GNST	Targeted in vitro photothermal destruction of breast cancer cells within 5 min of NIR irradiation (660 nm, cw laser,38 W/cm^2)	Van De Broek et al. (2011)
	SKBR3 cells	Bare non-targetted GNST	Photothermal destruction of breast cancer cells within 5 min of NIR irradiation (980 nm, cw laser,15 W/cm^2)	Yuan et al. (2012a, b)
Nanoaggregates	HeLa and B16F10 cells	pH-responsive 'smart' gold nanoparticle (SAN) system	In vitro photothermal ablation of the cancer cells using a NIR laser of 6.5 W/cm^2 or more for 10 min	Nam et al. (2009)
	B16F10 cells	Smaller sized SANs (<6 nm)	In vitro photothermal ablation of the cancer cells using a NIR laser of less than 20 W/cm^2	Hwang et al. (2014a, b)

Table 2 Chemo-photothermal therapy using different gold nanostructures

Gold nanostructure	Target cell line/ animal model	Nanoparticle design	Therapeutic outcome	References
Nanorods (GNR)	Huh-7 cells	Doxorubicin loaded hyaluronic acid-conjugated GNR-graphene oxide core/ shell nanocomposites (DOX@NGO-HA-GNR)	1.5–4 times more efficient killing of hepatoma cells in vitro as compared to chemotherapy or PTT alone	Xu et al. (2013)
	4T1 cells	Doxorubicin loaded DNA wrapped GNRs	Significant breast tumor growth inhibition and prevention of lung metastasis	Wang et al. (2014)
	4T1 cells	Doxorubicin loaded thermo- and pH- responsive polymer functionalized GNRs	Significant breast tumor growth inhibition and prevention of lung metastasis	Zhang et al. (2014)
Nanoshell (GNS)	HeLa cells	Doxorubicin and SPIO loaded cholesteryl succinylsilanenanomicelles, coated with GNS (CDF-GNS)	90% killing of cancer cells in vitro by synergistic effect of magnetic field guided chemo-photothermal therapy	Ma et al. (2012)
	SMMC-7721 cells	Doxorubicin loaded liposome/silica/GNS	Highly efficient killing of cancer cells in vitro by synergistic effect of chemo- and NIR photothermal therapy	Wu et al. (2011)
	Swiss mice bearing Hey tumors	EphB4 receptor targeted doxorubicin loaded hollow gold nanospheres (DOX@ HAuNs)	Significant tumour growth suppression	You et al. (2012)

(continued)

Table 2 (continued)

Gold nanostructure	Target cell line/ animal model	Nanoparticle design	Therapeutic outcome	References
Nanocages (GNC)	MDA-MB-435 cells	Doxorubicin loaded SV119-GNCconjugates	In vitro eradication of breast cancer stem cells	Sun et al. (2014a, b)
	HeLa cells	Doxorubicin loaded, thermoresponsive PNIPAM polymer functionalized GNC@mesoporous silica core/shell nanocomposites	Highly efficient killing of cancer cells in vitro by synergistic effect of chemo- and NIR photothermal therapy	Yang et al. (2013)
	MCF7 cells	pH responsive Fe_3O_4@ CaP-capped, Doxorubiicn loaded GNCs	47% killing of breast cancer cells in vitro by synergistic effect of chemo- and NIR photothermal therapy	Shi et al. (2012)
Nanostars (GNST)	MDA-MB-231 cells	Doxorubicin loaded, cyclic RGD peptide conjugated GNSTs	Targeted in vitro chemo-photothermal destruction of >90% breast cancer cells using NIR laser (765 nm, cw, 1 W/cm^2, 10 min)	Chen et al. (2013)
Nanoaggregates	Nu/nu nude mice bearing B16F10 melanoma cells	Doxorubicin loaded, pH-responsive 'smart' gold nanoparticle (SAN) system (SANDCs)	Significant accumulation of nanoaggregates in tumours with efficient reduction in tumor size as compared to SANs alone or a mixture of SANs and DOX	Nam et al. (2013a, b)

Table 3 Combined PDT/PTT and PTT/Immunotherapy using different gold nanostructures

Gold nanostructure	Target cell line/animal model	Nanoparticle design	Sensitizer	Therapeutic outcome	Ref
PTT+PDT					
Nanorods (GNRs)	CCRF-CEM cells	Sgc8 aptamer functionalized GNRs	Ce6	Combined PDT and PTT resulted in 60% cancer cell death as compared to 20% with only PDT and 36% with only PTT	Wang et al. (2012)
	Balb/c-nu mice bearing SCC7 squamous cell carcinoma tumours	RRLAC-peptide functionalized GNRs	$AlPcS_4$	Combined PDT and PTT resulted in 96% reduction in tumour volume as compared to 79% with only PDT and 25% with only PTT	Jang et al. (2011)
Nanocages (GNCs)	HeLa cells	Lipid coated GNCs	Hypocrellin B	Combined PDT and PTT resulted in 82.6% cancer cell death as compared to 35.4% with only PDT and 45.5% with only PTT	Gao et al. (2012)
	HeLa cells, mice bearing Ehrlich carcinoma tumors	Silica coated GNCs	Yb-2,4-dimethoxyhematoporphyrin	IR-luminiscence based detection of tumours in vivo along with combined PDT and PTT mediated killing of cancer cells both in vitro and in vivo	Khlebtsov et al. (2011)
Nanopopcorns (GNPs)	KB31 and SKBR3 cells	Iron oxide cluster core containing GNPs	SiNC	Magnetic field guided combined PTT (0.55 W/cm^2) and PDT mediated killing of cancer cells	Bhana et al. (2015)

PTT+Immunotherapy

Gold nanostructure	Target cell line/animal model	Nanoparticle design	Therapeutic outcome	Ref
Nanoshells (GNSs)	C57BL/6J mice bearing B16F10 tumours	PEGylated hollow GNSs	PTT combined with adoptive transfer of tumour directed activated T cells resulted in preventing tumour progression and inhibited outgrowth of lung metastasis	Bear et al. (2013)

Table 4 Combined photothermal therapy and multimodal imaging using different gold nanostructures

Gold nanostructure	Target cell line/ animal model	Nanoparticle design	Treatment strategy	References
Nanorods (GNRs)	Nude mice bearing MDA-MB-435 tumors	PEGylated GNRs	CT imaging guided photothermal destruction of tumours	Von et al. (2009)
	HOC-313 clone 8, HSC-3	Anti-EGFR conjugated GNR	Targeted in vitro PTT	Huang et al. (2006)
	Athymic nude mice bearing U87MG tumour	^{64}Cu radiolabelled, RGD peptide modified GNRs	PET imaging guided photothermal destruction of tumours; increase in tumour temperature by 27 °C and decrease in tumour size by 50%	Sun et al. (2014a, b)
	OSCC15, MDA-MB-231, MCF7 cells	Anti-EGFR antibody conjugated, polydopamine coated GNRs	OCT guided targeted PTT	Black et al. (2013)
	Atherosclerotic plaques	Silica GNRs	PA imaging and PTT	Yeager et al. (2014)
Nanoshells (GNS)	Nude mice bearing A431 tumors	C225 antibody conjugated, SPIONs encapsulated GNSs	Targeted magnetic resonance imaging (MRI) guided photothermal ablation of tumours at a laser power of 36 W/cm^2	Melancon et al. (2011)
	Balb/c nude mice bearing HT-1080 tumours	PFOB and SPIONs encapsulated PLA nanocapsules coated with PEGylated GNS	Bimodal US/MRI guided photothermal ablation of tumours in nude mice with a 82.2% reduction in tumour growth as compared to control	Ke et al. (2014)
	C57BL/6 mice bearing B16F10 tumours	Fe_3O_4/Ag core encapsulated hollow GNSs	MRI based detection of GNS loaded tumours in vivo; intratumoral injection of nanoparticles resulted in complete photothermal ablation of tumours after 13 days and intravenous injection only halted growth	Lin et al. (2014)

(continued)

Table 4 (continued)

Gold nanostructure	Target cell line/ animal model	Nanoparticle design	Treatment strategy	References
Nanorattles	MCF-7 cells	Estrogen receptor alpha antibody functionalized STHGNRs	SERS based detection and photothermal ablation using NIR laser (785 nm, 5.24 W/cm^2, 5 min)	Chen et al. (2014)
Nanopopcorns (GNPs)	LNCaP cells	A9RNA aptamer and Anti-PSMA antibody conjugated GNPs	SERS-based monitoring of targeted photothermal ablation of prostate cancer cells	Lu et al. (2010)
	SKBR3 cells	S6 aptamer conjugated GNP-SWCNT hybrids	SERS based highly specific detection of breast cancer cells followed by PTT (785 nm, 1.5 W/cm^2, 10 min)	Beqa et al. (2011)
Nanoaggregates	B16F10 mouse melanoma cells	MBA-conjugated SANs	Simultaneous SERS based imaging/diagnosis of cancer cells along with photothermal ablation of detected cells	Jung et al. (2013)
	HeLa cells	SANs	PT-OCT guided PTT	Xiao et al. (2013)

increase in aspect ratio show more absorption in the NIR region making them suitable for molecular imaging and PPTT. Huang et al. carried out an in-vitro study to determine both photothermal as well as molecular imaging properties of gold nanorods. They synthesized gold nanorods using seed-mediated growth method and conjugated it with anti-epidermal growth factor receptor monoclonal antibodies (anti-EGFR) by using the mechanism of electrostatic physisorption interaction. These anti-EGFR antibody-conjugated nanorods were incubated with malignant (HOC 313 clone 8 and HSC 3) as well as non-malignant epithelial cell line (HaCat). Due to overexpression of EGFR in malignant cells, anti-EGFR antibody-conjugated nanorods bound specifically to malignant cells and strongly scattered orange to red light in the dark field which enabled the diagnosis and visualization of the malignant cells from the non-malignant cells. They also noticed that, when exposed to a continuous red laser at 800 nm, EGFR overexpressing malignant cells required less than half of the laser energy as compared to normal cells to induce cell death (Huang et al. 2006).

Pissuwan et al. designed a CD11b antibody functionalized GNR and showed targeted in-vitro photothermal destruction of RAW 264.7 cells upon exposure to laser (5 mW Cw diode laser, 650 nm, 10 min) which resulted in 81% cancerous cell death in comparison to only 0.9% death of normal cells (Pissuwan et al. 2007). Wang et al. demonstrated targeted NIR mediated photothermal ablation of prostate cancer cells (DU145) along with cancer stem cells using aptamer (CSC1 and

CSC13) functionalized gold nanorods. CSC1 specifically targeted prostate cancer cells, while CSC13 targeted the cancer stem cells. They observed highly efficient cell destruction following laser irradiation. Okuno et al. performed an in-vivo study to evaluate the photothermal property of GNRs to treat lymph node tumors. They injected KM-Luc/GFP cells into the proper axillary lymph nodes (proper-ALNs) of MXH10/Mo–lpr/lpr mice to induce the formation of malignant fibrous histiocytoma like tumors. Upon direct injection of gold nanorods into the proper-ALNs tumors followed by irradiation with a NIR laser (1064 nm, cw, 1.5 W/cm^2, 3 min), they observed an increase in temperature upto 50 °C in the proper-ALNs tumors with significant inhibition of tumor growth and metastasis after day 3 (Okuno et al. 2013). Xu et al. developed DOX loaded hyaluronic acid conjugated GNR-graphene oxide core/shell nanocomposites and showed synergistic chemo-photothermal activity for killing of Huh-7-cells in-vitro. They observed 1.5-fold increase in killing of hepatoma cells as compared to chemotherapy alone and a fourfold increase in cancer cell death as compared to PTT alone (Fig. 4) (Xu et al. 2013). Yang et al. designed aptamer functionalized mesoporous silica coated gold nanorods and loaded them with DOX. They observed 97% killing of MCF 7 breast cancer cells in-vitro by synergistic effect of chemo-photothermal therapy upon irradiation with NIR laser (808 nm, CW, 1.2 W/cm^2, 10 min) as compared to 40% cell death by chemotherapy alone and 5% cell death PTT alone (Yang et al. 2012). Zhang et al. and Wang et al. demonstrated significant in-vivo inhibition of breast tumor growth along with prevention of lung metastasis by using DOX loaded thermoresponsive polymer functionalized GNRs and DOX loaded DNA wrapped GNRs respectively (Zhang et al. 2014; Wang et al. 2014). In addition to PTT alone and combined chemo-photothermal therapy, GNRs have also been used for combined PTT/PDT (Wang et al. 2012; Jang et al. 2011) and multimodal imaging guided PTT (Von et al. 2009; Sun et al. 2014a, b; Black et al. 2013; Yeager et al. 2014).

4.3 Gold Nanoshells (GNSs)

In 2003, Hirsch et al. were the first to report GNSs mediated NIR thermal ablative therapy under the guidance of magnetic resonance. In their work they used NIR active silica-gold nanoshells having diameter of inner core 110 ± 11 nm covered in a 10 nm gold shell which enabled its peak position at 820 nm in absorption spectrum. Further, they combined thiolated PEG (SH-PEG) on GNSs surface to make it stable in the physiological environments. Then they incubated it with Human breast epithelial carcinoma SK-BR-3 cells at 37 °C. After incubation, cells were exposed to NIR light (coherent, 820 nm, 35 W/cm^2, 7 min) to induce plasmonic photothermal ablation. To verify cell viability, cell membrane damage and nanoshell binding of the NIR light exposed SK-BR-3 cells, they stained it with three different dyes. (i) Calcein AM which reveals about cell viability, (ii) for cell membrane damage they used aldehyde-fixable fluorescein dextran dye (iii) silver enhancement stain used to determine nanoshell binding by using phase contrast microscopy.

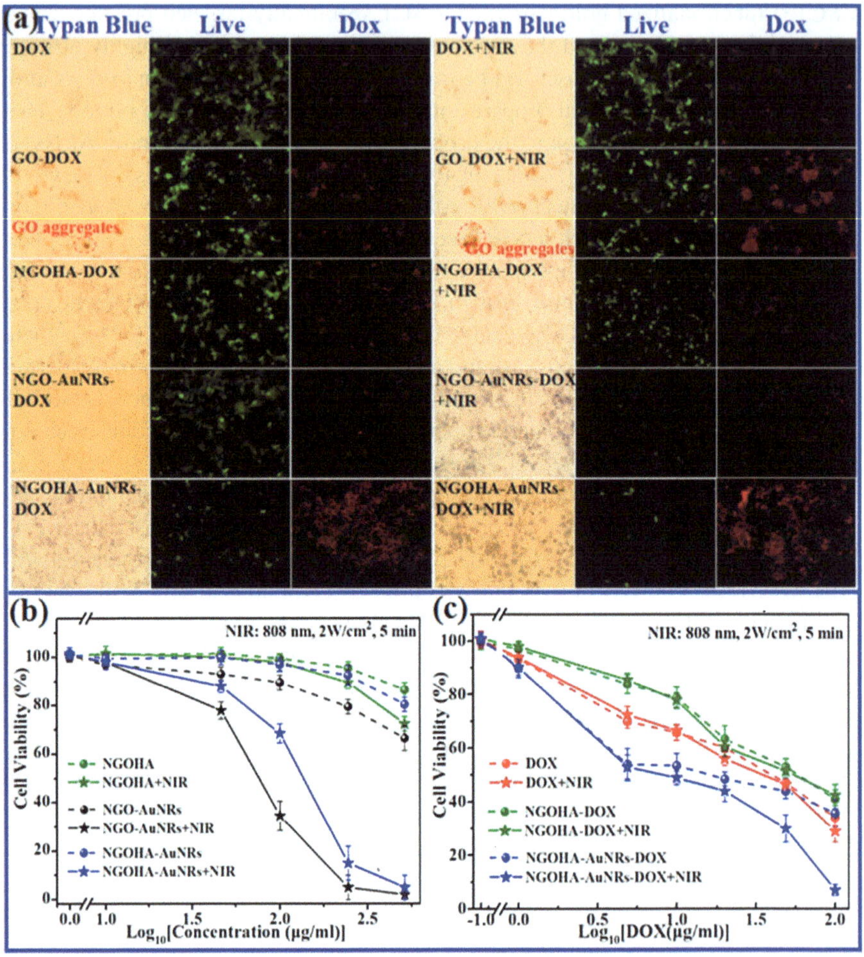

Fig. 4 Cytotoxicity of hepatoma Huh-7 cells under different treatments. (**a**) Fluorescence microscopy images of Huh-7 cells treated with chemotherapy or chemophototherapy. The dead cells were stained with trypan blue and the live cells with calcein. (**b**) Cytotoxicity of NGOHA, NGO-AuNRs, and NGOHA-AuNRs to Huh-7 cells with or without NIR irradiation. (**c**) Cytotoxicity of NGOHA-DOX, NGOHA-AuNRs-DOX, and free DOX to Huh-7 cells with or without NIR irradiation. The relative percentage of control cells not exposed to the delivery system (nontreated) was used to represent 100% cell viability. Data represent mean values for $n = 3$, and the bars are standard deviations for the means. (Reproduced with permission from Xu et al. 2013. Copyright 2013 American Chemical Society)

They also exploited these PEG-passivated NIR-absorbing gold–silica nanoshells to explain in-vivo photothermal ablation therapy in CB17-Prkd c SCID/J mice bearing Canine TVT tumours in the right and left hind leg. The PEG-passivated NIR-absorbing gold–silica nanoshells were injected interstitially into the tumour sites followed by exposure to NIR light (820 nm, 4 W/cm^2, 5-mm spot diameter, <6 min).

The temperature was monitored by using phase-sensitive, fast-spoiled gradient-echo MRI. By using Magnetic resonance temperature imaging (MRTI), they found that the average increase in temperature was 37.4 ± 6.6 °C on NIR exposure for 4–6 min. Histological evaluation of NIR irradiated nanoshell treated tumours revealed coagulation, cell shrinkage, and loss of nuclear staining which are common markers of thermal damage (Hirsch et al. 2003). Wu et al. reported a synergistic effect of chemo-photothermal therapy by using Doxorubicin loaded liposome/SiO_2/Au nanoshells. They demonstrated that DOX-loaded Au nanoshells are very much efficient to kill liver cancer cell (SMMC-7721) when exposed to NIR laser at a wavelength of 808 nm (Wu et al. 2011). You et al. investigated the in-vitro and in-vivo chemo-photothermal activity of DOX loaded hollow nanospheres by targeting EphB4 receptors in tumours and observed significant tumor growth suppression (You et al. 2012). Ma et al. demonstrated 90% killing of HeLa cells in-vitro by synergistic effect of magnetic field guided chemo-photothermal therapy using DOX-SPIO loaded cholesteryl succinyl silane nanomicelles coated with GNSs (Ma et al. 2012).

Stern et al. showed that the NIR active GNSs were very much effective for the treatment of prostate cancer. In their report, they showed ablation of two human prostate cancer cell lines PC-3 and C4–2 with the help of NIR active GNSs (Stern et al. 2008). Lowery et al. synthesized anti-HER2 conjugated GNSs and termed them as immunonanoshells which they further exploited for targeted tumor cell destruction with the help of NIR light. They used SK-BR-3 breast carcinoma cells for in-vitro photothermal study and revealed that immunonanoshells and NIR light can selectively induce cell death to breast carcinoma cells (Lowery et al. 2006). Liu et al. demonstrated targeted in-vitro photothermal destruction of BEL-7404 liver cancer cells using A54 peptide functionalized GNS. They used acridine orange staining to assess the viability of the BEL-7404 cells. The researchers irradiated the BEL-7404 cells with NIR laser for different time duration (1 min, 3 min and 5 min) and observed absence of cell organelles from the cytoplasm along with reduction in cell volume after 1 min of irradiation. But they found few microvilli on the cell surface and nucleolus was still present. After 3 min of irradiation microvilli disappeared from the cell surface and after 5 min they observed blebs in the cytoplasm of cell which indicated severe cell injury (Liu et al. 2009a, b). Sheikholeslami et al. used anti-HER2 antibody-tagged gold-silica nanoshells and examined its molecular imaging along with photothermal potency for the treatment of oral squamous cell carcinoma (KB cells). They irradiated KB cells with NIR laser (820 nm, 4 W/cm^2, 2 min) and observed more than 69.4% selective destruction of KB cells (Sheikholeslami et al. 2011). Ke et al. showed a bimodal US/MRI guided photothermal ablation of HT-1080 tumours in nude mice using PFOB and SPIONs (US and MRI agents) encapsulated PLA nanocapsules coated with PEGylated GNS followed by NIR laser irradiation (808 ± 10 nm Cw, 1.30 W cm^{-2}, 10 min) (Ke et al. 2014). In addition to this, MRI guided targeted photothermal ablation of tumors in-vivo using gold nanoshells have also been demonstrated (Melancon et al. 2011; Lin et al. 2014). Bear et al. demonstrated combined PTT/immunotherapy through adoptive transfer of tumor directed activated T cells into C57BL/6J mice bearing B16F10 tumors by using PEGylated hollow GNSs followed by NIR laser irradiation (coherent diode array laser, 808 nm, 3 W/cm^2, spot diameter = 8 mm, 3 min) that prevented further tumor progression and inhibited lung metastasis (Bear et al. 2013).

4.4 Gold Nanocages (GNCs) and Gold Nanorattles (GNRTs)

Hollow structure and porous thin wall of GNCs enable them to perform certain biomedical applications such as encapsulation of contrast enhancement agents for imaging, drugs for chemotherapy, surface functionalization with targeting ligands for cancer cell targeting and photothermal therapy. Chen et al. in their work synthesized GNCs with an edge length of 45 nm which enabled its peak position at 810 nm and demonstrated its targeted PTT activity on breast cancer cells (SK-BR-3) by conjugating GNCs with anti-HER-2 antibodies. To demonstrate the PTT action of GNCs, they incubated it with SKBR-3 cell line followed by NIR irradiation (Ti:sapphire laser with a 82 MHz) for 5 min. Their in-vitro study revealed that the immuno-GNCs were capable of selectively destructing cancer cells at a laser power density of 1.5 W/cm^2 which was much lower as compared to GNSs (35 W/cm^2) and GNRs (10 W/cm^2) (Fig. 5) (Chen et al. 2007; Au et al. 2008). Khan et al. synthesized gold nanocage-carbon nanotube hybrids conjugated with A9RNA aptamer for targeted in-vitro photothermal ablation of prostate cancer cells (LNCaP). Upon irradiation with Nd: YAG NIR laser (2 W cm^{-2}, 1064 nm, 10 min) they observed 95% LNCaP cells death (Khan et al. 2012).

Chen et al. further evaluated the potential of PEGylated GNCs for in-vivo photothermal treatment in U87MG-wtEFGR tumour bearing mice by irradiating with a NIR laser (λ = 808 nm) at a power density of 0.7 W/cm^2 for 10 min. Histological examination and PET scan revealed the irreversible tumour cell death by GNCs mediated PTT (Chen et al. 2010a, b). Shi et al. developed a novel nanocontainer using a hollow structure of GNCs capped with calcium phosphate (CaP) coated Fe_3O_4 nanoparticles (Fe_3O_4@CaP-capped GNCs) for drug delivery, drug targeting and PTT. DOX was loaded into the hollow nanocages and was selectively released into the tumour microenvironment upon solubilisation of CaP under acidic conditions. Using this nanocontainer, the group demonstrated 47% destruction of MCF7 breast cancer cells in vitro by synergistic effect of chemo- and NIR photothermal therapy which was observed only 23% without NIR irradiation (Shi et al. 2012). Gold nanocage mediated combined chemo-photothermal therapy has also been reported by other groups (Sun et al. 2014a, b; Yang et al. 2013). Khlebtsov et al. fabricated PDT sensitizer (Yb-2,4- dimethoxy-hematoporphyrin) with silica-coated GNCs and demonstrated photodynamic, luminescent and PTT properties of the nanoconjugate in vitro in HeLa cells and in vivo in mice bearing Ehrlich carcinoma tumors (Khlebtsov et al. 2011). In addition to this, Gao L et al. also demonstrated combined in-vitro PTT/PDT cancer therapy using hypocrellin-GNCs in HeLa cells. Firstly, they performed PTT and PDT experiments alone by using 790 nm two photon laser (85.5 pJ per pulse, 300 s) and observed that PDT and PDT alone lead to cell viability only up to 54.5% and 64.6% respectively. When they combined PDT

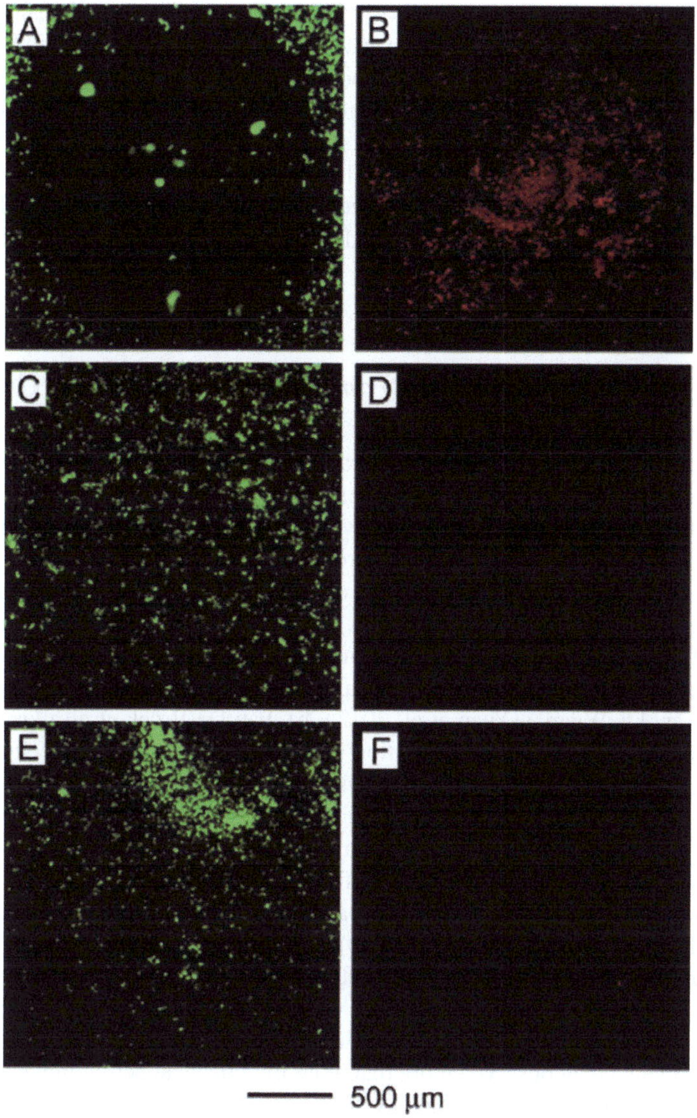

Fig. 5 SK-BR-3 breast cancer cells that were treated with immuno gold nanocages and then irradiated by an 810 nm laser at a power density of 1.5 W/cm^2 for 5 min showed a well-defined circular zone of dead cells as revealed by: (**a**) calcein AM assay (where green fluorescence indicates the cells were live), and (**b**) ethidium homodimer-1 (EthD-1) assay (where red fluorescence indicates the cells were dead). In the control experiment, cells irradiated under the same conditions but without immuno gold nanocage treatment maintained viability, as indicated by (**c**) calcein fluorescence assay and (**d**) the lack of intracellular EthD-1 uptake. Cells treated with immuno gold nanocages but irradiated at a lower power density (0.5 W/cm^2) for 5 min remained alive, as shown by (**e**) calcein fluorescence assay and (**f**) the lack of intracellular EthD-1 uptake. (Reproduced with permission from Chen et al. 2007. Copyright 2007 American Chemical Society)

and PTT and found cell viability up to 17.4% which was much better than the alone PDT and PTT (Gao et al. 2012). Chen et al. demonstrated a SERS guided PTT strategy using estrogen receptor-alpha antibody (ERa) functionalized SERS tag hidden gold nanorattles (STHGNRs). The researchers reported detection of MCF-7 breast cancer cells using SERS and also photothermal ablation of the cells using these STHGNRs upon irradiation at a 785 nm laser at a power density of 5.24 W/cm^2 for 5 min and cellular viability of MCF-7 cells was determined by using MTT assay and found that STHGNRs with laser irradiation are better theranostic agent for cancer treatment (Chen et al. 2014).

4.5 Gold Nanostars (GNSTs) and Gold Nanopopcorns (GNPs)

Yuan et al. demonstrated both in-vitro and in-vivo PEGylated GNSTs mediated photothermal ablation of breast cancer cell line (SKBR3) and dorsal window chambers in female CD-1 nude mice respectively. Their in-vitro study revealed that the PEGylated GNSTS were capable of producing localized photothermal ablation along with particle tracking within 5 min of a 980 nm continuous wave laser irradiation at a power density of 15 W/cm^2. For in-vivo, when they systemically administered PEGylated GNSTs in the animal model photothermal ablation was observed at the dorsal window chamber within 10 min of laser irradiation (785-nm continuous-wave laser, 1.1 W/cm^2) (Yuan et al. 2012a, b). Van De Broek et al. illustrated the target specific photothermal destruction of HER-2-positive SKOV3 cells by anti-HER-2 conjugated GNSTs using a 660-nm continuous wave NIR laser at a power density of 38 W/cm^2 for 5 min (Van De Broek et al. 2011). Yuan et al. used TAT peptide-functionalized GNSTs for in-vitro targeted PTT. They used BT549 breast cancer cell line and incubated with TAT-GNSTs followed by irradiation with an ultra-low power NIR laser (≥0.2 W/cm^2) and observed a distinct photothermal ablation of BT549 cells (Yuan et al. 2012a, b). Chen et al. developed a tumour targeted nanoconstruct by conjugating cRGD with GNSTs along with an anticancer drug (DOX) and evaluated the synergistic effect of PTT and chemotherapy on MDA-MB-231 breast cancer cell line. Under laser irradiation for 10 min (765-nm cw, 1.0 W/cm^2) they observed more than 90% of anticancer activity on cells which was much higher than DOX alone or GNST alone (Chen et al. 2013).

Bhana et al. synthesized NIR active hybrid GNPs nanostructures using self-assembled iron oxide cluster core and coated with silicon 2,3-napthalocyannie dihydroxide (SiNC) photosensitizer. These GNPs showed excellent synergistic PTT (0.55 W/cm^2) and PDT activity in KB-3-1 and SKBR-3 breast cancer cell lines under the guidance of magnetic field (Bhana et al. 2015). Lu et al. showed a significant mortality of human prostate cancer cells (LNCap) after 30 min of laser irradiation (758-nm cw laser, 12.5 W/cm^2) using GNPs conjugated with A9 RNA aptamers and anti-prostate-specific membrane antigen (PSMA) antibodies (Fig. 6) (Lu et al. 2010). Beqa et al. synthesized a hybrid structure consisting of GNPs decorated with single wall carbon nanotubes (SWCNTs) that showed enhanced in-vitro photothermal destruction of breast cancer line (SKBR3) upon exposure of NIR laser (785 nm, 1.5 W/cm^2, 10 min) (Beqa et al. 2011).

4.6 Gold Nanoaggregates

In a colloidal solution, electrostatic interaction between nanoparticles may promote their aggregation. External agents can also be added to promote aggregation between gold particles. For example, pyridine is used to promote the aggregation between gold particles through nitrogen lone-pair electrons (Khandelia et al. 2014; Messina et al. 2011). This aggregation can shift the plasmon resonance band towards the higher wavelength i.e. red shift and sometimes a second absorption peak at a higher wavelength is also observed, which makes these nanoaggregates efficient NIR active PTT agents.

Gold nanospheres usually have their LSPR peak in the visible region of the electromagnetic spectrum and hence cannot be used for NIR mediated PTT. However, aggregates or assemblies of gold nanospheres have been found to show absorbance in the NIR region that can be exploited for PTT purposes. Nam et al. synthesized a new class of gold nanostructures which they named as "smart" gold nanoparticles (SANs). The SANs were designed to possess pH responsive ligands on their surface, which in the presence of an acidic environment (intracellular or tumour microenvironment) underwent hydrolysis to form positive and negative charges on the surface of the nanoparticles. As a result of this, the individual nanoparticles experienced electrostatic attraction among them which led to the formation of gold nanoaggregates. This pH stimulated aggregation induced plasmonic coupling among the nanoparticles which ultimately resulted in shifting the absorbance of SANs to the NIR region. The researchers used these SANs (60 nm and 100 nm size) to demonstrate NIR mediated PTT on HeLa and B16F10 cancer cell lines. First they incubated SANs (60 nm) with cancer cell lines and irradiated with the laser power densities of 5, 6.5, 8 and 10 W/cm^2 (spot size, radius = 500 μm) for 10 min and observed cell death at a laser power of 6.5 W/cm^2 or more. Cell death was not observed outside the laser spot. With 100 nm SANs they observed cell mortality at a laser power more than 8 W/cm^2 (Fig. 7) (Nam et al. 2009). They further demonstrated the synergistic (chemo and thermo effect) anticancer effect on Nu/nu nude mice bearing B16F10 melanoma cells by conjugating Doxorubicin with SANs (SANDCs) and observed significant accumulation of SANDCs in tumours with efficient reduction in tumour size as compared to SANs alone or a mixture of SANs and DOX (Nam et al. 2013a, b). Jung et al. further exploited these pH responsive SANs as SERS active substrates for combined SERS based imaging/diagnosis and PTT. For simultaneous SERS diagnosis and PTT they combined SANs with Raman reporter surface molecule i.e. 4- mercaptobenzoic acid (MBA) and named this conjugate a MBA/SANs. They incubated MBA/SANs with B16 F10 cells and upon irradiation with laser light (785 nm, 25 mW) for 60 s they observed an enhanced Raman signal with an enhancement factor of 1.3×10^4. Irradiation with a NIR laser

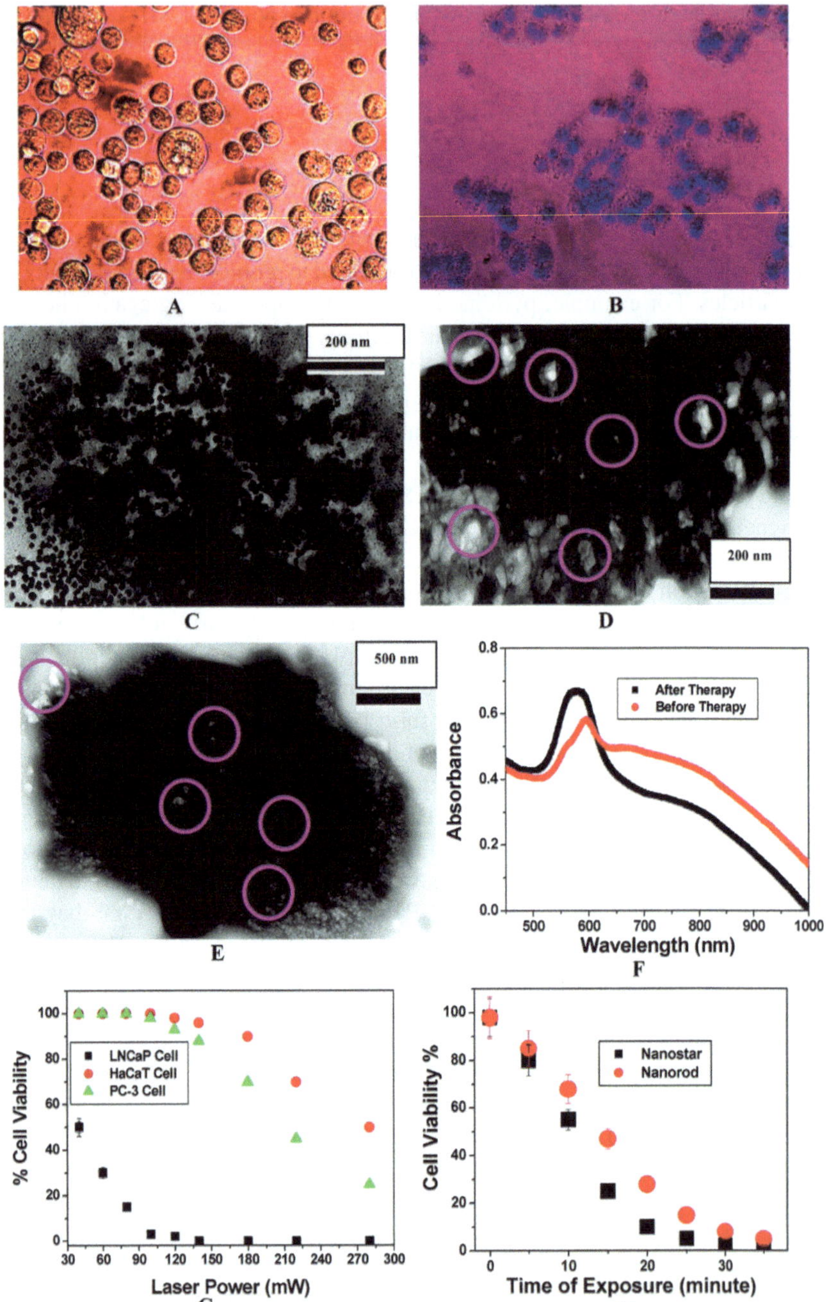

Fig. 6 (**a, b**) Bright-field inverted microscope images of multifunctional popcorn-shaped gold nanoparticle-conjugated LNCaP prostate cancer cells (**a**) before therapy and (**b**) after therapy for 30 min and after being stained with Trypan Blue. (**c**) TEM image showing deformation of

(785 nm, 19.5 W/cm^2, 10 min) resulted in significant damage to B16 F10 cells (Fig. 8) (Jung et al. 2013). Xiao et al. investigated the properties of SANs and showed photothermal-OCT (PT-OCT) guided detection of cancer cells combined with PTT using SANs (Xiao et al. 2013). Nam et al. increased the systemic circulation along with tumour accumulation of the pH-responsive SANs by encapsulating them in a PEG-grafted liposomal shell (Nam et al. 2013a, b). Hwang et al. synthesized smaller sized pH responsive SANs (<6 nm) and demonstrated photothermal ablation of B16F10 mouse melanoma cells in-vitro at a laser power of less than 20 W/cm^2. These smaller sized SANs were found to be more effective in accumulating in tumours which resulted in enhanced PTT (Hwang et al. 2014a, b).

5 Summary and Outlook

In summary, NIR active gold based nanomaterials of different structures such as nanorods, nanoshells, nanocages, nanorattles, nanopopcorns, nanostars and nanoaggregates have emerged as a novel class of potential plasmonic therapeutic agent for cancer treatment- from detection to cure. Tuneability in the SPR bands of nanoscale gold materials from the visible to NIR region makes them an ideal non-invasive cancer theranostic agent. It is highly desirable to understand the effect of size, shape and surface modification on the intrinsic properties of gold nanomaterials. For the efficient plasmonic photothermal treatment, researchers have optimized the photothermal conversion efficiencies of nanoscale gold materials by tuning in their optical properties and have demonstrated the in vitro and in vivo photothermal ablation as well as bioimaging. In this chapter, we presented a detailed overview on the different morphologies of gold nanostructures, their optical properties and NIR responsiveness. We further discussed the mechanism of heat generation by these plasmonic nanostructures and also described in detail about the possible mechanism of cell

Fig. 6 (continued) nano-popcorn structure after popcorn-shaped gold nanoparticle-conjugated LNCaP cells were exposed to 100 mW, 785 nm NIR continuous-wave radiation for 10 min. (**d**) TEM image showing structure deformation and irreparable damage of cancer cell surfaces after 20 min of radiation; purple circles show bubble formation. (**e**) TEM image demonstrating irreparable damage of cancer cell surfaces when multifunctional popcorn-shaped gold nanoparticle-conjugated LNCaP cells were exposed to 100 mW, 785 nm NIR continuous-wave radiation for 30 min; purple circles show bubble formation. (**f**) Absorption profile demonstrating nanoparticle structural changes after nanotherapy process. (**g**) Plot showing cell viability measured by MTT test after popcorn-shaped gold nanoparticle-conjugated LNCaP cells, PC-3 cells, and HaCaT cells were exposed to 785 nm NIR continuous-wave radiation at different power doses. (**h**) Plot comparing photothermal therapy response between well-characterized gold nanorods and popcorn-shaped gold nanoparticles when multifunctional nanoparticle-conjugated LNCaP cells were exposed to 100 mW, 785 nm NIR continuous-wave radiation for different times. (Reproduced with the permission from Lu et al 2010. Copyright 2010 American Chemical Society)

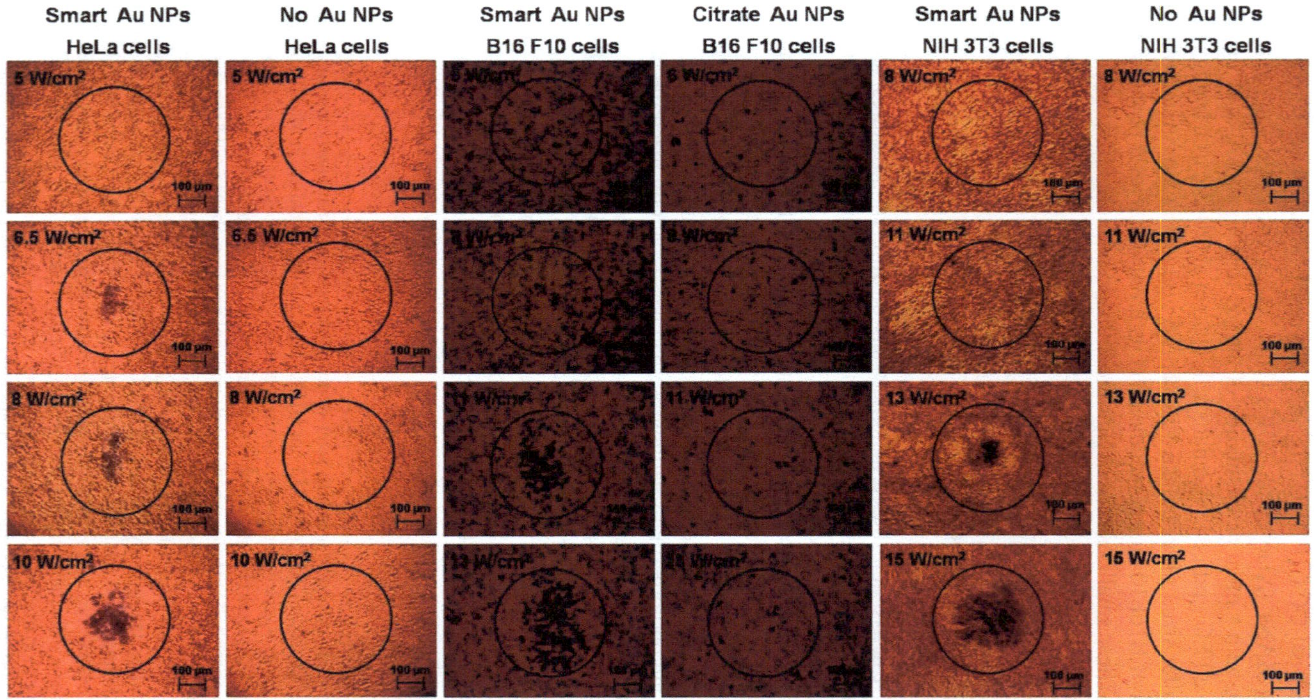

Fig. 7 Photothermal destruction of cells. HeLa cells were incubated with “smart” gold nanoparticles (first column) and with no nanoparticles (second column). B16 F10 cells were incubated with “smart” gold nanoparticles (third column) and with citrate gold nanoparticles (fourth column). NIH 3T3 cells were incubated with “smart” gold nanoparticles (fifth column) and with no nanoparticles (last column). Laser fluence rates are 5, 6.5, 8, and 10 W/cm^2 for HeLa cells; 6, 8, 11, and 13 W/cm^2 for B16 F10 cells; and 8, 11, 13, and 15 W/cm^2 for NIH 3T3 cells, respectively, from top to bottom row. Circles denote the position of laser spot. (Reproduced with the permission from Nam et al. 2009. Copyright 2009 American Chemical Society)

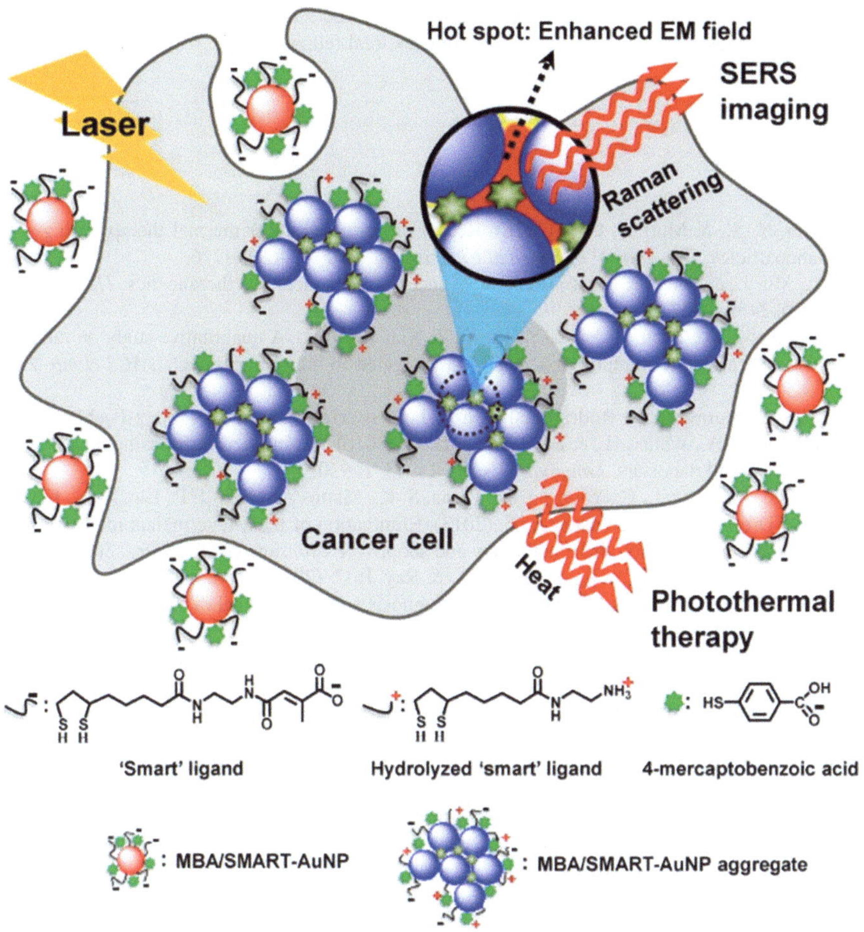

Fig. 8 Schematic illustration of the working mechanism of MBA/SMART-AuNP in a cancer cell (Reproduced with the permission from Jung et al. 2013. Copyright 2013 American Chemical Society)

death induced by PPTT. Finally, we highlighted the application of different gold nanostructures in cancer phototherapy. But, some issues are still unexplained about the fate of the gold nanomaterials inside the body and clearance of its degradant products from the body. Present researches are mainly focussed on the pharmacodynamics activity of gold nanomaterials and there are very few reports on its pharmacokinetics profile. Pharmacokinetics profile of the gold nanomaterials needs detailed investigation.

Acknowledgements AJ gratefully acknowledge the financial support from Indian Institute of Technology Mandi, Department of Science and Technology (DST) under project number: SERB/F/5627/2015-16 and Department of Biotechnology (DBT), Government of India, under project number: BT/PR14749/NNT/28/954/2015. SR would like to acknowledge DST INSPIRE Fellowship programme [IF160513] for providing doctoral fellowship.

References

Abadeer, N. S., & Murphy, C. J. (2016). Recent progress in cancer thermal therapy using gold nanoparticles. *The Journal of Physical Chemistry C, 120*(9), 4691–4716.

Ai, X., Mu, J., & Xing, B. (2016). Recent advances of light-mediated theranostics. *Theranostics, 6*(13), 2439–2457.

Au, L., Zheng, D., Zhou, F., Li, Z.-Y., Li, X., & Xia, Y. (2008). A quantitative study on the photothermal effect of immuno gold nanocages targeted to breast cancer cells. *ACS Nano, 2*(8), 1645–1652.

Barbosa, S., Agrawal, A., Rodríguez-Lorenzo, L., Pastoriza-Santos, I., Alvarez-Puebla, R. A., Kornowski, A., Weller, H., & Liz-Marzán, L. M. (2010). Tuning size and sensing properties in colloidal gold nanostars. *Langmuir, 26*(18), 14943–14950.

Bear, A. S., Kennedy, L. C., Young, J. K., Perna, S. K., Mattos Almeida, J. P., Lin, A. Y., Eckels, P. C., Drezek, R. A., & Foster, A. E. (2013). Elimination of metastatic melanoma using gold nanoshell-enabled photothermal therapy and adoptive T cell transfer. *PLoS One, 8*(7), e69073.

Beqa, L., Fan, Z., Singh, A. K., Senapati, D., & Ray, P. C. (2011). Gold nano-popcorn attached SWCNT hybrid nanomaterial for targeted diagnosis and potothermal therapy of human breast cancer cells. *ACS Applied Materials & Interfaces, 3*(9), 3316–3324.

Bhana, S., Lin, G., Wang, L., Starring, H., Mishra, S. R., Liu, G., & Huang, X. (2015). Near-infrared-absorbing gold nanopopcorns with iron oxide cluster core for magnetically amplified photothermal and photodynamic cancer therapy. *ACS Applied Materials & Interfaces, 7*(21), 11637–11647.

Black, K. C., Yi, J., Rivera, J. G., Zelasko-Leon, D. C., & Messersmith, P. B. (2013). Polydopamine-enabled surface functionalization of gold nanorods for cancer cell-targeted imaging and photothermal therapy. *Nanomedicine, 8*(1), 17–28.

Brunetaud, J. M., Mordon, S., Maunoury, V., & Beacco, C. (1995). Non-PDT uses of lasers in oncology. *Lasers in Medical Science, 10*(1), 3–8.

Calderwood, S. K., & Ciocca, D. R. (2008). Heat shock proteins: Stress proteins with Janus- like properties in cancer. *International Journal of Hyperthermia, 24*, 31–39.

Chen, J., Wang, D., Xi, J., Au, L., Siekkinen, A., Warsen, A., Li, Z.-Y., Zhang, H., Xia, Y., & Li, X. (2007). Immuno gold nanocages with tailored optical properties for targeted photothermal destruction of cancer cells. *Nano Letters, 7*(5), 1318–1322.

Chen, H., Shao, L., Ming, T., Sun, Z., Zhao, C., Yang, B., & Wang, J. (2010a). Understanding the photothermal conversion efficiency of gold nanocrystals. *Small, 6*(20), 2272–2280.

Chen, J., Glaus, C., Laforest, R., Zhang, Q., Yang, M., Gidding, M., Welch, M. J., & Xia, Y. (2010b). Gold nanocages as photothermal transducers for cancer treatment. *Small, 6*(7), 811–817.

Chen, H., Zhang, X., Dai, S., Ma, Y., Cui, S., Achilefus, S., & Gu, Y. (2013). Multifunctional gold nanostar conjugates for tumor imaging and combined photothermal and chemo-therapy. *Theranostics, 3*(9), 633–649.

Chen, Z., Yu, D., Huang, Y., Zhang, Z., Liu, T., & Zhan, J. (2014). Tunable SERS-tags-hidden gold nanorattles for theranosis of cancer cells with single laser beam. *Scientific Reports, 4*, 6709.

Choi, W. I., Kim, J.-Y., Kang, C., Byeon, C. C., Kim, Y. H., & Tae, G. (2011). Tumor regression in vivo by photothermal therapy based on gold-nanorod-loaded, functional nanocarriers. *ACS Nano, 5*(3), 1995–2003.

Chu, Z., Zhang, S., Zhang, B., Zhang, C., Fang, C.-Y., Rehor, I., Cigler, P., Chang, H.-C., Lin, G., Liu, R., & Li, Q. (2014). Unambiguous observation of shape effects on cellular fate of nanoparticles. *Scientific Reports, 4*, 4495.

Ciocca, D. R., & Calderwood, S. K. (2005). Heat shock proteins in cancer: Diagnostic, prognostic, predictive, and treatment implications. *Cell Stress & Chaperones, 10*(2), 86–103.

Dickerson, E., Dreaden, E., Huang, X., ElSayed, I., Chu, H., Pushpanketh, S., Mcdonald, J., & ElSayed, M. (2008). Gold nanorod assisted near-infrared plasmonic photothermal therapy (PPTT) of squamous cell carcinoma in mice. *Cancer Letters, 269*(1), 57–66.

El-Sayed, I. H., Huang, X., & El-Sayed, M. A. (2006). Selective laser photo-thermal therapy of epithelial carcinoma using anti-EGFR antibody conjugated gold nanoparticles. *Cancer Letters, 239*(1), 129–135.

Faraday, M. (1857). The Bakerian lecture: Experimental relations of gold (and other metals) to light. *Philosophical Transactions of the Royal Society of London, 147*(0), 145–181.

Fink, S. L., & Cookson, B. T. (2005). Apoptosis, pyroptosis, and necrosis: Mechanistic description of dead and dying eukaryotic cells. *Infection and Immunity, 73*(4), 1907–1916.

Gao, L., Fei, J., Zhao, J., Li, H., Cui, Y., & Li, J. (2012). Hypocrellin-loaded gold nanocages with high two-photon efficiency for photothermal/photodynamic cancer therapy in vitro. *ACS Nano, 6*(9), 8030–8040.

Gao, Y., Gu, J., Li, L., Zhao, W., & Li, Y. (2016). Synthesis of gold nanoshells through improved seed-mediated growth approach: Brust-like, in situ seed formation. *Langmuir, 32*(9), 2251–2258.

Ghosh, S. K., & Pal, T. (2007). Interparticle coupling effect on the surface plasmon resonance of gold nanoparticles: From theory to applications. *Chemical Reviews, 107*(11), 4797–4862.

Henglein, A. (1993). Physicochemical properties of small metal particles in solution: "Microelectrode" reactions, chemisorption, composite metal particles, and the atom-to-metal transition. *The Journal of Physical Chemistry, 97*(21), 5457–5471.

Hirsch, L. R., Stafford, R. J., Bankson, J. A., Sershen, S. R., Rivera, B., Price, R. E., Hazle, J. D., Halas, N. J., & West, J. L. (2003). Nanoshell-mediated near-infrared thermal therapy of tumors under magnetic resonance guidance. *Proceedings of the National Academy of Sciences, 100*(23), 13549–13554.

Hu, M., Chen, J., Li, Z.-Y., Au, L., Hartland, G. V., Li, X., Marquez, M., & Xia, Y. (2006). Gold nanostructures: Engineering their plasmonic properties for biomedical applications. *Chemical Society Reviews, 35*(11), 1084–1094.

Huang, X., & El-Sayed, M. A. (2011). Plasmonic photo-thermal therapy (PPTT). *Alexandria Journal of Medicine, 47*(1), 1–9.

Huang, X., El-Sayed, I. H., Qian, W., & El-Sayed, M. A. (2006). Cancer cell imaging and photothermal therapy in the near-infrared region by using gold nanorods. *Journal of the American Chemical Society, 128*(6), 2115–2120.

Huang, X., Jain, P. K., El-Sayed, I. H., & El-Sayed, M. A. (2007). Plasmonic photothermal therapy (PPTT) using gold nanoparticles. *Lasers in Medical Science, 23*(3), 217–228.

Huang, X., Kang, B., Qian, W., Mackey, M. A., Chen, P. C., Oyelere, A. K., El-Sayed, I. H., & El-Sayed, M. A. (2010). Comparative study of photothermolysis of cancer cells with nuclear-targeted or cytoplasm-targeted gold nanospheres: Continuous wave or pulsed lasers. *Journal of Biomedical Optics, 15*(5), 058002, 1–7.

Hwang, S., Nam, J., Jung, S., Song, J., Doh, H., & Kim, S. (2014a). Gold nanoparticle-mediated photothermal therapy: Current status and future perspective. *Nanomedicine, 9*(13), 2003–2022.

Hwang, S., Nam, J., Song, J., Jung, S., Hur, J., Im, K., Park, N., & Kim, S. (2014b). A sub 6 nanometer plasmonic gold nanoparticle for pH-responsive near-infrared photothermal cancer therapy. *New Journal of Chemistry, 38*(3), 918–922.

Jaiswal, A., Tian, L., Tadepalli, S., Liu, K., Fei, M., Farrell, M. E., Pellegrino, P. M., & Singamaneni, S. (2014). Plasmonic nanorattles with intrinsic electromagnetic hot-spots for surface enhanced Raman scattering. *Small, 10*(21), 4287–4292.

Jung, Y. J., Govindaiah, P., Park, T., Lee, S. J., Ryu, D. Y., Kim, J. H., & Cheong, I. W. (2010). Luminescent gold–poly(thiophene) nanoaggregates prepared by one-step oxidative polymerization. *Journal of Materials Chemistry, 20*, 9770–9774.

Jang, B., Park, J. Y., Tung, C. H., Kim, I. H., & Choi, Y. (2011). Gold nanorod– photosensitizer complex for near-infrared fluorescence imaging and photodynamic/photothermal therapy in vivo. *ACS Nano, 5*(2), 1086–1094.

Jung, S., Nam, J., Hwang, S., Park, J., Hur, J., Im, K., Park, N., & Kim, S. (2013). Theragnostic pH-sensitive gold nanoparticles for the selective surface enhanced Raman scattering and photothermal cancer therapy. *Analytical Chemistry, 85*(16), 7674–7681.

Ke, H., Wang, J., Tong, S., Jin, Y., Wang, S., Qu, E., Bao, G., & Dai, Z. (2014). Gold nanoshelled liquid perfluorocarbon magnetic nanocapsules: A nanotheranostic platform for bimodal ultrasound/magnetic resonance imaging guided photothermal tumor ablation. *Theranostics, 4*(1), 12–23.

Kelly, K. L., Coronado, E., Zhao, L. L., & Schatz, G. C. (2003). The optical properties of metal nanoparticles: The influence of size, shape, and dielectric environment. *The Journal of Physical Chemistry B, 107*(3), 668–677.

Khan, S. A., Kanchanapally, R., Fan, Z., Beqa, L., Singh, A. K., Senapati, D., & Ray, P. C. (2012). A gold nanocage–CNT hybrid for targeted imaging and photothermal destruction of cancer cells. *Chemical Communications, 48*(53), 6711–67113.

Khandelia, R., Jaiswal, A., Ghosh, S. S., & Chattopadhyay, A. (2014). Polymer coated gold nanoparticle–protein agglomerates as nanocarriers for hydrophobic drug delivery. *Journal of Materials Chemistry B, 2*(38), 6472–6477.

Khlebtsov, N. G., & Dykman, L. A. (2010). Optical properties and biomedical applications of plasmonic nanoparticles. *Journal of Quantitative Spectroscopy and Radiative Transfer, 111*(1), 1–35.

Khlebtsov, B., Panfilova, E., Khanadeev, V., Bibikova, O., Terentyuk, G., Ivanov, A., Rumyantseva, V., Shilov, I., Ryabova, A., Loshchenov, V., & Khlebtsov, N. G. (2011). Nanocomposites containing silica-coated gold–silver nanocages and Yb–2, 4-dimethoxyhematoporphyrin: Multifunctional capability of IR-Luminescence detection, photosensitization, and photothermolysis. *ACS Nano, 5*(9), 7077–7089.

Khoury, C. G., & Vo-Dinh, T. (2008). Gold nanostars for surface-enhanced Raman scattering: Synthesis, characterization and optimization. *The Journal of Physical Chemistry C, 112*(48), 18849–18859.

Kono, H., & Rock, K. L. (2008). How dying cells alert the immune system to danger. *Nature Reviews Immunology, 8*(4), 279–289.

Lee, K.-S., & El-Sayed, M. A. (2005). Dependence of the enhanced optical scattering efficiency relative to that of absorption for gold metal nanorods on aspect ratio, size, end-cap shape, and medium refractive index. *The Journal of Physical Chemistry B, 109*(43), 20331–20338.

Li, J.-L., & Gu, M. (2010). Surface plasmonic gold nanorods for enhanced two-photon microscopic imaging and apoptosis induction of cancer cells. *Biomaterials, 31*(36), 9492–9498.

Lin, A. Y., Young, J. K., Nixon, A. V., & Drezek, R. A. (2014). Encapsulated Fe3O4/Ag complexed cores in hollow gold nanoshells for enhanced theranostic magnetic resonance imaging and photothermal therapy. *Small, 10*, 3246–3251.

Link, S., & El-Sayed, M. A. (1999). Size and temperature dependence of the plasmon absorption of colloidal gold nanoparticles. *The Journal of Physical Chemistry B, 103*(21), 4212–4217.

Link, S., & El-Sayed, M. A. (2000). Shape and size dependence of radiative, non-radiative and photothermal properties of gold nanocrystals. *International Reviews in Physical Chemistry, 19*(3), 409–453.

Link, S., & El-Sayed, M. A. (2005). Simulation of the optical absorption spectra of gold nanorods as a function of their aspect ratio and the effect of the medium dielectric constant. *The Journal of Physical Chemistry B, 109*(20), 10531–10532.

Link, S., Mohamed, M. B., & El-Sayed, M. A. (1999). Simulation of the optical absorption spectra of gold nanorods as a function of their aspect ratio and the effect of the medium dielectric constant. *The Journal of Physical Chemistry B, 103*(16), 3073–3077.

Liu, M., Yang, P.-H., & Cai, J.-Y. (2009a). Optical properties and biomedical application of gold nanorods*. *Progress in Biochemistry and Biophysics, 36*(11), 1402–1407.

Liu, S.-Y., Liang, Z.-S., Gao, F., Luo, S.-F., & Lu, G.-Q. (2009b). In vitro photothermal study of gold nanoshells functionalized with small targeting peptides to liver cancer cells. *Journal of Materials Science: Materials in Medicine, 21*(2), 665–674.
Liu, Y., Xu, M., Chen, Q., Guan, G., Hu, W., Zhao, X., Qiao, M., Hu, H., Liang, Y., Zhu, H., & Chen, D. (2015). Gold nanorods/mesoporous silica-based nanocomposite as theranostic agents for targeting near-infrared imaging and photothermal therapy induced with laser. *International Journal of Nanomedicine, 10*, 4747–4761.
Liz-Marzán, L. M. (2004). Nanometals. *Materials Today, 7*(2), 26–31.
Lowery, A. R., Gobin, A. M., Day, E. S., Halas, N. J., & West, J. L. (2006). Immunonanoshells for targeted photothermal ablation of tumor cells. *International Journal of Nanomedicine, 1*(2), 149–154.
Lu, W., Singh, A. K., Khan, S. A., Senapati, D., Yu, H., & Ray, P. C. (2010). Gold nano-popcorn-based targeted diagnosis, nanotherapy treatment, and in situ monitoring of photothermal therapy response of prostate cancer cells using surface-enhanced Raman spectroscopy. *Journal of the American Chemical Society, 132*(51), 18103–18114.
Ma, Y., Liang, X., Tong, S., Bao, G., Ren, Q., & Dai, Z. (2012). Gold nanoshell nanomicelles for potential magnetic resonance imaging, light-triggered drug release, and photothermal therapy. *Advanced Functional Materials, 23*(7), 815–822.
Mahmoud, M. A. (2014). Optical properties of gold nanorattles: Evidences for free movement of the inside solid nanosphere. *The Journal of Physical Chemistry C, 118*(19), 10321–10328.
Melamed, J. R., Edelstein, R. S., & Day, E. S. (2015). Elucidating the fundamental mechanisms of cell death triggered by photothermal therapy. *ACS Nano, 9*(1), 6–11.
Melancon, M. P., Lu, W., Zhong, M., Zhou, M., Liang, G., Elliott, A. M., Hazle, J. D., Myers, J. N., Li, C., & Jason Stafford, R. (2011). Targeted multifunctional gold-based nanoshells for magnetic resonance-guided laser ablation of head and neck cancer. *Biomaterials, 32*(30), 7600–7608.
Messina, E., Cavallaro, E., Cacciola, A., Iatì, M. A., Gucciardi, P. G., Borghese, F., Denti, P., Saija, R., Compagnini, G., Meneghetti, M., Amendola, V., & Maragò, O. M. (2011). Plasmon-enhanced optical trapping of gold nanoaggregates with selected optical properties. *ACS Nano, 5*(2), 905–913.
Mulvaney, P. (1996). Surface plasmon spectroscopy of nanosized metal particles. *Langmuir, 12*(3), 788–800.
Nam, J., Won, N., Jin, H., Chung, H., & Kim, S. (2009). PH-induced aggregation of gold nanoparticles for photothermal cancer therapy. *Journal of the American Chemical Society, 131*(38), 13639–13645.
Nam, J., La, W.-G., Hwang, S., Ha, Y. S., Park, N., Won, N., Jung, S., Bhang, S. H., Ma, Y.-J., Cho, Y.-M., Jin, M., Han, J., Shin, J.-Y., Wang, E. K., Kim, S. G., Cho, S.-H., Yoo, J., & Kim, B.-S. (2013a). PH-responsive assembly of gold nanoparticles and "Spatiotemporally Concerted" drug release for synergistic cancer therapy. *ACS Nano, 7*(4), 3388–3402.
Nam, J., Ha, Y. S., Hwang, S., Lee, W., Song, J., Yoo, J., & Kim, S. (2013b). PH-responsive gold nanoparticles-in-liposome hybrid nanostructures for enhanced systemic tumor delivery. *Nanoscale, 5*(21), 10175–10178.
Neeves, A. E., & Birnboim, M. H. (1989). Composite structures for the enhancement of nonlinear-optical susceptibility. *Journal of the Optical Society of America B, 6*(4), 787–796.
Nie, S., Xing, Y., Kim, G. J., & Simons, J. W. (2007). Nanotechnology applications in cancer. *Annual Review of Biomedical Engineering, 9*(1), 257–288.
Okuno, T., Kato, S., Hatakeyama, Y., Okajima, J., Maruyama, S., Sakamoto, M., Mori, S., & Kodama, T. (2013). Photothermal therapy of tumors in lymph nodes using gold nanorods and near-infrared laser light. *Journal of Controlled Release, 172*(3), 879–884.
Oldenburg, S. J., Averitt, R. D., Westcott, S. L., & Halas, N. J. (1998). Nanoengineering of optical resonances. *Chemical Physics Letters, 288*(2–4), 243–247.
Papavassiliou, G. C. (1979). Optical properties of small inorganic and organic metal particles. *Progress in Solid State Chemistry, 12*(3–4), 185–271.

Pérez-Hernández, M., del Pino, P., Mitchell, S. G., Moros, M., Stepien, G., Pelaz, B., Parak, W. J., Gálvez, E. M., Pardo, J., & de la Fuente, J. M. (2015). Dissecting the molecular mechanism of apoptosis during photothermal therapy using gold nanoprisms. *ACS Nano, 9*(1), 52–61.

Pissuwan, D., Valenzuela, S. M., Killingsworth, M. C., Xu, X., & Cortie, M. B. (2007). Targeted destruction of murine macrophage cells with bioconjugated gold nanorods. *Journal of Nanoparticle Research, 9*(6), 1109–1124.

Pitsillides, C. M., Joe, E. K., Wei, X., Anderson, R. R., & Lin, C. P. (2003). Selective cell targeting with light-absorbing microparticles and nanoparticles. *Biophysical Journal, 84*(6), 4023–4032.

Ramos, M., & Sanchez. (2012). Induction of cell death in a glioblastoma line by hyperthermic therapy based on gold nanorods. *International Journal of Nanomedicine, 7*, 1511–1523.

Rodríguez-Oliveros, R., & Sánchez-Gil, J. A. (2011). Gold nanostars as thermoplasmonic nanoparticles for optical heating. *Optics Express, 20*(1), 621–626.

Schwartzberg, A. M., Olson, T. Y., Talley, C. E., & Zhang, J. Z. (2006). Synthesis, characterization, and tunable optical properties of hollow gold nanospheres. *The Journal of Physical Chemistry B, 110*(40), 19935–19944.

Sheikholeslami, F., Fekrazad, R., Rasaee, & Ardestani, S. (2011). Treatment of oral squamous cell carcinoma using anti-hER2 immunonanoshells. *International Journal of Nanomedicine, 6*, 2749–2755.

Shi, P., Qu, K., Wang, J., Li, M., Ren, J., & Qu, X. (2012). PH-responsive NIR enhanced drug release from gold nanocages possesses high potency against cancer cells. *Chemical Communications, 48*(61), 7640–7642.

Skrabalak, S. E., Chen, J., Sun, Y., Lu, X., Au, L., Cobley, C. M., & Xia, Y. (2008). Gold nanocages: Synthesis, properties, and applications. *Accounts of Chemical Research, 41*(12), 1587–1595.

Stern, J. M., Stanfield, J., Kabbani, W., Hsieh, J.-T., & Cadeddu, J. A. (2008). Selective prostate cancer thermal ablation with laser activated gold nanoshells. *The Journal of Urology, 179*(2), 748–753.

Sultan, R. A. (1990). Tumour ablation by laser in general surgery. *Lasers in Medical Science, 5*(2), 185–193.

Sun, X., Huang, X., Yan, X., Wang, Y., Guo, J., Jacobson, O., Liu, D., Szajek, L. P., Zhu, W., Niu, G., Kiesewetter, D. O., Sun, S., & Chen, X. (2014a). Chelator-Free64Cu-integrated gold nanomaterials for positron emission tomography imaging guided photothermal cancer therapy. *ACS Nano, 8*(8), 8438–8446.

Sun, T., Wang, Y., Xu, J., Zhao, X., Vangveravong, S., Mach, R. H., & Xia, Y. (2014b). Using SV119-gold nanocage conjugates to eradicate cancer stem cells through a combination of photothermal and chemo therapies. *Advanced Healthcare Materials, 3*(8), 1283–1291.

Turkevich, J., Stevenson, P. C., & Hillier, J. (1951). A study of the nucleation and growth processes in the synthesis of colloidal gold. *Discussions of the Faraday Society, 11*, 55–75.

Van de Broek, B., Devoogdt, N., D'Hollander, A., Gijs, H.-L., Jans, K., Lagae, L., Muyldermans, S., Maes, G., & Borghs, G. (2011). Specific cell targeting with nanobody conjugated branched gold nanoparticles for photothermal therapy. *ACS Nano, 5*(6), 4319–4328.

Vigderman, L., & Zubarev, E. R. (2013). High-yield synthesis of gold nanorods with longitudinal SPR peak greater than 1200 nm using hydroquinone as a reducing agent. *Chemistry of Materials, 25*(8), 1450–1457.

Von Maltzahn, G., Park, J., Agrawal, A., Bandaru, N. K., Das, S. K., Sailor, M. J., & Bhatia, S. N. (2009). Computationally guided photothermal tumor therapy using long-circulating gold nanorod antennas. *Cancer Research, 69*(9), 3892–3900.

Wang, J., Zhu, G., You, M., Song, E., Shukoor, M. I., Zhang, K., Altman, M. B., Chen, Y., Zhu, Z., Huang, C. Z., & Tan, W. (2012). Assembly of aptamer switch probes and photosensitizer on gold nanorods for targeted photothermal and photodynamic cancer therapy. *ACS Nano, 6*(6), 5070–5077.

Wang, J., Sefah, K., Altman, M. B., Chen, T., You, M., Zhao, Z., Huang, C. Z., & Tan, W. (2013). Aptamer-conjugated nanorods for targeted photothermal therapy of prostate cancer stem cells. *Chemistry – An Asian Journal, 8*(10), 2417–2422.

Wang, D., Xu, Z., Yu, H., Chen, X., Feng, B., Cui, Z., Lin, B., Yin, Q., Zhang, Z., Chen, C., Wang, J., Zhang, W., & Li, Y. (2014). Treatment of metastatic breast cancer by combination of chemotherapy and photothermal ablation using doxorubicin-loaded DNA wrapped gold nanorods. *Biomaterials, 35*(29), 8374–8384.

Wu, C., Yu, C., & Chu, M. (2011). A gold nanoshell with a silica inner shell synthesized using liposome templates for doxorubicin loading and near-infrared photothermal therapy. *International Journal of Nanomedicine, 6*, 807–813.

Wust, P., Hildebrandt, B., Sreenivasa, G., Rau, B., Gellermann, J., Riess, H., Felix, R., & Schlag, P. (2002). Hyperthermia in combined treatment of cancer. *The Lancet Oncology, 3*(8), 487–497.

Xiao, P., Li, Q., Joo, Y., Nam, J., Hwang, S., Song, J., Kim, S., Joo, C., & Kim, K. H. (2013). Detection of pH-induced aggregation of "smart" gold nanoparticles with photothermal optical coherence tomography. *Optics Letters, 38*(21), 4429–4432.

Xu, C., Yang, D., Mei, L., Li, Q., Zhu, H., & Wang, T. (2013). Targeting chemophotothermal therapy of hepatoma by gold nanorods/graphene oxide core/shell nanocomposites. *ACS Applied Materials & Interfaces, 5*(24), 12911–12920.

Yang, X., Liu, X., Liu, Z., Pu, F., Ren, J., & Qu, X. (2012). Near-infrared light-triggered, targeted drug delivery to cancer cells by aptamer gated nanovehicles. *Advanced Materials, 24*(21), 2890–2895.

Yang, J., Shen, D., Zhou, L., Li, W., Li, X., Yao, C., Wang, R., El-Toni, A. M., Zhang, F., & Zhao, D. (2013). Spatially confined fabrication of Core–Shell gold Nanocages@Mesoporous silica for near-infrared controlled photothermal drug release. *Chemistry of Materials, 25*(15), 3030–3037.

Yavuz, M. S., Cheng, Y., Chen, J., Cobley, C. M., Zhang, Q., Rycenga, M., Xie, J., Kim, C., Song, K. H., Schwartz, A. G., Wang, L. V., & Xia, Y. (2009). Gold nanocages covered by smart polymers for controlled release with near-infrared light. *Nature Materials, 8*(12), 935–939.

Yeager, D., Chen, Y.-S., Litovsky, S., & Emelianov, S. (2014). Intravascular photoacoustics for image-guidance and temperature monitoring during plasmonic photothermal therapy of atherosclerotic plaques: A feasibility study. *Theranostics, 4*(1), 36–46.

You, J., Zhang, R., Xiong, C., Zhong, M., Melancon, M., Gupta, S., Nick, A. M., Sood, A. K., & Li, C. (2012). Effective photothermal chemotherapy using doxorubicin-loaded gold nanospheres that target EphB4 receptors in tumors. *Cancer Research, 72*(18), 4777–4786.

Yuan, H., Khoury, C. G., Wilson, C. M., Grant, G. A., Bennett, A. J., & Vo-Dinh, T. (2012a). In vivo particle tracking and photothermal ablation using plasmon-resonant gold nanostars. *Nanomedicine: Nanotechnology, Biology and Medicine, 8*(8), 1355–1363.

Yuan, H., Fales, A. M., & Vo-Dinh, T. (2012b). TAT peptide-functionalized gold nanostars: Enhanced intracellular delivery and efficient NIR photothermal therapy using ultralow irradiance. *Journal of the American Chemical Society, 134*(28), 11358–11361.

Zharov, V. P., Kim, J. W., Curiel, D. T., & Everts, M. (2005). Self-assembling nanoclusters in living systems: Application for integrated photothermal nanodiagnostics and nanotherapy. *Nanomedicine: Nanotechnology, Biology and Medicine, 1*(4), 326–345.

Zhang, Z., Wang, J., Nie, X., Wen, T., Ji, Y., Wu, X., Zhao, Y., & Chen, C. (2014). Near infrared laser-induced targeted cancer therapy using thermoresponsive polymer encapsulated gold nanorods. *Journal of the American Chemical Society, 136*(20), 7317–7326.

Nanomaterials-Based siRNA Delivery: Routes of Administration, Hurdles and Role of Nanocarriers

Nitin Gupta, Divya Bharti Rai, Ashok Kumar Jangid, Deep Pooja, and Hitesh Kulhari

Abstract Ribonucleic acid interference (RNAi) is a potential alternative therapeutic approach to knock down the overexpression of genes in several disorders especially cancers with underlying genetic dysfunctions. For silencing of specific genes involved in cell cycle, small/short interfering ribonucleic acids (siRNAs) are being used clinically. The siRNA-based RNAi is more efficient, specific and safe antisense technology than other RNAi approaches. The route of siRNA administration for siRNA therapy depends on the targeted site. However, certain hurdles like poor stability of siRNA, saturation, off-target effect, immunogenicity, anatomical barriers and non-targeted delivery restrict the successful siRNA therapy. Thus, advancement of an effective, secure, and long-term delivery system is prerequisite to the medical utilization of siRNA. Polycationic nanocarriers mediated targeted delivery system is an ideal system to remove these hurdles and to increase the blood retention time and rate of intracellular permeability. In this chapter, we will mainly discuss the different biocompatible, biodegradable, non-toxic (organic, inorganic and hybrid) nanocarriers that encapsulate and shield the siRNA from the different harsh environment and provides the increased systemic siRNA delivery.

Keywords Naked siRNA · Overexpression of genes · Hurdles/barriers · Systemic delivery · Cationic nanocarriers · Targeted delivery

N. Gupta · D. B. Rai · A. K. Jangid · H. Kulhari (✉)
School of Nano Sciences, Central University of Gujarat, Gandhinagar, Gujarat, India
e-mail: hitesh.kulhari@cug.ac.in

D. Pooja
Applied Biology Division, CSIR-Indian Institute of Chemical Technology, Hyderabad, Telangana, India

S. Singh, P. K. Maurya (eds.), *Nanotechnology in Modern Animal Biotechnology*, https://doi.org/10.1007/978-981-13-6004-6_3

1 Introduction

In recent years, one of the most important transformations in the area of biology has occurred as RNA molecular biology. In 1990, RNA interference (RNAi) phenomenon was firstly discovered in plants and *Caenorhabditis elegans* (nematode) in the form of single-stranded RNA (ssRNA) by Carolyn Napoli and Richard Jorgensen. Later, in 1998, Andrew Fire and Craig Mello found RNAi as a double-stranded RNA (dsRNA) in mammalian cells; pursue the development of an innovative therapeutic tool in the biomedical field. Andrew Fire and Craig Mello demonstrated that dsRNA is more effective for down-regulation of gene expression than ssRNA (Xu and Wang 2015; Fire et al. 1998). The interference mechanism of siRNA is involved to control many cellular phenomena including an intranuclear arrangement of heterochromatin and development, multiplication, differentiation and proliferation of cells (Berezikov 2011). The core of mechanism is RNA molecules; microRNA (miRNA), siRNA, and germline-directed Piwi-interacting RNA (piRNA) which actually directs degradation of the protein-encoding messenger RNA (mRNA) through a complex of certain enzymes. While piRNA and microRNA intrinsically generate from the genome, the siRNA perform as an exogenously designed fragment for interference therapy (Wilson and Doudna 2013) Thus, post-transcriptional gene silencing regulates the outcome of the targeted gene by repressing translation.

siRNA has several merits over conventional therapeutic agents like synthetic drugs or bioflavonoids. The key advantages are remarkable target-specific gene knockdown/silencing property, lesser side effects, potent gene regulation, thermodynamic stability, ease of synthesis with low production costs than protein or antibodies and there is no need of cellular expression system and complex protein purification (Liu et al. 2004; Chougule 2012). One of the major advantages of RNAi is the capability to access "non-druggable" targets; for instance, certain targeted proteins which become inaccessible to drug due to alteration in enzymatic function or conformation. For such cases, synthetic RNA can be sequenced to target almost any desired gene of human genome (Aagaard and Rossi 2007). This advancement is associated with the emergence of short noncoding RNA stretch known as siRNA. Because of its unique property to regulate the targeted genes expression (in-vivo and in-vitro both), currently siRNA has obtained more attention as a potent therapeutic agent in several biological and genetic abnormalities such as neurodegenerative diseases, Huntington's disease, hematological diseases, inherited genetic disorders, dominant genetic disorders, cardiovascular disorders, various cancers, viral infections, autoimmune diseases, ocular diseases and many other illnesses caused by action from one or several genes, which can be regulated via RNAi phenomenon (Gavrilov and Saltzman 2012; Dorn et al. 2004; Jagannath and Wood 2007). The genetic expression may be altered at various genetic levels, involving RNA processing, RNA stability, chromatin structure, chromosome segregation transcription, and translation (Carthew and Sontheimer 2009) siRNA is a better therapeutic agent than conventionally used therapeutic agents. For the effective therapeutic therapy, the first and most essential step is to ensure the delivery of siRNA to the targeted/site-specific tissue and cells from the administrated site. Although, the systemic circulation of naked RNA molecule is challenged with several hurdles/problems such as fast enzymatic degradation in the physiological and biological media, entrapment by phagocytes, extravasation from blood to tumor tissues, recognition by the immune system, renal clearance, poor tissue penetration, incompetent endocytosis, immunostimulation, and off-target effects. These several drawbacks prevent the delivery of siRNA to the active site (Lee et al. 2016; Gewirtz 2007). So, viral and non-viral carriers developed as carrier systems. Initially, the viral vectors were used as delivery platforms where the viruses deliver the therapeutic siRNA into the targeted cells through the mechanism of transduction. They are suitable for both transient and long-term gene silencing, depending on the type of carrier viruses. Moreover, viral vectors have the ability to enter the nucleus of the cells (Qiu et al. 2016a). However, these conventional vectors have a risk of causing immunogenicity, toxicity, and insertional mutagenesis. The smaller size of viral particles with small size also limits the size of a therapeutic gene that can be inserted and carried. Some of the carriers and conjugated with various molecules are used to deliver siRNA such as aptamer oligonucleotide, sterol or antibody conjugates with siRNA. But these conjugations and carriers are not enough for the systemic delivery of siRNA.

Therefore, for overcoming the associated hurdles during systemic delivery and to successfully deliver siRNA-based drugs in-vivo with easy and improved extensive practical applications, various carriers or vectors are required that can convey the siRNA to the desired site of activity. Non-viral based biocompatible, biodegradable, non-toxic, nano-sized carrier systems extensively employ nanomaterials (NMs) as compact carriers for in-vivo delivery of siRNA. NCs have the diameter/size <100 nm. Due to small size, NCs have some advanced and unique physiochemical properties than micro or macro carriers such as they possess a greater ratio of surface area to volume with the increased number of surface-active molecules. These nanocarriers (NCs) form complexes with the siRNA or encapsulate it to develop an intact and inert molecule or particle which that is capable of transfecting the targeted cells and inducing the genetic interference (Dana et al. 2017). Davis et al. (2010) showed the site-specific transport of siRNA using NCs in humans via systemic injection in earlier work, which developed as a foundation for later research related to the medical application of siRNA (Davis et al. 2010). After this, several researchers and scientists have developed/synthesized several NCs such as cationic polymeric nanoparticles (polyplex), liposomal nanocarriers (lipoplex), carbon nanomaterials, lipid-based nanomicelles etc. These NCs protect the siRNA from rapid enzymatic degradation in biological and physiological media, reduce immunogenicity, increase blood retention time and cellular uptake, and increase the targeted delivery. These advantages are responsible for improving the therapeutic effects of siRNA during systemic circulation.

This chapter presents the basic mechanism of siRNA interference, its advantages, various routes of administration, hurdles/challenges during systemic delivery naked siRNA, and overcoming of these hurdles using various nanocarriers.

1.1 Mechanism of siRNA Interference

siRNA interference is an intrinsic process of post-transcriptional gene regulation where a non-coding stretch of RNA sequence down-regulate the activity of several coding genes and silence the gene expression (Draz et al. 2014). Mostly a short dsRNA fragment known as siRNA mediates homology-based regulation of gene expression in eukaryotic cells (Fig. 1) (Almeida and Allshire 2005). This process can be compared to another approach based on miRNA knockdown and replacement. For inhibition, a synthetic ssRNA known as antagomirs/anti-miRs employed as the antagonist of intrinsic miRNA through an antisense mechanism. The replacement process involves synthetic miRNAs which are introduced in-vivo to mimic the function of the endogenous miRNAs. This RNA mimicking leads to gene silencing and degradation/inhibition of mRNA consequently. This chapter elaborates the clinical application of RNA interference in medical treatment and exclusively focuses on the siRNA therapy (Aagaard and Rossi 2007).

The first step of RNAi is the initiation phase which involves the production of effectors molecules. There are two specific nucleotides at the 3′-overhang of dou-

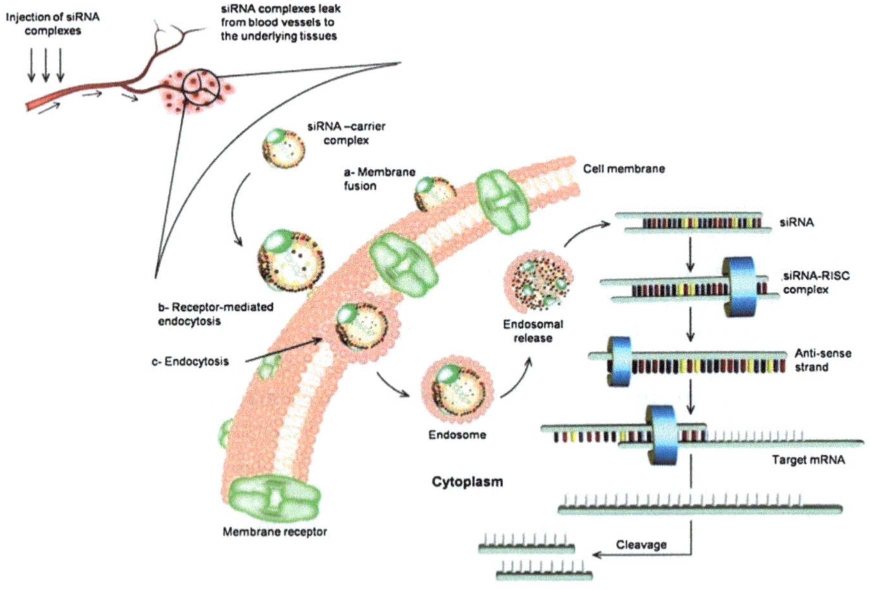

Fig. 1 Mechanism of siRNA interference in-vivo. (Draz et al. 2014)

ble-stranded RNA molecules approximately 21–22 base pair long which is targeted by an enzyme complex known as Dicer, in homology-dependent mechanism to cleave recognized mRNA. The RNase III-like enzyme termed as Dicer, degrades the longer dsRNA, generating short interfering RNAs in the homology-dependent mechanism (Gavrilov and Saltzman 2012). This processing of dsRNA into siRNA constructs by endonuclease Dicer occurs in the cell cytoplasm.

After the formation of effector molecules, the execution phase begins when a multiprotein complex bound the effector molecules to stimulate gene silencing. siRNAs are bound by a multi-subunit assembly of certain proteins forming RNA-induced silencing complex (RISC) (Hammond et al. 2000). The RISC complex includes three proteins, namely Dicer, TRBP, and Argonaute 2, each with a distinct role; (1) Dicer joins siRNAs to the RISC (2) TAR-RNA binding protein (TRBP) assists the siRNA-RISC bounding (3) Ago-2 is the slicing protein of RISC complex that acts at specific cleavage site between bases 10 and 11 from the 5′-terminal of the antisense siRNA strand, in the target mRNA molecules (Aagaard and Rossi 2007). The mRNA strands which are almost complementary of guide RNA strand recognized and sliced by Ago-2 protein. In the second instance, this complementary siRNA-mRNA pair sometimes resembles miRNA bounded to target sites. miRNAs are processed within the nucleus by Drosha-DGCR8 complex, thereby repressing translation or degrading the transcript.

The Argonaute (Ago) proteins are the sole member of RISC and the catalytic domain of Ago-2 enables the cleavage of targeted mRNA region (Meister et al. 2004). The siRNAs embedded within RISC are duplex, Ago-2 proteins detach and remove one of the strands from RISC bounded double-stranded siRNA referred to as the passenger strand. The other single strand, known as guide RNA, remains embedded in the RISC complex (Fig. 2). The functionally active complex with guide RNA recognizes the target due to specific intermolecular base pair interaction (Tang 2005). Ago-2 protein is composed of three functional subunits; namely PIWI, PAZ, and MID. Among these, the PIWI domain possesses the structural folding similar to RNase H and provides thermodynamic enthalpy to the RISC complex for

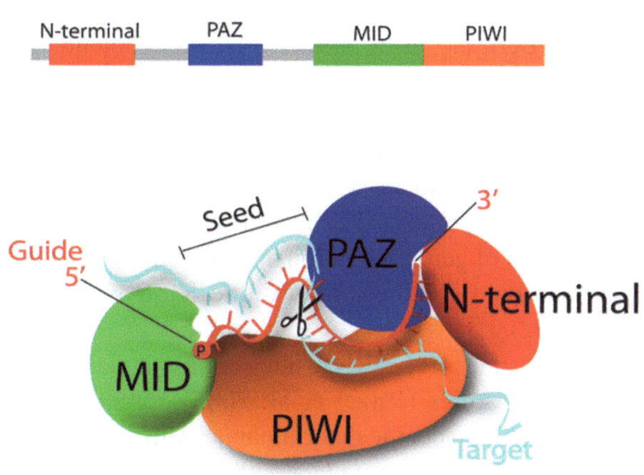

Fig. 2 Formation of the RISC complex. (Gavrilov and Saltzman 2012)

cleavage (Song et al. 2004). The PAZ subunit of the Ago-2 protein binds two terminal 3′-nucleotide specifically for connecting it to the RISC complex. The staggered ends of the siRNA consisting of the specific nucleotide lie in the core domain in a hydrophobic cavity composed of various aromatic residues where the base of the terminal nucleotide is stacked (Song et al. 2003). The 5′-phosphate group of siRNA is bounded to the carboxyl-end of the protein in the middle of the MID and PIWI domains through a magnesium ion (Ma et al. 2005). It is observed that the Argonaute family also contributes while guide-strand selection. Argonaute conserves the strand which has less thermodynamically stable 5′-terminal as the guide strand and cleaves another one to release the passenger strand (Schwarz et al. 2003).Certain basic principles such as thermodynamic stability of the siRNA end regulate the selection of the strand which remains anchored to the RISC complex (Schwarz et al. 2003). The strand with less stability at the terminals usually converts to form mRNA molecules which can be recognized and later on degraded by Ago-2. The strand complementary to the targeted mRNA act as guide RNA.

2 Advantages of siRNA

RNAi has several extra benefits and advantages compared to others therapeutic agents (already discussed in Sect. 1). However here, we are discussing some of the significant advantages of siRNA in details including its therapeutic efficacy/potency for suppressing of gene expression, self-targeted/selective delivery on infected cells, and safety.

2.1 Potency

Contrary to other bioactive molecules applied so far including antisense DNA oligonucleotides and RNA-enzymes, RNAi has a greater potential in antisense technology (Bertrand et al. 2002). The high potential of RNAi approach observed because of the fact that effector molecules may function at a lower concentration than in the case of ribozymes or antisense oligos. This is a major advantage in a clinical scenario. The underlying factor for the above success of RNAi over other antisense-based practices for treatment purpose is that it autonomously target the complementary transcripts by using the host cellular machinery, often leading to suppression of gene expression up to much extend. Of note, small antisense ssRNAs may also bind complementarily into RISC and conduct target mRNA (Martinez et al. 2002).

The efficiency of siRNA interference is decided by many parameters such as structural features (Patzel et al. 2005), target mRNA accessibility (Heale et al. 2005), position-specific determinants (Amarzguioui and Prydz 2004), and differential end stability (Schwarz et al. 2003). The most important parameter among is the thermodynamic stability of terminal points which depends on asymmetry and holds an important function in the selection of mRNA. This efficiency is usually measured by half of the maximum inhibition value known as IC_{50} values.

2.2 Selectivity

The high degree of specificity of siRNA-based approach is a major quality which enables the recognition of targeted and non-targeted sequences. The basis of the selectivity is the complementarity of mRNA bases, also referred to as Watson-Crick base pairing interactions. So, the antisense strand is designed based on targeted mRNA sequences and hypothetically. It is possible to treat all kind of genetic disorder through RNAi method once we know the nucleotide sequence of the associated genes. This unlimited capability of antisense-mediated suppression has supported systems for mutating nearly all the known genes of protein production comprised in the human genome at an extensive range. The high specificity may even permit distinguishing between abnormal alleles of a gene which arises dysfunction from the native allele which may differ in just a single or few nucleotide bases. This is obviously helpful for focusing on predominant mutants, as an example in the case of a few oncogenes (Aagaard and Rossi 2007).

Furthermore, certain malignancies such as cancer can result in different etiology such as abnormal chromosomal number, multiple gene mutations, or epigenetic changes (Clark 2007). So, this disease may be significantly different in the case of each patient and requires personalized treatment. Moreover, cases not responding to chemotherapy method due to resistance also may get the advantage of siRNA therapy (Gottesman 2002).

2.3 Safe and Cheap Alternative

The RNAi effector molecule executes the cleaving of mRNA at the later stage of translational gene expression, unlike other antisense molecules. So there is no chance of interaction of siRNA with DNA and thereby deleterious mutation and pre-birth abnormalities due to gene therapy (Xu and Wang 2015). As well as the process is used to synthesized siRNA is easy and cheaper. For the synthesis of RNAi, no any costly instruments and hazardous chemical are used.

3 Routes of Administration of siRNA

There have been various clinical trials reported in RNAi therapy by local and systemic administration of siRNA to various organs such as kidney, eye, liver, and skin which is dictated by the types of cells and tissues at the targeted site. For localized delivery, siRNA can be administered specifically to the skin surface, eye, lungs etc. On the other hand, oral, intravenous (IV) or intraperitoneal (IP) routes of systemic delivery are suitable for those genetic dysfunctions which are not confined to a particular site or not easily amenable to other clinical therapies. But the systemic siRNA administration has many hurdles such as less specificity and bioavailability to the targeted sites, rapid clearance or systemic toxicity (Durcan et al. 2008).

3.1 Localized Delivery

Administration at the local sites has turned into better and compelling option, permitting the utilization of lower dosages and diminishing the undesirable effects. siRNA specifically administered through different routes such as topically, gastrointestinal tract (GIT), respiratory tract, nervous system, ocular etc. The proficient administration of naked siRNA has been done through the topical application at the anterior chamber or by injecting within the vitreous at the posterior chamber. Particularly as a localized manner, delivery of siRNA to through the topical and ocular surface is easy compared to other organs, tissues, and cells. Continuous progress has been made towards its use as a therapeutic procedure for eye diseases. Currently, certain siRNA-based drugs for topical and ocular defects have passed the preclinical phase and various clinical trials studies are going on (Nguyen et al. 2012a, b; Pañeda 2013).

3.1.1 Topical Administration

The skin is the outermost covering of the body with the largest surface area, hence prone to numerous skin diseases including some genetic defects as well such as dermatitis, rheumatoid arthritis, psoriasis, pachyonychia congenital, alopecia areata, melanoma, wounds etc. Topical application is most beneficial for the treatment of these disorders due to easy accessibility at the affected site, decreased systemic side effects, elimination of the first-pass metabolism, thereby decreasing the therapeutic dose. Being a non-invasive and easy-to-administer method, the topical application also enhances patient's compliance to treatment (Prausnitz et al. 2004).

siRNA module has been tried for the treatment of psoriasis which is one of the most common skin disorders and associated with overgrowth in keratinocytes due to declination in apoptosis that makes a plaque in addition to the inflammatory component of the disease. Conventional treatment of psoriasis topically applies corticosteroids, which leads to weight gain, itching dryness, stinging, fluid retention, irritation, hypopigmentation, atrophy, hypertension, and other side effects. It is highly desirable to find drug molecules with fewer side effects. Therefore, targeting the genetic cause i.e. upregulation of the pro-inflammatory cytokines is a more suitable way to treat psoriasis. The overproduction of cytokine can be regulated by suppressing tumor necrosis factor alpha protein. Animal studies show that silencing of gene encoding tumor necrosis factor-alpha resulted in the improvement in the psoriatic lesion.

Another successful example of the clinical utility of siRNA over drugs is observed in the case of allergic contact dermatitis (ACD). Only a short-term treatment with topical corticosteroids and calcineurin inhibitors were available earlier. However, the development of siRNA molecules in recent years has enabled more specific targeting of genetic factors. ACD is genetically characterized by the elevated expression of CD86 in the affected skin area. The reduction of CD86 expression by RNAi silencing decrease the inflammatory response of the immune system and hypersensitivity (Aldawsari et al. 2015).

Despite the above advantages, there are several anatomical barriers due to composition and cellular distribution of the superficial surface of the skin. The outer stratum corneum is impermeable to the therapeutic RNA and acts as a barrier in topical administration (Madison 2003). The issue of low penetration overcome by using carrier systems for delivering which can encapsulate therapeutic siRNA molecule and penetrate through stratum corneum to reach the target site within the dermal layers First, the carrier system should be able to encapsulate or form a complex with siRNA; second, the carrier system must possess the ability to cross the stratum corneum to reach the target cells in the skin. More importantly, the carrier system should be able to bypass endolysosomal degradation before reaching the nucleus to obtain the clinical impact of RNAi after topical application (Aldawsari et al. 2015).

3.1.2 Ocular Administration

Ophthalmology is the earliest field where the efficiency of RNAi-based therapeutics was studied through clinical trials. Gene therapy has proven to be most successful in the case of ocular complexities due to certain unique properties. It is an immunogenically-privileged and easily assessable site. Moreover, it is comparatively a distinct part from the rest of the body (Conley and Naash 2010). Hence, the low therapeutic dose is required and the probability of systemic toxicity also confines within the local area by subsequently enabling targeted delivery. Thus, the RNAi method can be employed as a proficient treatment for different visual impairments such as irreversible blindness (Conley and Naash 2010). siRNA-based therapy has introduced advancement to resolve visual impairments results due to age-dependent macular degeneration, glaucoma and photoreceptor degeneration been utilized to distinguish qualities that advance harm in the eye and could be the premise of new medications for some, sicknesses, including glaucoma, age-related macular degeneration, and photoreceptor degeneration (Campochiaro 2006).

For example, major ophthalmic disorders in humans such as anterior ischemic optic neuropathy and glaucoma-related blindness are the consequences of Retinal ganglion cell (RGC) destruction is efficiently treated by intravitreal administration of siRNA. The studies on rat models suggest that loss of RCG is due to cleavage by a caspase-2 nuclease, which is unregulated in case of optic nerve damage. Intravitreal delivery of a chemically-modified synthetic siRNA for inhibiting the expression caspase-2 expression enhanced RGC survival remarkably. The persistence of exogenously delivered siRNA has a longer span of up to a month and a safer RNA interference without prompting interferon-mediated inflammation in the retinal region which is immunogenically-privileged, contrary to the conventional ophthalmic drug formulations which sometimes cause ocular irritation. The trials of RNAi resulted in suppression of caspase-2 enzyme and thus siRNA are clinically proven as the latent neuroprotective component for intervention in ocular diseases concerning to RGC loss (Ahmed et al. 2011).

Despite being a suitable site for targeted siRNA release, few problems have been observed in ocular administration. The first physical barrier is protective viscous fluid secrete by the lacrimal gland and known as tear film. It continually washes of the applied formulation before complete absorption. The second hindrance is produced by conjunctival and corneal epithelial cells which have low penetrability (de la Fuente et al. 2010). Similar hurdle is produced by a blood-retinal barrier (Bodor and Buchwald 2005). Because of the mentioned physiological constraints, ocular administration of interfering RNA suffers from low bioavailability. Therefore, delivery of RNA via ocular site depends on the efficient delivery platform.

3.1.3 Pulmonary Administration

Lungs are the most vital organ of the respiratory system and attribute to special morphological features such as large surface area and vascularized structure. These features enable the site-specific administration to lungs through aerosol formulation or intranasal instillation and provide novel methods for curing respiratory related disease (Thomas et al. 2007). Also, siRNA is prevented from degradation by nuclease during pulmonary administration through the respiratory tract. However, there are certain physical hurdles, for example, beating moment of cilia and flow of mucus for clearance of respiratory route epithelial cells and the negatively charged cell membrane surface (Gutbier et al. 2010). These hindrances influence the proficiency of in-vivo delivery of siRNA (Griesenbach et al. 2006). Numerous investigations have examined in spite of the above troubles, the scope of localized administration for site-specific application of siRNA to the organs of the respiratory system. Such investigations discover new medical strategies for major respiratory-associated disorders, for example, influenza, respiratory syncytial infection, cancer and severe acute respiratory syndrome (SARS), atypical mycobacterial infections, lung cancer, respiratory alphaherpesvirus infection, pulmonary fibrosis, asthma, chronic obstructive pulmonary disorder (COPD), emphysema, respiratory syncytial virus (RSV), parainfluenza virus (PIV), rhinoviruses coronavirus (SCV), ischemic reperfusion injury and other endogenously accessible disease in the lungs by confined delivery at peculiar site. Pulmonary administration can be done by inhalation, intranasal route, and intra-tracheal route.

Localized delivery of siRNA in lung disorders provided an alternative when the systemic administration does not efficiently suppress the targeted proteins. Most common example of commercial siRNA formulation with pulmonary administration method is ALN-RSV01, used in case of upper respiratory tract infection which occurs due to the growth of RSV. This formulation acts on the N gene and inhibit the nucleocapsid synthesis in RSV, thereby control the replication of RSV in nasal passages and trachea.

Cystic fibrosis is another genetic disorder which involves the chronic inflammation mediated through IkB/protein complex that is involved in transcriptional regulation of DNA, cytokinesis and cellular development. This protein is produced by cystic fibrosis trans membrane conductance regulator (CFTR) gene and epithelial Na^+ transporter (ENaC) gene which founds to be mutated in the case of cystic fibrosis. Pulmonary siRNA delivery decreases the expression of epithelial Na^+ channel (ENaC) gene or down-regulate its activity to normalize the Na^+-Cl^- transport across the cell membrane and prevent the excess mucus release, luminal hydration, pulmonary infection and epithelial cell destruction (Khatri et al. 2012).

3.1.4 Gastrointestinal Administration

siRNA particularly could be administered within the gastrointestinal system by means of endoscopic infusion for the treatment of infections. The surface of the GI tract is covered by a layer of fluid called mucus. This viscous fluid stimulates the release of RNAi formulation straightforwardly to its site of activity, thereby eliminating any undesirable effects and systemic toxicity to adjacent tissues and organs (Pellish et al. 2008). Moreover to the application of siRNA for treatment GI-associated diseases, this strategy also has been utilized for illustrating the course of different gastrointestinal infections and the determination of possible targets for RNA interference, for example, tumour necrosis factor alpha (TNF-α) in the case of ulcerative colitis, Crohn's disorder and inflammatory bowel diseases (IBD) (Zhang et al. 2006). IBD earlier treated with infliximab, a chimeric IgG-1 monoclonal antibody against (TNF-α). But these antibodies can trigger anti-antibodies production and elicit immune hypersensitivity. On contrary, localized silencing of TNF-α through siRNA molecules may not trigger any immune response.

Traditional treatment of oesophageal cancer involves aggressive surgical and adjuvant chemoradiation therapy. One common etiology observed in various types of oesophageal malignancies is hyperactivity of Bcl-X_L, the gene responsible for inhibiting apoptosis. Therefore, by interfering with the Bcl-X_L expression via siRNA-based approach, the oesophageal squamous cell cancer line was destructed in the apoptotic programme.

Similar to oesophageal malignancies, pancreatic cancer has a delayed diagnosis and chemotherapy practices. The common reason observed for adenocarcinoma is an elevated level of serine-arginine protein kinases (SRPK) which led to erroneous splicing of pre-mRNA is the causative in certain malignant cancer including pancreatic cancer. Because serine-arginine protein kinases (SRPK) are the class of proteins with a regulatory role in pre-mRNA splicing, and elevation in the expression of SRPK1, trials were done by Hayes et al. utilized SRPK1-specific siRNA to block SRPK1 activity. Consequently, growth in pancreatic and colonic cancer cell lines declined and apoptotic death was much higher as compared to common anticancer drugs such as gemcitabine and cisplatin (Pellish et al. 2008).

3.1.5 Central Nervous System Administration

The brain is the most complex organ of the body, not only regarding anatomy and function but also in terms of medical therapies. Due to its complexity contributed by various types of cells and vivid functions, the conventional treatment methods do not prove to be as effective in the case of neuron-associated problems. In this specific situation, therapeutics based on RNAi could identify the genetic basis and comprehend the course of neural disorders as a part of prognosis. Later, it could evaluate the remedial capability of targeted RNAi therapy for silencing genes involved in neurological disorders. For example, PTEN-induced putative kinase 1 (PINK1) is targeted for treatment of Parkinson's disease (Deng et al. 2005); Huntington gene for Huntington's disease (Harper et al. 2005); purinoceptor 3 (P2X3) for neuropathic pain (Dorn et al. 2004) and EGFR for intracranial tumors (Boado 2005). Certain neurodegenerative diseases such as Alzheimer's disease-BACE1 (Singer et al. 2005); amyotrophic lateral sclerosis-SOD1 (Ralph et al. 2005) and other dysfunctions mediated by hypoxic/ischemic events could also be treated through siRNA-mediated silencing. Contrary to the potential specificity of the RNAi method, the efficiency is compromised due to the highly impermeable blood-brain barrier that confines the availability of RNAi-based formulations across the brain from peripheral blood vessels (Cazzin and Ring 2010).

Alzheimer's disease (AD) is a protein misfolding disorder caused by defective extracellular amyloid β-peptide (Aβ) disease and leads to progressive cognitive decline and memory loss. The researcher has found that γ-secretase which cleaves γ-site of the amyloid precursor protein (APP) is a key component that generates Aβ from through sequential proteolytic cleavage events. BACE1 is the gene expressing for γ-secretase, hence the therapeutic siRNA molecule targeting BACE1 considered as a potential gene therapy of AD. Similarly, a targeted decrease in APP expression through the same mechanism also serves the purpose.

Huntington's disease (HD) is another neurodegenerative disease benefited by this new siRNA module. The disease is associated with the upregulation of Huntingtin (Htt) gene, which adds extra polyglutamine stretch in exon1 region of the HTT protein product. This mutant protein type misfolds and forms inclusion bodies. The striated neurons are the most degenerated cell types, hence CNS is severely affected in the case of HD. Silencing mutant Htt mRNA through RNAi the approach has been seen for clinical utilization (Gherardini et al. 2014).

3.1.6 Vaginal Administration

The vaginal route is the primary site of invasion for some microbial agents causing infections especially sexually transmitted diseases, cancer, and inflammatory problems. In such instance, the topical administration of RNA-based product is the most ideal approach to anticipate sexually transmitted diseases. This type of application provides benefits of lower therapeutic dose and maximum bioavailability at the site infection (Rossi 2009). Therefore, the studies are on-going in context of targeted-siRNA therapy using identified gene targets such as CCR5 for human immunodeficiency infection (HIV); E6, E7, and Grb10 for cervical tumor and UL-29ANDUL-27 for HSV-2 infection. However, the most usual drawback of vaginal application is patient compliance.

HIV is one of the initial viral diseases, where targeting by RNAi, was effective in treating the viral infection as well apart from genetic disorders. The siRNA molecule developed to target different viral genes including gag, tat, nef, vif, rev and env involved in the viral cycle of HIV. The RNA may also act on host cellular cofactors such as CD4, CXR4, NFkB, and CCR5 as well as reverse transcriptase and control HIV replication effectively.

The vaginal administration has been more successful in case of cervical cancer result due to overexpression of HPV oncogene and triggers unregulated proliferation of cells. Therefore, inhibiting the activity of HPV gene via siRNA is very effective in slowing the proliferation of cancerous cervical cells. The siRNA therapy can specifically target human HPV 16/18 E6/E7 genes which have a significant effect in stopping the multiplication of cervical tumors (Wu et al. 2011).

3.2 Systemic Delivery

Comparatively, systemic delivery is more suitable in the clinical situation of blood-borne disorders and malignancies. Thus, the recent researches are more focusing on the alternatives to bring out the systemic delivery of siRNA as a more efficient and safe procedure of gene silencing. The chapter further discusses the role of NCs to overcome the pitfalls of systemic siRNA administration. Various nanoparticles can be used as non-viral vectors to facilitate delivery rather than naked siRNA molecule.

3.2.1 Oral Administration

The oral route is the most ideal mode of administration due to patient compliance. Moreover, other benefits such as cost-effectiveness and facile administration by oral route attract attention for drug delivery and are also preferred for local administration of intestinal therapeutics. But the acidic environment of stomach readily degrades the naked oligonucleotide. Therefore, the stable administration via the oral route achieved by nano carrier-mediated delivery (Akhtar 2009).

3.2.2 Intravenous Administration

Although IV administration is a successful mode of administration for many drug formulations, it has been observed that the delivery of siRNA by this method is hampered by the very short lifespan kidney filtration, serum protein aggregation and enzymatic degradation of therapeutic RNA after injecting. It has been exhibited in mice, rats, human and other primates that bare siRNAs are rapidly cleared from the body circulation after IV infusion with the half-life as short as a few minutes or approximately 0.5 h also, a large accumulation of therapeutic oligonucleotide occurs in the kidney after IV administration (Huang et al. 2016).

3.2.3 Intraperitoneal Administration

In addition to less stability of therapeutic RNA administered through IV route, other common after-effects of repeated IV administration are phlebitis or loss of veins and risk of embolism. In this regard, the IP delivery provides a better option, especially in the case of cancers of ovaries, digestive system, and peritoneum. The IP administration moreover found to have enhanced the in-vivo stability of siRNA and enables administration of a large volume of formulations at low concentrations so as to avoid particle aggregation (Singhania et al. 2011).

Many malignant tumors which arise from the stimulation of nuclear factor-kappa B (NF-kB), for instance, gastric tumor shows poor prognosis and resistance to chemotherapy alone. Thus, NF-kB is a common target for RNA silencing. A study demonstrates efficient intraperitoneal administration of NF-kB p65 siRNA along with a common chemotherapeutic agent (paclitaxel) for treating peritoneal prevailing of gastric tumor. Lower expression of NF-kB p65 expression was observed in western blot analysis and apoptotic destruction of cells transfected with NF-kB p65 siRNA as compared to treatment with paclitaxel alone (Inoue et al. 2008).

4 Major Hurdles to the Therapeutic Delivery of siRNA

The efficacy of RNA interference may differ for various siRNAs oligonucleotides targeting same mRNA sequence but at distinct nucleotide. Full efficiency of siRNA is not obtained as only some of the delivered siRNAs reach the action site without distortion in mammalian cells and successfully perform cleaving action (Naito and Ui-Tei 2013). Even though, approximately 58–78% of arbitrary siRNAs with non-specific sequences, mediate mRNA silencing with an efficacy more than half of the proficiency (>50%). Out of this, most of the (90–95%) proficiency is accounted by just 11–18% of the non-specific siRNA oligonucleotides (Chalk et al. 2004).

The decreased efficacy of siRNA may have many causes including instability, unspecific targeting to diseased-cells, and competition with endogenous RNAs, saturation of RNAi/miRNA complex, immune response activation and off-target impact. The problems become more prominent in applying the systemic mode of administration for delivering siRNA which imposes some more disadvantages compared to localized administration. The main problem is to conserve the activity of siRNA during circulation in the systemic blood vessels. The systemic circulation of siRNA is further hindered due to many anatomical and physiological protective barriers present in the human body such as renal filtration, vascular endothelium, extracellular matrix, serum protein aggregation, cellular internalization, phagocytes uptake, endosomal degradation, and RISC loading. siRNA need to overcome these barriers before it reaches its site of action, to exhibit clinically significant impact with considerable proficiency.

4.1 *Transient Effect*

RNAi mediated gene silencing can be stimulated by either direct or indirect mechanism. The first involves an RNA-mediated process where an exogenous synthetic siRNAs make a complex with a 21- base pair long duplex. This effector molecule is then targeted towards the specific cell. Meanwhile, the second approach involves an endogenous miRNA-based strategy where the longer RNA hairpin transcripts are processed intracellularly to produce the effector siRNA. While the direct method is more straightforward and efficient for gene cleavage, the impact of externally introduced, synthetically directly generated siRNA is short term. Thus, the synthetic siRNA requires to be administered repeatedly with each dose being expensive and time-consuming. The miRNA-based RNAi drugs, on the contrary, shows high stability and long-time effect. Hence a single treatment is not enough for effective gene therapy (Aagaard and Rossi 2007).

4.2 Stability

Another great pitfall of siRNA is related to the in-vivo stability. The exposed siRNA are liable to hydrolysis in the extracellular compartment due to the action of various enzymes present in serum and target tissues. The half-life of naked siRNAs in serum as reported so far is several minutes up to an hour (Rettig and Behlke 2012). The siRNA molecule structurally contains a ribose sugar backbone. On contrary to a hydrogen atom as in DNA, it is the hydroxyl group at 2′ position of the pentose ring which is targeted by serum nucleases (Czech et al. 2011). These nucleases break the phosphodiester backbone of RNA and degrade this nucleic acid. Thus, it is a challenge to deliver intact siRNA with conserved therapeutic activity through the circulation. Meanwhile, siRNAs must not only survive in the serum but should also reach their target cells in the specific tissues for internalization to occur appropriately and to achieve required RNAi effect (Guo et al. 2010a).

4.3 Physiological Barriers

As the siRNA administered and introduced into the systemic circulation, its route towards the site of action is obscured by many structural and anatomical defensive barriers (Fig. 3). The step of the clearance through the kidney and reticuloendothelial system (RES) of other organs such as lung, liver, bone marrow and spleen is the primary obstruction. It entraps all the particles with a diameter larger than 100 nm. Thus, the ideal RNA molecules to bypass renal and RES clearance should have a diameter ranging from 20 to 100 nm approximately. Additionally, therapeutic material within this size range shows an enhanced uptake at tumor sites passively through leaky capillaries of a pore size of 100–800 nm (Song et al. 2005).

After passing through the RES, as the siRNA enters the targeted tissue, it is hindered by local barriers present in the extracellular matrix and the layer of endothelial cells. Typically, the endothelial lining of tissues have a pore size from 4.5 to 25 nm, hence the smallest pore can allow particles of 4–5 nm only (Rettig and Behlke 2012). In such circumstances, only naked siRNA molecules or the one conjugated with small molecular ligands easily penetrate through the endothelial cells. Also, the naked or complexed siRNA must squeeze out of the dense extracellular space to reach the surface of the target cell.

4.4 Cellular Uptake and Endosomal Engulfing

Even after bypassing the nuclease activity in serum and extracellular matrix, the RNAi pathway still hindered by various obstacles. After reaching proximity to the target cells, uptake of siRNA by the target cells is also crucial, prior the therapeutic

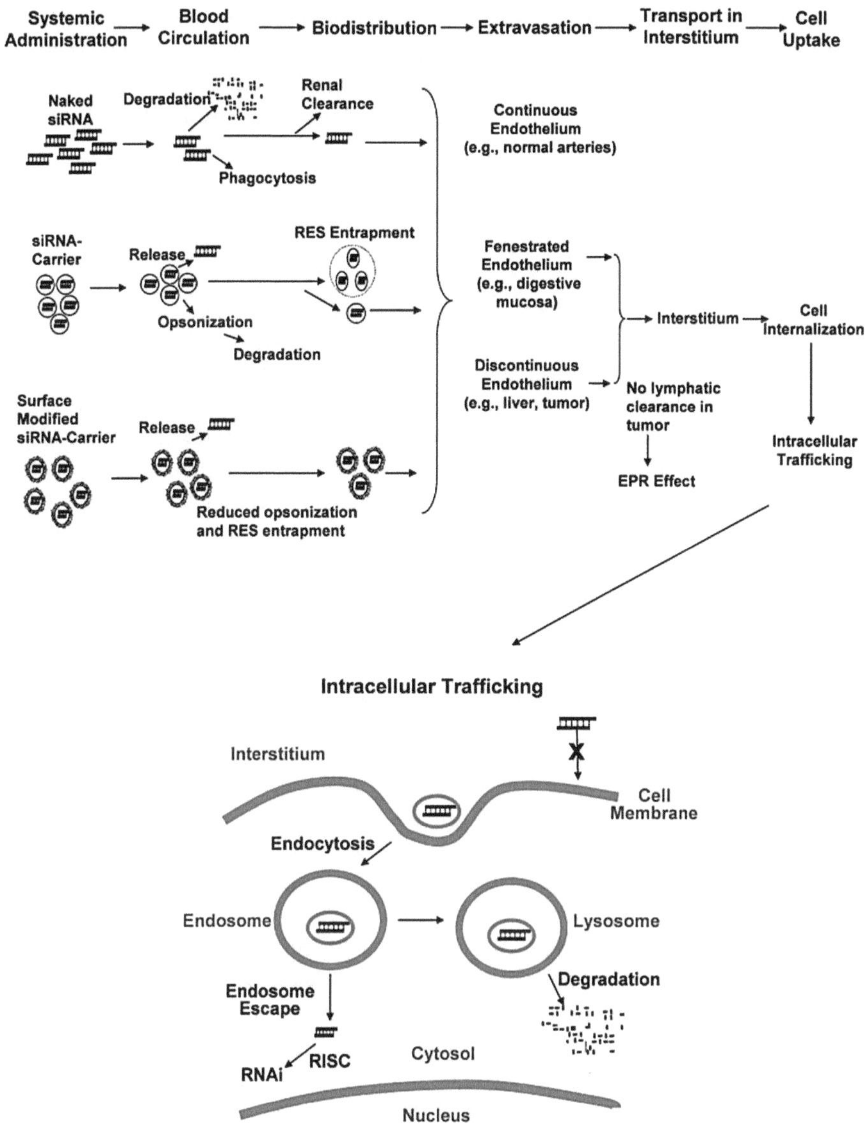

Fig. 3 Barriers to systemic siRNA delivery. (This figure is produced after reprint permission under license no. 4445840089493). (Wang et al. 2010)

RNA can perform their gene silencing function. The exposed siRNA molecule being negatively charged and large, constrain its own passage through the cell membrane and anticipates aggregation inside the cell.

The second barrier to overcome is endolysosomal compartments within the cytoplasmic space. These vesicles also contain various intracellular RNAase which may degrade the siRNA molecule once internalized i.e. endosomal engulfing. Strategies for designing and delivering siRNA are therefore essential for enabling endocytosis and subsequent endosomal escape of siRNA molecule. Additionally, siRNAs being susceptible to hydrolysis by endosomal nuclease enzymes can be incorporated into RISC with improved efficacy, and to compel the clinical utility (Gavrilov and Saltzman 2012).

4.5 Off-Target Effect

Recent studies conducted by Birmingham et al. (2006), Mouldy (2006), and Wilson and Doudna. (2013) have shown that the specificity of RNAi is sometimes compromised. The evidence from these studies shows that most siRNA molecules are probably not as specific as it was presumed and that siRNA treatment can bring about off-target gene silencing. The off-target effect refers to the suppression of genetic region other than the causative gene target. Such a phenomenon may produce undesirable genetic alteration and cellular dysfunction. The most common reason of "off-target" silencing is the seed region of the siRNA sequence. This region is characterized by partial homology of 6–7 nucleotide long sequence particularly inside the 3′ untranslated sequence (3′UTR) which bound with a different mRNAs rather than the desired target mRNA (Birmingham et al. 2006). In the condition, a seed region siRNA acquiring complementarity with certain regulatory miRNAs may distort the gene expression (Mouldy 2006). Moreover, failure of discrimination between the guide strand and the passenger strand by RISC may prompt the chances of complementary interaction between siRNA and non-targeted mRNA. The off-target effect can also arise as an after effect of the immune activation. RNA if perceived by immunoreceptors sometimes, such as Toll-like receptors (TLRs) and trigger signalling biomolecules such as cytokines which may alter gene expression (Wilson and Doudna 2013).

Off-target cleavage can't be overlooked as it prompts adverse mutations at the genetic level and results in unexpected consequences. Thus, the selection and sequencing of candidate siRNAs should be done specifically. Two basic processes can be followed for determining the potent RNAi sequence specific to the targeted gene. Predictive bioinformatic approaches implemented at the initial stage of siRNA to develop the most suitable sequences of the siRNA molecule. Later, exclusive trials on all potent therapeutic siRNA sequences should be performed to observe normal protein expression profiles. Any inadvertent off-target effect may result in an abnormal interpretation of data and potential toxicity.

4.6 *Saturation of RNAi Machinery*

The bioactive material is usually analogues to the cellular biomolecules and works similarly to the in-vivo biomolecules. Externally delivered therapeutic siRNAs almost resembles intrinsic siRNAs and depend on the endogenous machinery for inducing RNA interference. Hence, there are significant risks of saturating such pathways and hence disturb the native biological systems that based on cellular processing to execute their role.

4.7 *Stimulation of Immune System by siRNAs*

While in general, the active immune mechanism serves for defence mechanism and is advantageous. However, sometimes, it evokes undesired immune sensitization, which may lead to adverse effects (Marques and Williams 2005). The siRNA might be discriminated as a foreign particle through various defence mechanisms and components of the immune system. For example, siRNA may be recognized by cytoplasmic and endosomal Toll-like receptors expressed on the surface of different immunocompetent cells. Consequently, the release of cytokines and type-I interferon generates inflammatory reactions. The immune stimulation property of siRNA is directly related to its size. siRNA oligonucleotide up to 30 base pairs usually doesn't confer cytotoxicity. A cell could safely be transfected with 30-nucleotide bp longer dsDNA without any cellular toxicity. On contrary, longer RNA strands elicit an immune response as it bounds to either TLRs or to certain enzymes such as 2′,5′-oligoadenylate synthetase-RNase L or protein kinase to trigger interferon. There are specific GU-rich motifs which are actually determined by TLR7, refer as "danger motif". For example, in the endosome vesicle of plasmacytoid dendritic cells 5′-GUCCUUCAA-3′ sequence is recognized by the TLR7 and stimulates innate mechanisms to release inflammatory cytokines further. Another example is of TLR9 activity is countering CpG motifs in antisense oligonucleotides. Thus the immune actions can be controlled by mode of delivery and passage across various intracellular compartments of the siRNA. RNA oligonucleotides larger than 27–29 base pairs and highly potent Dicer substrates siRNAs need to be safely administered and monitored for immunogenic side-effects (Rose et al. 2005).

5 Overcoming the Hurdles to siRNA Delivery Using Nanocarriers

To overcome the above-discussed therapeutics hurdles, and to achieve efficient and potent systemic delivery of unstable siRNA, several methods have been developed that can increase the stability in serum, improve in-vivo delivery and efficacy of gene silencing tendency. Some of them are-

(i) Designing artificially chemically modified siRNA structures which are prepared by substituting various molecules or atoms at different positions for examples-substituting phosphodiester (PO_4) linkages with phosphorothioate (PS) at the 3′-end, or introducing the new molecules that resist the enzymatic degradation like adding O-methyl (2′-O-Me), fluoro (2′-F) groups or methoxyethyl (2′-O-MOE) group significantly prolonged circulation half-life in systemic route and improved RNAi efficiency (Czauderna 2003; Chiu and Rana 2003).
(ii) By complexation with a molecule or carrier, for example, the cholesterol-siRNA complex. The cholesterol-siRNA complexes improve pharmacokinetics (increases half-life of siRNA) (Wolfrum et al. 2007).
(iii) By encapsulating siRNA in a carrier such as viral vectors. Various natural viral vectors have already in use for systemic transport of siRNA to treat various carcinomas. Jung et al. used an adenoviral vector encoding siRNA significantly suppressed the outgrowth of the pituitary tumor by transforming gene overexpressed in both in-vivo and in-vitro hepatocellular carcinoma cells (Jung et al. 2006). Sabbioni et al. (2006) showed type 1-based amplicon vectors that based on herpes simplex virus, inhibited in-vivo tumor formation of human polyomavirus BK-transformed cells (pRPc cells) (Sabbioni et al. 2006).

Even though the above-mentioned methods have several advantages and provide a good platform to improve delivery of naked siRNA. However, the frequent application of these methods is limited because they are also having numerous drawbacks such as safety issues (insertion mutagenesis and immunogenicity) low loading efficiency and only laboratory/small-scale production. Furthermore, siRNA is more liable to nuclease degradation than circular plasmids. Therefore, we need a stable delivery system to improve the systemic delivery of siRNA.

Now a day, an alternative approach, synthesis of nonviral vectors (synthetic carriers) is inflowing to resolve or reduce these drawbacks and to enhance in-vivo siRNA delivery. Nonviral vectors have various sizes, shapes, dimensions, and consist of different biocompatible and biodegradable materials. These carriers can load or entrap the high amount of siRNA and shielding from various hurdles or barriers such as rapid elimination, enzymatic degradation, reduced potency in-vitro and in-vivo both etc. Currently, different dimensions, shapes, nanosized (at least one dimension <100 nm) nonviral vectors are applied for the systemic delivery of genes, drugs, or co-delivery of both genes and drugs together. These nanosized, nonviral vectors enhance pharmacokinetics (increase half-life) with EPR (enhanced permeability and retention) effect, decrease rapid renal filtration/clearance, and protect from extracellular enzymatic degradation and support in escape from endosomes intracellular. These NCs have some beneficial properties like high biocompatibility, non-immunogenicity etc. Polymer, lipid-based nanomicelles liposomes, and dendrimers are mostly used nonviral NCs.

6 The Aid of Nanocarriers for siRNA Delivery

6.1 Enhanced Blood Retention Time

The size of naked siRNAs is about 7 in length (Hansen et al. 2001; Smith et al. 1996) and 2 nm in diameter (Sinden et al. 1998) and with a molecular weight of about 13 kDa (Whitehead et al. 2009). The kidney is made up of numerous glomeruli, and the main function of glomeruli is the natural filtration barrier of undesired by-products in the blood. The size of pore the glomeruli membrane is approximately 8 nm (Wartiovaara et al. 2004). Any molecules that have a molecular weight and a size below 50 kDa and 8 nm, respectively are typically rapidly expelled from the kidney (Rappaport et al. 1995). After systemic delivery, siRNA is rapidly expelled out from the body system through either renal clearance and filtration through the renal system or the entrapment in the RES in the liver, spleen, lungs and bone marrow. For increasing the retention time of siRNA in the body, siRNA is entrapped indifferent NCs that have the size range between 20 and 100 nm in diameter. This optimal size range helps to avoid potency for avoiding both renal clearance and RES entrapment and allows NCs to remain in the circulation for a longer period and to increase the rate of intracellular penetration in tumor cells and tissues via passive targeting (Jang et al. 2003; Pecot et al. 2011).

6.2 Enhanced Stability and Cell Penetration Property

The specific physiochemical properties of naked siRNA are their anionic charge due to the presence of phosphate groups in the structure and hydrophilicity. As the plasma membrane is made up of lipid bilayers, it is lipophilic in nature and having anionic charge. Therefore, the intracellular penetration of siRNA across the plasma membrane is quite limited by passive transportation due to the electrostatic repulsion. When the siRNA is administrated in the bloodstream, it is immediately covered by a major component of blood plasma protein i.e. serum albumin and initiates our immune system to engulf albumin coated siRNA by phagocytes NCs entraps or covered the siRNA, avoid the engulfment by the phagocytes cells and increase the cell penetration of siRNA. Polycationic (positively charged) NCs are the best suited for high siRNA loading, shielding from degradation during bloodstream circulation, and systemic delivery of siRNA, especially to metastatic tumors. The most commonly used polymers are polyethyleneimine (PEI) (Illum and Davis 1984; Norman et al. 1993). According to Bonnet et al. (2008) linear PEI-siRNA (lPEI/siRNA) conjugate showed better cytocompatibility and non-immunogenic than branched PEI-siRNA (bPEI/siRNA) conjugate (Bonnet et al. 2008). However, the NCs that are prepared from a single PEI polymer is not stable during systemic delivery of siRNA and agglomerate in the bloodstream due to slat medium and serum protein. The prominent solution to overcome the problem is to use a copolymer/mixed polymer (di or triblock copolymers) which are more stable, for example, PEI can be mixed with poly (ethylene glycol) (PEG) to prepare a PEGylated PEI/siRNA. PEG polymer covers at the outside of the PEI/siRNA conjugate by forming a hydrophilic layer that sufficiently reduces the agglomeration tendency of NCs with increased EPR effect (Malek et al. 2008; Merkel et al. 2009). Kim et al. (2012) prepared mannosylated and pegylated polyethyleneimine/siRNA (PEI-PEG-MAN/siRNA) polyplex NCs with an approximate size of 357.33 nm for increased cellular uptake of siRNA. In this conjugate, PEI entrapped a high amount of negatively charged siRNA, PEG provided outer stable hydrophilic membrane, and mannose was used as a cell binding ligand for macrophages (Kim et al. 2012).

Some other polycationic peptides, like poly-histidine polymers, and lipids NCs have also used for the systemic and improved delivery of siRNA (Midoux et al. 2009).

6.3 Site-Specific Delivery

Although after entrapment of siRNA by different positively charged NCs enhance the therapeutic efficacy of siRNA, the delivery systems are not enough to deliver a high amount of siRNA to the specified tissues/cells by passive diffusion transport. Due to this, NCs are not capable to differentiate infected (cancer) cells from normal/healthy cells. Also, positively charged NCs agglomerate in different cell media and biological fluids. To increase the concentration of siRNA on the targeted site, the nanocarriers-mediated targeted delivery system is a new and novel approach to increase the therapeutic efficacy of siRNA. Receptor-mediated endocytosis is an active transport system that improves the rate of cellular uptake of siRNA loaded NCs. The active, targeted transport system can be achieved by surface modification of NCs using different targeting ligands. These specific ligands bind to a site-specific receptor which is overexpressed on targeted cells and tissues (Li et al. 2016; Akinc et al. 2010). Examples of some commonly used ligands are (poly)saccharides, (Zhang et al. 2016) vitamins, (Duan et al. 2016) antibodies, (Gu et al. 2017) peptides, aptamers, transferrin (Tf) etc.

6.4 pH-Sensitive Trigger Release

The development of pH-sensitive NCs is the most commonly used approach to improve delivery of siRNA after systemic administration, especially in cancer. The pH of blood, extracellular matrix and fluid are mildly basic (pH ~7.4), whereas the pH of the intracellular compartment is 6.8. However, cancer cells have even more acidic (pH 4–6) for both extracellularly and intracellularly, caused by the higher rate of glycolysis and tendency to produce higher CO_2 and lactic acid. High lactate levels indicate metastases, tumor recurrence, and prognosis in some cancer patients. Basically, there are three different materials are used to develop pH-sensitive NCs: protonizable, acid-labile, and destabilizing compounds (Torchilin 2014; Sawant and Torchilin 2012).

Poly(2-(diisopropylamino)ethyl methacrylate) (PDPA) is a smart polymer, that is used to form NCs. These NCs bypass the endosomal compartment and improve the efficacy of siRNA delivery (Qian and Berkland 2015). It is a pH-sensitive polymer that is hydrophobic in nature at neutral and physiological pH medium. However, in mild acidic medium (pH <6.2) it changes from hydrophobic to hydrophilic in nature (Licciardi et al. 2005). This polymer can easily encapsulate the desired amount of siRNA and shield the siRNA from enzymatic degradation at physiological pH. The encapsulated siRNA is released in the cytoplasm after dissociation of NCs in early endosomes or lysosomes (pH <6) via protonation of the PDPA and display an outstanding 'proton sponge' effect (PSE) due to tertiary amines at the terminal surface (Yu et al. 2011).

Numerous NCs have been developed that are stable at blood pH and give the protection of siRNA during circulation in the bloodstream but they degrade at low pH in tumor tissue environment and release the siRNA (Kim et al. 2012).

PSE is the effect that is used for protecting the degradation of therapeutics agents loaded NCs from different cell organelles such as early endosome and lysosome using various aminated polymers like PDPA, PEI, (Benjaminsen et al. 2013) poly-L-Lysine (PLL) etc. These polymers are destroyed to cell organelles upon acidification in low or acidic pH and safely deliver the genetic material in the nucleus. The PSE produces from numerous slightly basic cojugates (between pH 5 and 6), promoting proton (H^+) absorption in cellular organelles with lower pH and an osmotic pressure generated around the organelle membrane. Due to the generation of osmotic pressure, acidic endosome burst and release the materials into the cytoplasm. The advantage of the PSE is that we could easily control this effect by partially altering the groups from carboxylic acid (-COOH) to tertiary (3°) amines. Whenever both the functional groups are attached on the surface of NCs, form electrostatic and steric interactions that are extremely reactive to the acidic organelles and siRNA is appropriately attached and cellular entry (Yezhelyev et al. 2008).

6.5 *Avoid Intracellular Endosomal Engulfing*

As discussed earlier, naked siRNA cannot efficiently across the plasma membrane due to unfavorable physicochemical properties of siRNAs. A very less amount of siRNA is passed through from the semi-permeable plasma membrane. After the intracellular delivery, siRNA is less stable in the bloodstream and has a very less half-life (<5 min) due to rapid enzymatic degradation by endogenous serum nucleases (Tseng et al. 2009). The intracellular trafficking of siRNAs starts in early endosomal vesicles where siRNA fuse with sorting endosomes and finally transfer their contents to the late endosomes. A high amount of siRNA is entrapped by the late endosomes and lysosomes and very less amount of siRNA is reached into the cytoplasm to come into play regardless of the release mechanism. To avoid the rapid degradation and shielding from engulfing by the late endosomes and lysosomes, various NCs have been developed to shield the naked siRNA such as cationic lipid, (Lu et al. 2009) polymers and di or triblock copolymers NCs, (Qian and Berkland 2015) nanomicelles (Zhou et al. 2016) etc.

7 Classification of Nanocarriers Used for Systemic Delivery of siRNA

The NC that is applied to shield and to improve in-vitro and in-vivo delivery of siRNA must be safe and non-toxic for human health because they come in direct contact with humans. Naturally obtained molecules are safer as compared to chemically synthesized molecules. Because of the synthesis of NCs several hard acids and bases, and others hazardous chemical are used. These acids and hazardous chemicals give several toxic effects on human health and the environment also. Mostly NCs have the diameter >100 nm. However, the long chain polymeric NCs can have the diameter >500 nm. NCs developed for the systemic delivery of siRNA can be broadly classified into three different types:

7.1 *Organic Nanocarriers*

Organic nanocarriers (ONCs) are prepared from natural or synthetic organic molecules and the mixture of both. ONCs are typically soft, biocompatible, biodegradable, less toxic, non-immunogenic, highly stable, and can shield the siRNA. From the last few decades, several scientists and researchers have already used numerous organic NCs for protecting the drugs and nucleic acids from the harsh environment, enzymatic degradation and blood plasma proteins during systemic circulation inside the body. The most commonly used ONCs are nanomicelles, liposomes, polymeric nanoparticles, dendrimers, and carbon nanomaterials (graphene oxide sheets, nanotubes, spherical shape fullerene). These NCs have some additional benefits. The synthesis procedures for ONCs are quite easy and they can encapsulate more drugs or nucleic acids molecules with a controlled delivery. The surfaces of ONCs can be simply decorated with targeting biomolecules to develop targeted carrier systems. These advantages make ONCs the most appealing systems for drugs and genes delivery and other biomedical applications.

7.1.1 Polymeric Nanocarriers/Polyplexes

Currently, several natural and synthetic polymeric nanocarriers (PNCs) (Fig. 4c) offers a great platform to deliver drugs and gene with improved efficacy. These polymers are highly biocompatible, biodegradable, non-toxic in nature and have outstanding controlled release character. Generally, PNCs are cationic, anionic and non-ionic in nature. The polycationic polymer has emerged as one of the most promising candidates for developing efficient gene delivery vector.

Polymers can be broadly classified into two-types based on their sources: natural and synthetic. The most commonly used natural polymer is chitosan (CS), poly(lactic acid-co-glycolic acid) (PLGA), (Zhou et al. 2012) atelocollagen, (Minakuchi et al. 2004) inulin (Cavallaro et al. 2017). CS is a linear, natural, cationic, FDA (Food and Drug Administration) approved polysaccharide (a long chain of monosaccharide carbohydrate) co-block polymer of glucosamine and N-acetylated glucosamine. Polymer and has been used for both in-vitro and in-vivo systemic delivery of siRNA because it has nontoxicity, biocompatibility, and biodegradability properties (Malhotra et al. 2013; Vauthier et al. 2013). Currently, CS is the highly used polysaccharide polymer for delivering of therapeutic agents (drugs and genetic agents) due to its high permeability capability across the semipermeable cell membrane (des Rieux et al. 2006). In one of the study, prepared an novel amino acid-functionalized Arg-Gly-Asp (RGD)-chitosan nanoparticle (RGD-CHNP) that significantly increased selective intratumoral delivery of siRNA for regulation of many growth-promoting genes (PLXDC1, FAK, and POSTN) accompanied by therapeutic efficacy in the A2780, HeyA8, and SKOV3ip1 orthotopic animal studies of ovarian cancer models (Han et al. 2010). Besides natural polymers, several biocompatible, biodegradable, non-toxic synthetic PNCs have also been used for systemic delivery of siRNA such as PLL (Cun et al. 2011), PEI (Nimesh and Chandra 2009), PEG, dimethylamino ethyl methacrylate, polyfluorene, and cyclodextrin-based polycations. These polymers are linear and branched and similarly the efficient way to shield and efficient delivery of siRNA. Both types of PNCs can encapsulate a high amount of siRNA and deliver it via the passive route. For increasing the delivery concentration of siRNA on specific infected cells/tissues by escaping the normal cells and tissues, different actively targeted molecules are attached to the surfaces of PNCs (Table 1).

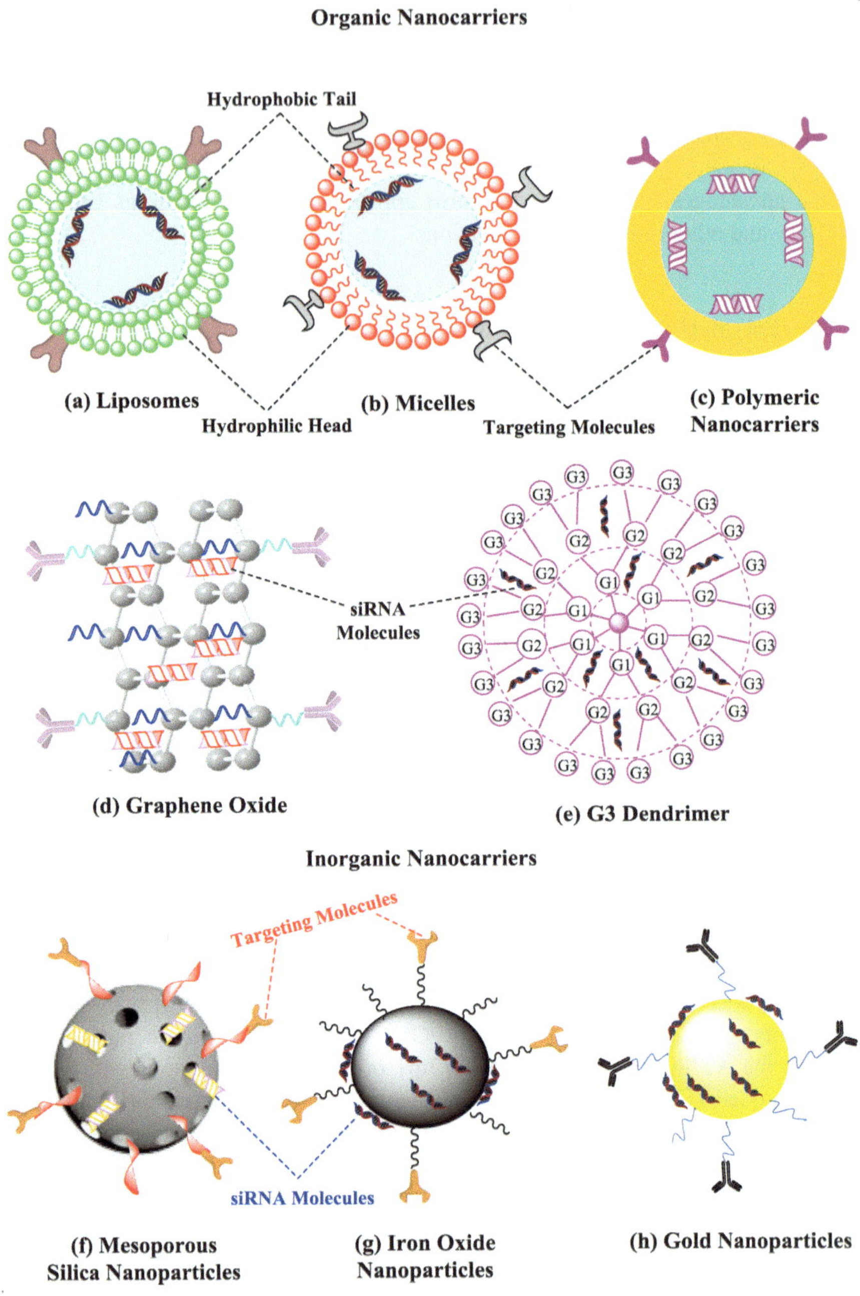

Fig. 4 Schematic illustration of various nanocarriers

Table 1 Application of various polymeric nanoparticles for systemic delivery of siRNA

Nanocarriers	Shape	Size	ζ (mV)	Targeted/siRNA molecule	Cell lines	Type of study	Aid	References
PLGA-PLL-PEG	Spherical	~150 nm	–	iRGD	HEK293T and A549	In-vitro	Targeted delivery, controlled release, enhanced siRNA potency, endosomal escape	Zhou et al. (2012)
PEI-HYD	Spherical	150–250 nm	Low or neutral surface charge (±10 mV)	CGRGDS peptides/siAldh1a2	HUVECs cells	In-vitro, in-vivo	Site-specific delivery	Wang et al. (2016)
PLGA	Spherical	218.3–257.9 nm	−45.5 to −37.5 mV	–	H1299	In-vitro	High encapsulation efficiency (~70%), sustained release, highly stable, protection from nuclease degradation	Cun et al. (2011)
Chitosan-mPEG	Spherical	263.6 ± 13.5 nm	−7.59 ± 0.94 mV	CP15	Mouse xenograft model of colorectal cancer	In-vivo	Targeted delivery, enhanced cell permeability, and enhanced retention time in blood	Malhotra et al. (2013)

PLGA Poly (DL-lactic-co-glycolic acid), *PLL* Poly-l-lysine, *PEI-HYD* Polyethyleneimine modified with hydrazide groups, *HUVECs* human umbilical vein endothelial cells, A549- human lung cancer, H1299- human non-small cell lung carcinoma cell line

7.1.2 Nanomicelles-Based Delivery

The term "micelles" was given by McBain in 1913. Micelles are formed by supramolecular self-assembly of surfactants, lipids, aqueous insoluble polymers (Euliss et al. 2006). Amphiphilic molecules are the molecules that have hydrophobic nonpolar "tail" at one end and hydrophilic polar "head" to the other end (Fig. 4b). Micelles are the colloidal suspension of the amphiphilic molecules. When amphiphilic molecules are mixed in the aqueous medium, the nonpolar hydrophobic tail is arranged inside the molecules and letting the polar side i.e. "head" to remain an outer side in direct contact with the aqueous medium and a hollow spherical or cylindrical structure is formed by the process of self -assembly. For example, Cao et al. (2011) mixed cationic PCL and PEI which self-assembled, forming biodegradable micelles. These micelles were utilized to carry anti-apoptotic Bcl-2 specific siRNA and along with an anti-cancer drug, docetaxel (DOX). Further, folic acid was surface-conjugated as a targeting ligand for hepatic cancer cells. This nano-assembly simultaneous delivered drug and siRNA to yield a combined RNAi-chemotherapeutic benefit against hepatic cancer (Cao et al. 2011). Similarly, Zhu et al. (2014) synthesized PEG-pp-PEI-PE nanomicelles for the delivery of siRNA against survivin or GFP along with paclitaxel (PTX). Efficient down-regulation of GFP gene and survivin were observed in PTX-resistant non-small cell lung cancer cells after the micelles-mediated siRNA delivery compared to free drug (Zhu et al. 2014). In another study, prepared a PEGylated PEI-siRNA micellar nanoparticles that shown high entrapment efficiency and long-term blood circulation of siRNA loaded nanoparticles with reduced aggregation, opsonization, and inflammation response. In this study, polycationic PEI was used for high loading capacity of negatively charged siRNA and PEG was used for others advantages (Wu et al. 2017) (Table 2).

7.1.3 Carbon-Based Nanomaterials

Carbon nanomaterials (CNMs) are synthesized from allotropes of carbon such as graphene, graphene oxide (GO), nanotubes, fullerene. For example, a study conducted by Varkouhi et al. (2011), carbon nanotubes functionalized with PEI and pyridinium were explored for siRNA delivery. Both functionalized cationic CNTs complexed anionic siRNA and provide the silencing efficiency of 10–30% and 10–60% of cytotoxicity (Varkouhi et al. 2011). Another study by Ren et al. (2017) was done to prepare and characterize a GO-based carrier, surface-functionalized with PLL and Arg-Gly-Asp-Ser oligonucleotides to actively target tumors, These nanocarriers were then used for loading VEGF-siRNA and observed to perform a slow and sustained release of siRNA along with low toxicity as compared to bare GO (Ren et al. 2017).

Figure 4d showed the two-dimensional, sheet-like GO structure that has numerous oxygen groups like carboxylic acids, hydroxyl, epoxy, and alkoxy on its edges and both side of surfaces. The average thickness of GO is about 1–2 nm and length/

Table 2 Application of various lipid-based nanocarriers/lipoplexes for systemic delivery of siRNA

Lipid-based NPs	Shape	Size	ζ (mV)	Targeted molecule	Cell lines	Type of study	Aid	References
DOPE, DOTMA, DPPE-PEG 2000	Spherical	112–122 nm	42–56 mV	Targeting peptides (Y, ME27)	BALB/c adult mice	In-vivo	Receptor-mediated targeted delivery	Yu-Wai-Man et al. (2016)
AtuFECT01	Spherical	117.8 nm	46.4 mV	Anti-CD31 antibody	HeLa cell and heart, lung, liver, spleen of mice	In-vitro and In-vivo	Targeted delivery, stable in the bloodstream	Santel et al. (2006)
Toc-siRNA	Spherical	–	–		Hepa 1–6 cells mice liver	In-vivo	Improved stability and cleaving efficiency of siRNA, targeted delivery, no side effects	Nishina et al. (2008)
DLinDAP, DLinDMA, DLinKDMA, or DLinKC2-DMA	Spherical	75–85 nm			Raw 264.7 cells	In-vitro		Lin et al. (2013)

DOPE 1,2-dioleoyl-sn-glycero-3-phosphoethanolamine, *DOTMA* 1,2-di-O-octadecenyl-3-trimethylammonium propane, *DPPE-PEG 2000* 1,2-dipalmitoyl-*sn*-glycero-3-phosphoethanolamine N-[methoxy(polyethylene glycol)-2000], *HTFs* Primary human Tenon's fibroblasts, *DLinKC2-DMA* 2,2-dilinoleyl-4-(2-dimethylaminoethyl)-[1,3]-dioxolane, *DLinKDMA* 2,2-dilinoleyl-4-dimethylaminomethyl-[1,3]-dioxolane, *DLinDAP* 1,2-dilinoleyloxy-3-dimethylaminopropane, *DLinDAP* 1,2-dilinoleoyl-3-dimethylaminopropane, *α Toc-siRNA* α-tocopherol (vitamin E)- siRNA, *GFP* green fluorescent protein width is between 50 and 200 nm depending on the biomedical application. GO has a high surface area and capability of holding molecules on both sides of sheets due to this GO has high loading capacity.

Carbon nanotubes (CNTs) are hollow one-dimension nanostructures. The CNTs may be cylindrical, tubular or needle in shape. Depending on the number of layers/walls, CNTs are of three-types: single-walled carbon nanotubes (SWCNTs), double-walled carbon nanotubes (DWCNTs) and multi-walled carbon nanotubes (MWCNTs). The typical thickness of SWCNTs is about 0.4–1 nm and that of DWCNTs is 1.4–20 nm (Schedin et al. 2007; Bae et al. 2010). MWCNTs are 20–100 nm thick (Bartelmess et al. 2015) and 50–500 nm in length.

Fullerene has zero-dimensional foot-ball like hollow structure with a diameter below 100 nm. Fullerene has a different number of carbon atoms on its structure like C_{60}, C_{70} etc. C_{60} is the most commonly used fullerene, having 60 carbons in its structure and is composed of 20 hexagonal rings and 12 pentagonal rings and is called as Buckyballs.

Pristine CNMs are less biocompatible and have less siRNA loading capacity due to electrostatic repulsion between them. CNMs are modified or coated with several cationic polymers like PEG, PEI, chitosan, PAH (Poly (allylamine hydrochloride)), PLGA (Poly (lactic acid-co-glycolic acid)) are the frequently used polymers. After modification or coating with the polymer, CNMs can hold a high concentration of siRNA due to the simple electrostatic attraction between of them. For improved delivery of siRNA, polymer coated CNMs are further modified with suitable ligands (Table 3)

7.1.4 Liposomes/Lipoplexes

Liposomes are well-ordered, self-assembled vesicles of amphiphilic lipids molecules that form a minute spherical shape nanoparticles having single-lamellar or multi-lamellar lipid bilayer structure of phospholipid molecules that have aqueous medium inside used to encapsulation and deliver drugs or genetic materials into a cell (Fig. 4a) (Akbarzadeh et al. 2013). Liposomes are highly biocompatible and biodegradable, small size, vesicles of amphiphilic lipid with the high surface area-to-volume ratio. They have some superior physiochemical properties like increased rate of biological membrane permeability, the flexibility of functionalization, tuning of surface charge. These properties make liposomes a better nanoformulation to deliver drugs and genes than others NCs and also have been successfully applied in the clinic (Hatakeyama et al. 2011). Polycationic liposomes as building blocks to entrap of negatively charged siRNA through electrostatic interactions and form lipid complexes systems called lipoplexes (Wolff and Rozema 2008). So far, many liposomal formulations had got the FDA approval and many more are currently in development or in clinical trials enabling features such as stable circulation and tumor-specific targeting versions of siRNA-liposome carriers (Noble et al. 2009; Drummond et al. 2010; Patil and Jadhav 2014). Liposomal NCs can be divided into

Table 3 Various carbon nanomaterials used for systemic and targeted delivery of siRNA

Type of CNMs	Shape and dimension	Grafted polymers	Targeted receptor and molecule	siRNA-molecule	Cell lines	Type of study	Type of therapies	Aid	References
Graphene Oxide (GO)	Sheet and 2D	PEG and PAH	Folate receptor and folic acid	Histone deacetylase 1 (HDAC1) and K-Ras siRNAs	MIA PaCa-2	In-vivo	Photothermal and gene therapy both	Non-toxic, targeted gene therapy and photothermal effects	Yin et al. (2017)
		PEG and PEI	Folic acid	MCF-7/ADR cells	MC3T3-E1	In-vitro	Photothermal, chemotherapy, and gene therapy both	Overcome drug-resistant cancer cell, high siRNA loading, targeted delivery	Zeng et al. (2017)
SWCNTs	Cylindrical/ tube like and 1D	PEI	Aptamer and piperazine		MCF-7 and MDA-MB-231	In-vitro	Gene therapy	Biocompatible, load high amount, targeted delivery	Mohammadi et al. (2015)
TPFE-fullerene	Spherical and 0D		TLR4 and EGFP	siGFP and siNEG	Mice lung	In-vivo	Targeted lung delivery	Easily accumulation and clearance from the lung	Minami et al. (2014)

PAH Poly(allylamine hydrochloride), *PEG* Polyethylene glycol, *0D* Zero Dimension, *1D* One Dimension, *2D* Two Dimension, *EGFP* enhanced green fluorescent protein, *TLR4-* mouse toll-like receptor 4, *TPFE* tetra(piperazino)fullerene epoxide, *MC3T3-E1* murine osteoblast cell line, *MIA PaCa-2* Pancreatic cancer cells four main types based on their physical structures- nucleic-acid-lipid particles, lipoplexes, lipopolyplexes, and membrane/core nanoparticles.

Lipoplexes are composed of several bilayer of lipids in spherical form and siRNAs are entrapped between the consecutive lipid bilayers. Stable nucleic-acid-lipid particles have the outer surface is almost neutral but inner surface are highly positively charged due to this inner positive charge siRNAs are entrapped in the interior of liposomes. Only pH-sensitive lipid can be synthesized by these types of liposomes. These liposomes showed cationic behaviour at acidic pH (pH-4.0) and no-ionic behaviour at neutral pH (pH-7.4). Lipopolyplexes are synthesized by coating the liposomes from oppositely charged polymers. If lipid NCs has cationic charged then they are coated with anionic charged polymers and vice versa. A membrane/core NCs has some different structure than other liposomes NCs. This type of NCs is similar core-shell-like entities in which single or multiple inorganic NPs are embedded in the core and surrounded by a lipid bilayer form as the outer shell (Xia et al. 2016).

Landen et al. (2005) siRNA was encapsulated into 1,2-dioleoylsn-glycero-3-phosphatidylcholine (DOPC, a neutral lipid) lipid NCs and this NCs were decorated with EphA2 (oncoprotein, a targeting ligand) for improved systemic siRNA delivery. These liposomal NCs was tested on two different orthotopic mouse model ovarian malignant cell types HeyA8 and SKOV3ip1. EphA2-siRNA-DOPC NCs have significantly reduced the tumor growth compared to non-silencing siRNA. Whenever paclitaxel was also loaded on this NCs conjugate the cancer cell growth was drastically decreased on comparison to individual chemotherapy with paclitaxel and RNAi therapy with a non-silencing siRNA (Landen et al. 2005).

7.1.5 Dendrimers

Dendrimers are the nanosized, spherical, monodispersed, highly branched, three-dimensional, synthetic macromolecules. A typical dendrimer molecule is composed of three structural regions: *the innermost core*, comprising of atomic or molecular entity with at least two identical functional groups; *branching generations* with inner cavities, and *specific functional groups* present on the surface (Fig. 4e) (Verma et al. 2012; Pooja et al. 2014). Numerous functional groups (-COOH or -OH) are suitable for the attachment of various ligands that makes it more suitable for targeted delivery of therapeutic agents. Dendrimers are made up of repeated branching unit and form a shell of either same groups and different groups, this shell is known as generation "G" (Kulhari et al. 2015). When are mixed with dendrimers known as dendriplexes. Presently, dendrimers are used for the delivery of several hydrophobic and hydrophilic drugs, genes, bioimaging agents (Table 4). Depending on the type of central atom, different classes of dendrimers are available such as PAMAM (Polyamidoamine), PPI (Poly (propylene imine), PLL, and poly(2,2-bis(hydroxymethyl)propionic acid, etc. (Svenson and Tomalia 2012). Due to its branch structure, several voids/spaces are formed where drugs or nucleic acids are

Table 4 Application of dendrimers for systemic delivery of siRNA

Dendrimers	Generation (G)	Hybridized with	Targeted molecule	siRNA type	Cell type	Type of study	Aid	References
PAMAM	G_4	Dodecylated	–	Bcl-2 siRNA	HeLa and MDA-MB231 cells	In-vitro	Polymer G4-23C12 successfully transport siRNA, reduced Bcl-2 mRNA, and protein expressions	Chang et al. (2015)
	G_5	PEG	Peptides- B6 (TfR targeted), GE11 (EGFR targeted)	–	HeLa and LS174T cells	In-vitro	Two cell lines targeted using two targeting ligands, low toxicity	Urbiola et al. (2018)
	G_5	TEA-core	–	Sticky siRNA	PC-3	In-vitro	High siRNA silencing, endosomal escape promoters	Liu et al. (2012)
PPI	G_5	Dithiol containing PEG	LHRH peptide	–	Negative SKOV-3 and positive A2780 and A549	In-vitro	High lateral and steric stability in the blood environment, targeted delivery	Taratula et al. (2009)
Dendritic (PLL-KG6)	G_6			si–ApoB	Liver	In-vivo	Monodispersed structure, high plasmid DNA transfection ability, low cytotoxicity	Watanabe et al. (2009)

α-CDE α-cyclodextrin, *LHRH* Luteinizing Hormone-Releasing Hormone, *PAMAM* Poly(amidoamine), *PPI* Poly(propylene imine), *si–ApoB* siRNA-Apolipoprotein B, *TfR* transferrin, *EGFR* epidermal growth factor receptors, *LS174T* colorectal adenocarcinoma, *HeLa* cervix adenocarcinoma cells, *PC-3* Human Prostate adenocarcinoma cells, *TEA-core* Triethanolamine-Core, A2780- Positive human ovarian, A549- human lung cancer cell lines entrapped in these voids. Amine-terminated cationic dendrimers are good for high loading efficiency and controlled release of negatively charged siRNA.

Recently, for treating cancer researchers developed a tumor microenvironment-sensitive polypeptides-amphiphilic dendrimer as a carrier of siRNA/Paclitaxel (TMSP-ADENS/siRNA/PTX) complex for robust co-delivery of therapeutic agents. In this complex, siRNA and paclitaxel were encapsulated within the hydrophilic centre and internal hydrophobic layer, respectively. The outer PEG layer increased the half-life of dendrimers. This complex illustrated the highest rate while regulation of VEGF mRNA in A375 xenograft mice. Furthermore, the tumour-inhibiting and angiogeny-inhibiting activity were also observed in another study on HT-1080 xenograft (Li et al. 2018).

7.2 *Inorganic Nanocarriers*

Inorganic NCs are prepared synthetically from inorganic materials which are hard, water-insoluble, less biodegradable, toxic (Tatiparti et al. 2017). Due to these limitations, inorganic NCs are less used than ONCs to deliver the siRNA molecules. Frequently used inorganic NCs are mesoporous silica nanoparticles; different metal NPs such as gold NPs (AuNPs), selenium NPs (SeNPs), silver NPs (AgNPs); metal oxide NPs like super-paramagnetic iron oxide nanoparticles (SPIONs). Bare inorganic NCs are less biocompatible and have low siRNA loading capacity than ONCs. The reason is that these NCs are mainly solid and therefore they are unable to encapsulate any molecule. The molecules can only be adsorbed of any molecules on its outer most surface. For increasing their biocompatibility and loading capacity outer surfaces of these NCs are coated with several biocompatible, biodegradable, natural or synthetic polymers and lipids (Table 5).

7.2.1 Mesoporous Silica and Silicon-Based Nanoparticles

The use of mesoporous silica nanoparticles (MSNs) is increasing in biomedical and pharmacy field. MSNPs are not only having a spherical shape, high surface area, tunable small size, biocompatibility, and biodegradability, easily surface functionality but also have several ordered porous structural surfaces. Therapeutic hydrophobic drugs, bioimaging agents, genes, chemotherapeutic agents are trapped or stored in these pores (Fig. 4f). Due to this specific character, MSNs show some advantages of high loading efficiency and controlled release of encapsulated agents (Mekaru et al. 2015). Further, MSNs are small in size with a diameter ranging between 50 and 200 nm, therefore they are passively transported and across membrane barrier and give EPR effect through and leaky vasculature and accumulate in the tumor cells (Maeda 2001). Homogenous, monodispersed MSNs are mostly synthesized by the sol-gel method. For high loading and specific site of delivery of anionic siRNA,

Table 5 Inorganic nanocarriers used for delivery of siRNA

Type of NPs	Coated polymer	Shape, size and zeta potential	siRNA molecule	Targeted molecule	Animal/cell lines	Type of study	Aid	References
Mesoporous silica nanoparticles	PEI	Spherical	Pgp-siRNA		KB-31, KB-V1 cells	In-vitro	Overcome chemotherapeutic resistance in cancer cells	Meng et al. (2010)
SPIONs	PEG-g-PEI	Spherical, 63 ± 6.1 nm and +27.0 ± 1.1 mV	–	RGD	Bel-7402 cells	In-vitro and in-vivo	Targeted delivery, In-vitro and in-vivo MR imaging in mice	Wu et al. (2013)
Au NPs	–		Polyvalent siRNA		HeLa cell	In-vitro	Knockdown of luciferase expression	Giljohann et al. (2009)
Au nanostar	PEG	Star shape. 210.5 ± 10.3 nm, and 12.35 ± 2.28 mV	siCOX-2	2-amino-2-deoxy-D-glucose	HepG2 and SGC7901	In-vitro	Photothermal therapy, targeted delivery	Zhu et al. (2017)
Se NPs	–	Dendritic structures, 10 nm, and 12.7 mV	Anti-Oct4	RGDfC peptide	HepG2 cancer cell line	In-vitro and in-vivo both	Targeted improved delivery, biocompatible	Xia et al. (2017)

MDR Multidrug resistance; *Pgp gene* P-glycoprotein gene, *SPIONs* Superparamagnetic iron oxide nanoparticles, *PLKs* Polo-like kinases, *KB-31, KB-V1 cells* Human cervix carcinoma (a derivative of HeLa), *Se NPs* Selenium nanoparticles, *PEG-g-PEI* Polyethylene glycol-grafted polyethyleneimine, Bel-7402-human hepatocellular carcinoma cell line

MSNs are coated with cationic polymers, PEG, PEI, and amine-terminated PAMAM dendrimers (Meng et al. 2010; Ngamcherdtrakul et al. 2015).

7.2.2 Calcium Phosphate Core-Shell Nanocarriers

Calcium phosphate (CaP) particles are most commonly used biocompatible, biodegradable inorganic carrier for high loading of siRNA. Cap is a component that is also present in human bones. CaP particles have several demerits also they are obtained in micro-sized with polydispersed. Nano-sized monodispersed CaP NPs was obtained, when they were precipitated with PEG. CaP NPs are not shown any toxicity at a higher concentration than RNAi-induced concentrations and also shown better control over endogenous VEGF mRNA expression in cultured pancreatic cancer lines.

Researchers synthesized lipid/calcium/phosphate core-shell nanoparticles (LCP NPs) by microemulsion technology for improving systemic delivery of anti-luciferase siRNA. In this type of NPs, calcium phosphate core is surrounded by cationic lipid shell and decorated by an anisamide ligand complementary for sigma receptor on B16F10 melanoma cell surface. After IV injection of siRNA, the luciferase luminescence signal in metastatic B16F10 tumor cells was significantly diminished in C57BL/6 mice (Yang et al. 2012). Several others studies have also been done for systemic delivery of siRNA using CaP NPs inorganic nanoparticles (Qiu et al. 2016b; Pittella et al. 2011).

7.2.3 Metal and Metal Oxide Nanoparticles

From the last few decades, scientist and researchers have developed, tested and applied several types of metals (gold, silver, selenium, cobalt, nickel etc.) and metal oxides (iron oxide, silver oxide, manganese oxide etc.) NPs for the biomedical applications.

Initially, Au NPs were used only for drug delivery, biomolecular sensing, and hyperthermal therapy. However, currently, AuNPs are also being used in intracellular gene delivery and therapy due to their unique and controllable optical properties, and easy alteration of surfaces with thiolate molecules (Fig. 4h). Giljohann et al. (2009) prepared the polyvalent siRNA-gold nanoparticle conjugate (siRNA-AuNPs) which was more stable in 10% serum, exhibited a prolonged half-life and more knockdown of luciferase gene expression than to free RNA duplexes (Giljohann et al. 2009). There are several ways to load a sufficient amount of siRNA on Au NPs:

(i) Thiol bond- siRNA is directly attached on AuNPs surface via a gold-thiol (R-S-H) bond.
(ii) Ionic bond- Anionic siRNA can be attached on the surfaces of the cationic AuNPs due to the electrostatic interaction.

(iii) Polymer coated surface- siRNA adhered to the AuNPs surface coated with biocompatible and biodegradable with polymers (Lee et al. 2008).

Iron oxide NPs are other frequently used metal nanoparticles and have numerous biomedical applications. It is frequently used to deliver various therapeutic agents. Iron oxide NPs are also known as magnetic nanoparticles (MNPs) because they have high magnetic behaviour and this behaviour is used for remotely-controlled site-specific delivery of several therapeutic molecules (drugs and genes) by applying an external magnetic field (Fig. 4g). MNPs have a diameter between 10 and 100 nm range and are obtained in two different oxidation states- Fe_2O_3 (ferric oxide/hematite) and Fe_3O_4 (magnetite/lodestone) NPs. When the diameter of NPs are <30 nm, they show the property of paramagnetism in the hysteresis of the magnetization loop of the particles because of the negligible coercivity arising from the infinitely small energy barrier (Lian et al. 2003).

7.3 *Hybrid Nanocarriers*

Organic and inorganic NCs are having their individual benefits and have been applied to remove various hurdles in delivering genetic materials. However, both classes of NPs has some limitations. Inorganic NPs such as MSNs may have more stability, mechanical, chemical and imaging properties but accumulates as non-degradable substances and confers toxicity. On the other side, organic NCs such as polymers, liposomes, and micelles are biocompatible as well as biodegradable with high encapsulation efficiency both on the surface and within the core, but less stable in-vivo. For removing the drawbacks of both type of NPs, the hybrid NCs are the best solution. These hybrid NCs are prepared with the fusion of both organic and inorganic NCs. Numerous biocompatible and biodegradable polymers are coated or grafted on the surfaces of both NCs. These polymers increase biocompatibility, loading tendency of genes with reduced toxicity (Tatiparti et al. 2017). Meanwhile the inorganic constituents provide diagnostic or therapeutic properties. Therefore, the hybrid NCs have benefit to offers the theranostics application with high delivery efficiency (Sailor and Ji-Ho 2012; Young et al. 2016) (Table 6).

8 Conclusion

Conventional therapy is already being applied to treat several deadly diseases such as cancer, tuberculosis, AIDS etc. However, their use is quite limited due to several weaknesses like non-targeted delivery, high dose and dose frequency that gives numerous adverse effects and multidrug resistance. Currently, gene therapy acclaimed as a revolutionary option of new therapy. siRNA delivery is an alternative approach, that has been pursued actively because of their excellent safety, high

Table 6 Different hybrid nanocarriers applied for delivery of siRNA

Type of NPs	Hybrid with	Targeted molecule	siRNA molecule	Cell lines	Type of study	Aids	References
Mesoporous silica nanoparticles	PEI–PEG copolymer	Anti-HER2 monoclonal antibody (trastuzumab)		BT474, JIMT1, SKBR3, HCC1954, MCF7, MDA-MB-231, and MDA-MB-468; MCF-10a; HepG2; HEK-293	In-vitro	Protection from blood enzymatic degradation, blood compatible, non imunogenic	Ngamcherdtrakul et al. (2015)
Iron oxide nanoparticles	Lipidoid (oleic acid or oleic amine)	–		HeLa cell	In-vitro	High loading, enhanced permeability and retention (EPR) effect	Jiang et al. (2013)
Mesoporous silica nanoparticles	G2 amine-terminated PAMAM dendrimers	–	Bcl-2 siRNA	A2780/AD human ovarian cancer cells	In-vitro	Enhances efficacyof chemotherapy drugs, decrease multidrug-resistant	Chen et al. (2009)
Layer-by-layer assembly technique Au NPs	PEI/PAH-Cit/PEI/MUA			HeLa cell		Enhanced delivery, escape endosome, release siRNA into the cytoplasm	Guo et al. (2010b)
Mesoporous silicon nanoparticles	GO nanosheets	A peptide derived from RVG	Dy547-labeled siRNA	Neuro-2a cells injured brain cell	In-vitro and in-vivo studies both	Enhanced cellular uptake and gene silencing	Joo et al. (2016)

Human breast cancer cell lines (BT474, JIMT1, SKBR3, HCC1954, MCF7, MDA-MB-231, and MDA-MB-468), breast epithelial cells (MCF-10a), human liver carcinoma (HepG2), and human embryonic kidney cells (HEK-293), *PEI/PAH-Cit/PEI/MUA* polyethyleneimine/cis-aconitic anhydride functionalized poly(allylamine)/ polyethyleneimine/11-mercaptoundecanoic acid, *RVG* Rabies virus glycoprotein efficacy and specificity, easy modification, and unlimited therapeutic targets. But the therapeutic efficacy of naked siRNA is reduced due to systemic instability, before reaching to the target site. To get rid of these hurdles, the most vital solution is to develop biocompatible, biodegradable, and site-specific nano-sized carriers. Every NC has its own benefits and limitations depending on its physicochemical composition, offering a flexible and wide variety of effective agents based on siRNA. Organic NCs such as polymeric or lipid-based nanoparticles and micelles have excellent tendency to encapsulate the therapeutic molecules nevertheless they are unstable and easily degrade within the systemic circulation causing the premature release of therapeutic molecules before reaching the specific site. Mixed polymer micelles are quite stable compared to single polymer micelles. Inorganic naked metal and metal oxide NCs like Au NPs and MNPs are stable and can be easily functionalized. In spite of these advantages, several studies have proved that larger size and a higher concentration of these NCs are toxic. Moreover, naked NCs are also unstable in systemic delivery and rapidly agglomerated by blood plasma protein. Hence, these NCs require stabilizing materials. At last, the nanocarrier mediated delivery systems better solution to resolve issues related to the potent systemic delivery of siRNA. Some of the delivery systems also have proven their values and passed the initial clinical phases.

References

Aagaard, L., & Rossi, J. J. (2007). RNAi therapeutics: Principles, prospects and challenges. *Advanced Drug Delivery Reviews, 59*(2–3), 75–86.

Ahmed, Z., Kalinski, H., Berry, M., Almasieh, M., Ashush, H., Slager, N., Brafman, A., Spivak, I., Prasad, N., Mett, I., et al. (2011). Ocular neuroprotection by SiRNA targeting caspase-2. *Cell Death & Disease, 2*, e173.

Akbarzadeh, A., Rezaei-Sadabady, R., Davaran, S., Joo, S. W., Zarghami, N., Hanifehpour, Y., Samiei, M., Kouhi, M., & Nejati-Koshki, K. (2013). Liposome: Classification, preparation, and applications. *Nanoscale Research Letters, 8*(1), 1–8.

Akhtar, S. (2009). Oral delivery of SiRNA and antisense oligonucleotides. *Journal of Drug Targeting, 17*(7), 491–495.

Akinc, A., Querbes, W., De, S., Qin, J., Frank-Kamenetsky, M., Jayaprakash, K. N., Jayaraman, M., Rajeev, K. G., Cantley, W. L., Dorkin, J. R., et al. (2010). Targeted delivery of RNAi therapeutics with endogenous and exogenous ligand-based mechanisms. *Molecular Therapy, 18*(7), 1357–1364.

Aldawsari, M., Babu, R. J., & Chougule, M. B. (2015). Progress in topical SiRNA delivery approaches for skin disorders. *Current Pharmaceutical Design, 21*, 4594–4605.

Almeida, R., & Allshire, R. C. (2005). RNA silencing and genome regulation. *Trends in Cell Biology, 15*(5), 251–258.

Amarzguioui, M., & Prydz, H. (2004). An algorithm for selection of functional SiRNA sequences. *Biochemical and Biophysical Research Communications, 316*(4), 1050–1058.

Bae, S., Kim, H., Lee, Y., Xu, X., Park, J. S., Zheng, Y., Balakrishnan, J., Lei, T., Ri Kim, H., Song, Y., II, et al. (2010). Roll-to-roll production of 30-inch graphene films for transparent electrodes. *Nature Nanotechnology, 5*(8), 574–578.

Bartelmess, J., Quinn, S. J., & Giordani, S. (2015). Carbon nanomaterials: Multi-functional agents for biomedical fluorescence and Raman imaging. *Chemical Society Reviews, 44*(14), 4672–4698.

Benjaminsen, R. V., Mattebjerg, M. A., Henriksen, J. R., Moghimi, S. M., & Andresen, T. L. (2013). The possible "proton Sponge" effect of polyethylenimine (PEI) does not include change in lysosomal PH. *Molecular Therapy, 21*(1), 149–157.

Berezikov, E. (2011). Evolution of microRNA diversity and regulation in animals. *Nature Reviews. Genetics, 12*(12), 846–860.

Bertrand, J. R., Pottier, M., Vekris, A., Opolon, P., Maksimenko, A., & Malvy, C. (2002). Comparison of antisense oligonucleotides and SiRNAs in cell culture and in vivo. *Biochemical and Biophysical Research Communications, 296*(4), 1000–1004.

Birmingham, A., Anderson, E. M., Reynolds, A., Ilsley-Tyree, D., Leake, D., Fedorov, Y., Baskerville, S., Maksimova, E., Robinson, K., Karpilow, J., et al. (2006). 3′ UTR seed matches, but not overall identity, are associated with RNAi off-targets. *Nature Methods, 3*(3), 199–204.

Boado, R. J. (2005). RNA interference and nonviral targeted gene therapy of experimental brain cancer. *NeuroRx, 2*(1), 139–150.

Bodor, N., & Buchwald, P. (2005). Ophthalmic drug design based on the metabolic activity of the eye: Soft drugs and chemical delivery systems. *The AAPS Journal, 7*(4), E820–E833.

Bonnet, M. E., Erbacher, P., & Bolcato-Bellemin, A. L. (2008). Systemic delivery of DNA or SiRNA mediated by linear polyethylenimine (L-PEI) does not induce an inflammatory response. *Pharmaceutical Research, 25*(12), 2972–2982.

Campochiaro, P. A. (2006). Potential applications for RNAi to probe pathogenesis and develop new treatments for ocular disorders. *Gene Therapy, 13*(6), 559–562.

Cao, N., Cheng, D., Zou, S., Ai, H., Gao, J., & Shuai, X. (2011). The synergistic effect of hierarchical assemblies of SiRNA and chemotherapeutic drugs co-delivered into hepatic cancer cells. *Biomaterials, 32*(8), 2222–2232.

Carthew, R. W., & Sontheimer, E. J. (2009). Origins and mechanisms of MiRNAs and SiRNAs. *Cell, 136*(4), 642–655.

Cazzin, C., & Ring, C. J. A. (2010). Recent advances in the manipulation of murine gene expression and its utility for the study of human neurological disease. *Biochimica et Biophysica Acta, Molecular Basis of Disease, 1802*(10), 796–807.

Chalk, A. M., Wahlestedt, C., & Sonnhammer, E. L. L. (2004). Improved and automated prediction of effective SiRNA. *Biochemical and Biophysical Research Communications, 319*(1), 264–274.

Chang, H., Zhang, Y., Li, L., & Cheng, Y. (2015). Efficient delivery of small interfering RNA into cancer cells using dodecylated dendrimers. *Journal of Materials Chemistry B, 3*(41), 8197–8202.

Chen, A. M., Zhang, M., Wei, D., Stueber, D., Taratula, O., Minko, T., & He, H. (2009). Co-delivery of doxorubicin and Bcl-2 SiRNA by mesoporous silica nanoparticles enhances the efficacy of chemotherapy in multidrug-resistant cancer cells. *Small, 5*(23), 2673–2677.

Chiu, Y. L., & Rana, T. M. (2003). SiRNA function in RNAi: A chemical modification analysis. *RNA, 9*(9), 1034–1048.

Chougule, M. B. (2012). Current scene and prospective potentials of SiRNA in cancer therapy. *Journal of Pharmacogenomics Pharmacoproteomics, 03*(05), 3–5.

Clark, S. J. (2007). Action at a distance: Epigenetic silencing of large chromosomal regions in carcinogenesis. *Human Molecular Genetics, 16*(R1), R88–R95.

Conley, S. M., & Naash, M. I. (2010). Nanoparticles for retinal gene therapy. *Progress in Retinal and Eye Research, 29*(5), 376–397.

Cun, D., Krohn, D., Jonas, M., Bunker, M., Whiteside, P., Scurr, D., Foged, C., & Mørck, H. (2011). European journal of pharmaceutics and biopharmaceutics high loading efficiency and sustained release of SiRNA encapsulated in PLGA nanoparticles: Quality by design optimization and characterization. *European Journal of Pharmaceutics and Biopharmaceutics, 77*(1), 26–35.

Czauderna, F. (2003). Structural variations and stabilizing modifications of synthetic SiRNAs in mammalian cells. *Nucleic Acids Research, 31*(11), 2705–2716.
Czech, M. P., Aouadi, M., & Tesz, G. J. (2011). RNAi-based therapeutic strategies for metabolic disease 3121. *Nature Reviews. Endocrinology, 7*, 1759–5037. (Electronic)), 473–484.
Dana, H., Chalbatani, G. M., Mahmoodzadeh, H., Karimloo, R., Rezaiean, O., Moradzadeh, A., Mehmandoost, N., Moazzen, F., Mazraeh, A., Marmari, V., et al. (2017). Molecular mechanisms and biological functions of SiRNA. *International Journal of Biomedical Sciences, 13*(2), 48–57.
Davis, M. E., Zuckerman, J. E., Choi, C. H. J., Seligson, D., Tolcher, A., Alabi, C. A., Yen, Y., Heidel, J. D., & Ribas, A. (2010). Evidence of RNAi in humans from systemically administered SiRNA via targeted nanoparticles. *Nature, 464*(7291), 1067–1070.
de la Fuente, M., Raviña, M., Paolicelli, P., Sanchez, A., Seijo, B., & Alonso, M. J. (2010). Chitosan-based nanostructures: A delivery platform for ocular therapeutics. *Advanced Drug Delivery Reviews, 62*(1), 100–117.
Deng, H., Jankovic, J., Guo, Y., Xie, W., & Le, W. (2005). Small interfering RNA targeting the PINK1 induces apoptosis in dopaminergic cells SH-SY5Y. *Biochemical and Biophysical Research Communications, 337*(4), 1133–1138.
des Rieux, A., Fievez, V., Garinot, M., Schneider, Y.-J. J., & Préat, V. (2006). Nanoparticles as potential oral delivery systems of proteins and vaccines: A mechanistic approach. *Journal of Controlled Release, 116*(1), 1–27.
Dorn, G., Patel, S., Wotherspoon, G., Hemmings-Mieszczak, M., Barclay, J., Natt, F. J. C., Martin, P., Bevan, S., Fox, A., Ganju, P., et al. (2004). SiRNA relieves chronic neuropathic pain. *Nucleic Acids Research, 32*(5), e49–e49.
Draz, M. S., Fang, B. A., Zhang, P., Hu, Z., Gu, S., Weng, K. C., Gray, J. W., & Chen, F. F. (2014). Nanoparticle-mediated systemic delivery of SiRNA for treatment of cancers and viral infections. *Theranostics, 4*(9), 872–892.
Drummond, D. C., Noble, C. O., Guo, Z., Hayes, M. E., Connolly-Ingram, C., Gabriel, B. S., Hann, B., Liu, B., Park, J. W., Hong, K., et al. (2010). Development of a highly stable and targetable nanoliposomal formulation of topotecan. *Journal of Controlled Release, 141*(1), 13–21.
Duan, L., Yan, Y., Liu, J., Wang, B., Li, P., Hu, Q., & Chen, W. (2016). Target delivery of small interfering RNAs with vitamin E-coupled nanoparticles for treating hepatitis C. *Scientific Reports, 6*, 24867.
Durcan, N., Murphy, C., & Cryan, S. A. (2008). Inhalable SiRNA: Potential as a therapeutic agent in the lungs. *Molecular Pharmaceutics, 5*(4), 559–566.
Euliss, L. E., DuPont, J. A., Gratton, S., & DeSimone, J. (2006). Imparting size, shape, and composition control of materials for nanomedicine. *Chemical Society Reviews, 35*(11), 1095.
Fire, A., Xu, S., Montgomery, M. K., Kostas, S. A., Driver, S. E., & Mello, C. C. (1998). Potent and specific genetic interference by double-stranded RNA in *Caenorhabditis Elegans. Nature, 391*(6669), 806–811.
Gavrilov, K., & Saltzman, W. M. (2012). Therapeutic SiRNA: Principles, challenges, and strategies. *The Yale Journal of Biology and Medicine, 85*(2), 187–200.
Gewirtz, A. M. (2007). On future's doorstep: RNA interference and the pharmacopeia of tomorrow. *The Journal of Clinical Investigation, 117*(12), 3612–3614.
Gherardini, L., Bardi, G., Gennaro, M., & Pizzorusso, T. (2014). Novel SiRNA delivery strategy: A new "Strand" in CNS translational medicine? *Cellular and Molecular Life Sciences, 71*, 1–20.
Giljohann, D. A., Seferos, D. S., Prigodich, A. E., Patel, P. C., & Mirkin, C. A. (2009). Gene regulation with polyvalent SiRNA–nanoparticle conjugates. *Journal of the American Chemical Society, 131*(6), 2072–2073.
Gottesman, M. M. (2002). Mechanisms of cancer drug resistance. *Annual Review of Medicine, 53*(1), 615–627.
Griesenbach, U., Kitson, C., Garcia, S. E., Farley, R., Singh, C., Somerton, L., Painter, H., Smith, R. L., Gill, D. R., Hyde, S. C., et al. (2006). Inefficient cationic lipid-mediated SiRNA and antisense oligonucleotide transfer to airway epithelial cells in vivo. *Respiratory Research, 7*(1), 26.

Gu, J., Al-Bayati, K., & Ho, E. A. (2017). Development of antibody-modified chitosan nanoparticles for the targeted delivery of SiRNA across the blood-brain barrier as a strategy for inhibiting HIV replication in astrocytes. *Drug Delivery and Translational Research, 7*(4), 497–506.

Guo, P., Coban, O., Snead, N. M., Trebley, J., Hoeprich, S., Guo, S., & Shu, Y. (2010a). Engineering Rna for targeted Sirna delivery and medical application. *Advanced Drug Delivery Reviews, 62*(6), 650–666.

Guo, S., Huang, Y., Jiang, Q., Sun, Y., Deng, L., Liang, Z., Du, Q., Xing, J., Zhao, Y., Wang, P. C., et al. (2010b). Enhanced gene delivery and SiRNA silencing by gold nanoparticles coated with charge-reversal polyelectrolyte. *ACS Nano, 4*(9), 5505–5511.

Gutbier, B., Kube, S. M., Reppe, K., Santel, A., Lange, C., Kaufmann, J., Suttorp, N., & Witzenrath, M. (2010). RNAi-mediated suppression of constitutive pulmonary gene expression by small interfering RNA in mice. *Pulmonary Pharmacology & Therapeutics, 23*(4), 334–344.

Hammond, S. M., Bernstein, E., Beach, D., & Hannon, G. J. (2000). An RNA-directed nuclease mediates post-transcriptional gene silencing in drosophila cells. *Nature, 404*, 293.

Han, H. D., Mangala, L. S., Lee, J. W., Shahzad, M. M. K., Kim, H. S., Shen, D., Nam, E. J., Mora, E. M., Stone, R. L., Lu, C., et al. (2010). Targeted gene silencing using RGD-labeled chitosan nanoparticles. *Clinical Cancer Research, 16*(15), 3910–3922.

Hansen, K. M., Ji, H. F., Wu, G., Datar, R., Cote, R., Majumdar, A., & Thundat, T. (2001). Cantilever-based optical deflection assay for discrimination of DNA single-nucleotide mismatches. *Analytical Chemistry, 73*(7), 1567–1571.

Harper, S. Q., Staber, P. D., He, X., Eliason, S. L., Martins, I. H., Mao, Q., Yang, L., Kotin, R. M., Paulson, H. L., & Davidson, B. L. (2005). RNA interference improves motor and neuropathological abnormalities in a Huntington's disease mouse model. *Proceedings of the National Academy of Sciences, 102*(16), 5820–5825.

Hatakeyama, H., Akita, H., Ito, E., Hayashi, Y., & Oishi, M. (2011). Biomaterials systemic delivery of SiRNA to tumors using a lipid nanoparticle containing a tumor-Specific cleavable PEG-lipid. *Biomaterials, 32*(18), 4306–4316.

Heale, B. S. E., Soifer, H. S., Bowers, C., & Rossi, J. J. (2005). SiRNA target site secondary structure predictions using local stable substructures. *Nucleic Acids Research, 33*(3), 1–10.

Huang, Y., Cheng, Q., Ji, J.-L., Zheng, S., Du, L., Meng, L., Wu, Y., Zhao, D., Wang, X., Lai, L., et al. (2016). Pharmacokinetic behaviors of intravenously administered SiRNA in glandular tissues. *Theranostics, 6*(10), 1528–1541.

Illum, L., & Davis, S. S. (1984). The organ uptake of intravenously administered colloidal particles can be altered using a non-ionic surfactant (Poloxamer 338). *FEBS Letters, 167*(1), 79–82.

Inoue, M., Matsumoto, S., Saito, H., Tsujitani, S., & Ikeguchi, M. (2008). Intraperitoneal administration of a small interfering RNA targeting nuclear factor-kappa B with paclitaxel successfully prolongs the survival of xenograft model mice with peritoneal metastasis of gastric cancer. *International Journal of Cancer, 123*(11), 2696–2701.

Jagannath, A., & Wood, M. (2007). RNA interference based gene therapy for neurological disease. *Briefings in Functional Genomics & Proteomics, 6*(1), 40–49.

Jang, S. H., Wientjes, M. G., Lu, D., & Au, J. L. S. (2003). Drug delivery and transport to solid tumors. *Pharmaceutical Research, 20*(9), 1337–1350.

Jiang, S., Eltoukhy, A. A., Love, K. T., Langer, R., & Anderson, D. G. (2013). Lipidoid-coated Iron oxide nanoparticles for efficient DNA and SiRNA delivery. *Nano Letters, 13*(3), 1059–1064.

Joo, J., Kwon, E. J., Kang, J., Skalak, M., Anglin, E. J., Mann, A. P., Ruoslahti, E., Bhatia, S. N., & Sailor, M. J. (2016). Porous silicon-graphene oxide core-shell nanoparticles for targeted delivery of SiRNA to the injured brain. *Nanoscale Horizons, 1*(5), 407–414.

Jung, C. R., Yoo, J., Ye, J. J., Kim, S., Chu, I. S., Young, I. Y., Jong, Y. C., & Im, D. S. (2006). Adenovirus-mediated transfer of SiRNA against PTTG1 inhibits liver cancer cell growth in vitro and in vivo. *Hepatology, 43*(5), 1042–1052.

Khatri, N. I., Rathi, M. N., Kolte, A. A., Kore, G. G., Lalan, M. S., Trehan, S., & Misra, A. R. (2012). Patents review in SiRNA delivery for pulmonary disorders. *Recent Patents on Drug Delivery & Formulation, 6*, 45–65.

Kim, N., Jiang, D., Jacobi, A. M., Lennox, K. A., Rose, S. D., Behlke, M. A., & Salem, A. K. (2012). Synthesis and characterization of mannosylated pegylated polyethylenimine as a carrier for SiRNA. *International Journal of Pharmaceutics, 427*(1), 123–133.

Kulhari, H., Pooja, D., Singh, M. K., & Chauhan, A. S. (2015). Optimization of carboxylate-terminated poly(amidoamine) dendrimer-mediated cisplatin formulation. *Drug Development and Industrial Pharmacy, 41*(2), 232–238.

Cavallaro, G., Sardo, C., Scialabba, C., Licciardi, M., & Giammona, G. (2017). Smart inulin-based polycationic nanodevices for SiRNA delivery. *Current Drug Delivery, 14*(2), 224–230.

Landen, C. N., Chavez-Reyes, A., Bucana, C., Schmandt, R., Deavers, M. T., Lopez-Berestein, G., & Sood, A. K. (2005). Therapeutic EphA2 gene targeting in vivo using neutral liposomal small interfering RNA delivery. *Cancer Research, 65*(15), 6910 LP–6916918.

Lee, S. H., Bae, K. H., Kim, S. H., Lee, K. R., & Park, T. G. (2008). Amine-functionalized gold nanoparticles as non-cytotoxic and efficient intracellular SiRNA delivery carriers. *International Journal of Pharmaceutics, 364*(1), 94–101.

Lee, S. J., Kim, M. J., Kwon, I. C., & Roberts, T. M. (2016). Delivery strategies and potential targets for SiRNA in major cancer types. *Advanced Drug Delivery Reviews, 104*, 2–15.

Li, M., Zhang, W., Wang, B., Gao, Y., Song, Z., & Zheng, Q. C. (2016). Ligand-based targeted therapy: A novel strategy for hepatocellular carcinoma. *International Journal of Nanomedicine, 11*, 5645–5669.

Li, X., Sun, A. N., Liu, Y. J., Zhang, W. J., Pang, N., Cheng, S. X., & Qi, X. R. (2018). Amphiphilic dendrimer engineered nanocarrier systems for co-delivery of SiRNA and paclitaxel to matrix metalloproteinase-rich tumors for synergistic therapy. *NPG Asia Materials, 10*(4), 238–254.

Lian, S., Kang, Z., Wang, E., Jiang, M., Hu, C., & Xu, L. (2003). Convenient synthesis of single crystalline magnetic Fe3O4 nanorods. *Solid State Communications, 127*(9–10), 605–608.

Licciardi, M., Tang, Y., Billingham, N. C., Armes, S. P., & Lewis, A. L. (2005). Synthesis of novel folic acid-functionalized biocompatible block copolymers by atom transfer radical polymerization for gene delivery and encapsulation of hydrophobic drugs. *Biomacromolecules, 6*(2), 1085–1096.

Lin, P. J. C., Tam, Y. Y. C., Hafez, I., Sandhu, A., Chen, S., Ciufolini, M. A., Nabi, I. R., & Cullis, P. R. (2013). Influence of cationic lipid composition on uptake and intracellular processing of lipid nanoparticle formulations of SiRNA. *Nanomedicine: Nanotechnology, Biology and Medicine, 9*(2), 233–246.

Liu, J., Carmell, M. A., Rivas, F. V., Marsden, C. G., Thomson, J. M., Song, J. J., Hammond, S. M., Joshua-Tor, L., & Hannon, G. J. (2004). Argonaute2 is the catalytic engine of mammalian RNAi. *Science (80-), 305*(5689), 1437–1441.

Liu, X., Liu, C., Laurini, E., Posocco, P., Pricl, S., Qu, F., Rocchi, P., & Peng, L. (2012). Efficient delivery of sticky SiRNA and potent gene silencing in a prostate cancer model using a generation 5 triethanolamine-core PAMAM dendrimer. *Molecular Pharmaceutics, 9*(3), 470–481.

Lu, J. J., Langer, R., Chen, J. A., & Novel Mechanism, I. (2009). Involved in cationic lipid-mediated functional SiRNA delivery. *Molecular Pharmaceutics, 6*(3), 763–771.

Ma, J.-B. B., Yuan, Y.-R. R., Meister, G., Pei, Y., Tuschl, T., & Patel, D. J. (2005). Structural basis for 5′ -end-specific recognition of guide RNA by the *A. Fulgidus* piwi protein. *Nature, 434*(7033), 666–670.

Madison, K. C. (2003). Barrier function of the skin: "La raison d'Être" of the epidermis. *The Journal of Investigative Dermatology, 121*(2), 231–241.

Maeda, H. (2001). The enhanced permeability and retention (EPR) effect in tumor vasculature: The key role of tumor-selective macromolecular drug targeting. *Advances in Enzyme Regulation, 41*(00), 189–207.

Malek, A., Czubayko, F., & Aigner, A. P. E. G. (2008). Grafting of polyethylenimine (PEI) exerts different effects on DNA transfection and SiRNA-induced gene targeting efficacy. *Journal of Drug Targeting, 16*(2), 124–139.

Malhotra, M., Tomaro-Duchesneau, C., Saha, S., & Prakash, S. (2013). Systemic SiRNA delivery via peptide-tagged polymeric nanoparticles, targeting PLK1 gene in a mouse xenograft model of colorectal cancer. *International Journal of Biomaterials, 2013*, 1–13.

Marques, J. T., & Williams, B. R. G. (2005). Activation of the mammalian immune system by SiRNAs. *Nature Biotechnology, 23*(11), 1399–1405.

Martinez, J., Patkaniowska, A., Urlaub, H., Lührmann, R., & Tuschl, T. (2002). Single-stranded antisense SiRNAs guide target RNA cleavage in RNAi. *Cell, 110*(5), 563–574.

Meister, G., Landthaler, M., Patkaniowska, A., Dorsett, Y., Teng, G., & Tuschl, T. (2004). Human Argonaute2 mediates RNA cleavage targeted by MiRNAs and SiRNAs. *Molecular Cell, 15*(2), 185–197.

Mekaru, H., Lu, J., & Tamanoi, F. (2015). Development of mesoporous silica-based nanoparticles with controlled release capability for cancer therapy. *Advanced Drug Delivery Reviews, 95*, 40–49.

Meng, H., Liong, M., Xia, T., Li, Z., Ji, Z., Zink, J. I., & Nel, A. E. (2010). Engineered design of mesoporous silica nanoparticles to deliver doxorubicin and P-glycoprotein SiRNA to overcome drug resistance in a cancer cell line. *ACS Nano, 4*(8), 4539–4550.

Merkel, O. M., Beyerle, A., Librizzi, D., Pfestroff, A., Behr, T. M., Sproat, B., Barth, P. J., & Kissel, T. (2009). Nonviral SiRNA delivery to the lung: Investigation of PEG-PEI polyplexes and their in vivo performance. *Molecular Pharmaceutics, 6*(4), 1246–1260.

Midoux, P., Pichon, C., Yaouanc, J. J., & Jaffrès, P. A. (2009). Chemical vectors for gene delivery: A current review on polymers, peptides and lipids containing histidine or imidazole as nucleic acids carriers. *British Journal of Pharmacology, 157*(2), 166–178.

Minakuchi, Y., Takeshita, F., Kosaka, N., Sasaki, H., Yamamoto, Y., Kouno, M., Honma, K., Nagahara, S., Hanai, K., Sano, A., et al. (2004). Atelocollagen-mediated synthetic small interfering RNA delivery for effective gene silencing in vitro and in vivo. *Nucleic Acids Research, 32*(13), e109–e109.

Minami, K., Okamoto, K., Doi, K., Harano, K., Noiri, E., & Nakamura, E. (2014). SiRNA delivery targeting to the lung via agglutination-induced accumulation and clearance of cationic tetraamino fullerene. *Scientific Reports, 4*, 4916.

Mohammadi, M., Salmasi, Z., Hashemi, M., Mosaffa, F., Abnous, K., & Ramezani, M. (2015). Single-walled carbon nanotubes functionalized with aptamer and piperazine-polyethylenimine derivative for targeted SiRNA delivery into breast cancer cells. *International Journal of Pharmaceutics, 485*(1–2), 50–60.

Mouldy, S. (2006). Single-stranded small interfering RNA are more immunostimulatory than their double-stranded counterparts: A central role for 2′-hydroxyl uridines in immune responses. *European Journal of Immunology, 36*(5), 1222–1230.

Naito, Y., & Ui-Tei, K. (2013). Designing functional SiRNA with reduced off-target effects. *Methods in Molecular Biology, 942*, 57–68.

Ngamcherdtrakul, W., Morry, J., Gu, S., Castro, D. J., Goodyear, S. M., Sangvanich, T., Reda, M. M., Lee, R., Mihelic, S. A., Beckman, B. L., et al. (2015). Cationic polymer modified mesoporous silica nanoparticles for targeted SiRNA delivery to HER2 $^{+}$ breast cancer. *Advanced Functional Materials, 25*(18), 2646–2659.

Nguyen, Q. D., Schachar, R. A., Nduaka, C. I., Sperling, M., Basile, A. S., Klamerus, K. J., Chi-Burris, K., Yan, E., Paggiarino, D. A., Rosenblatt, I., et al. (2012a). Dose-ranging evaluation of intravitreal SiRNA PF-04523655 for diabetic macular edema (the DEGAS study). *Investigative Ophthalmology and Visual Science, 53*(12), 7666–7674.

Nguyen, Q. D., Schachar, R. A., Nduaka, C. I., Sperling, M., Basile, A. S., Klamerus, K. J., Chi-Burris, K., Yan, E., Paggiarino, D. A., Rosenblatt, I., et al. (2012b). Phase 1 dose-escalation study of a SiRNA targeting the RTP801 gene in age-related macular degeneration patients. *Eye, 26*(8), 1099–1105.

Nimesh, S., & Chandra, R. (2009). Polyethylenimine nanoparticles as an efficient in vitro SiRNA delivery system. *European Journal of Pharmaceutics and Biopharmaceutics, 73*(1), 43–49.

Nishina, K., Unno, T., Uno, Y., Kubodera, T., Kanouchi, T., Mizusawa, H., & Yokota, T. (2008). Efficient in vivo delivery of SiRNA to the liver by conjugation of α-tocopherol. *Molecular Therapy, 16*(4), 734–740.

Noble, C. O., Guo, Z., Hayes, M. E., Marks, J. D., Park, J. W., Benz, C. C., Kirpotin, D. B., & Drummond, D. C. (2009). Characterization of highly stable liposomal and immunoliposomal formulations of vincristine and vinblastine. *Cancer Chemotherapy and Pharmacology, 64*(4), 741–751.

Norman, M. E., Williams, P., & Illum, L. (1993). Influence of block copolymers on the adsorption of plasma proteins to microspheres. *Biomaterials, 14*(3), 193–202.

Pañeda, C. (2013). SYL040012, a SiRNA for the treatment of glaucoma. *Acta Ophthalmologica, 91*(s252), 0–0.

Patil, Y. P., & Jadhav, S. (2014). Novel methods for liposome preparation. *Chemistry and Physics of Lipids, 177*, 8–18.

Patzel, V., Rutz, S., Dietrich, I., Köberle, C., Scheffold, A., & Kaufmann, S. H. E. (2005). Design of SiRNAs producing unstructured guide-RNAs results in improved RNA interference efficiency. *Nature Biotechnology, 23*(11), 1440–1444.

Pecot, C. V., Calin, G. A., Coleman, R. L., Lopez-Berestein, G., & Sood, A. K. (2011). RNA interference in the clinic: Challenges and future directions. *Nature Reviews Cancer, 11*(1), 59–67.

Pellish, R. S., Nasir, A., Ramratnam, B., & Moss, S. F. (2008). Review article: RNA interference – potential therapeutic applications for the gastroenterologist. *Alimentary Pharmacology & Therapeutics, 27*(9), 715–723.

Pittella, F., Zhang, M., Lee, Y., Kim, H. J., Tockary, T., Osada, K., Ishii, T., Miyata, K., Nishiyama, N., & Kataoka, K. (2011). Enhanced endosomal escape of SiRNA-incorporating hybrid nanoparticles from calcium phosphate and PEG-block charge-conversional polymer for efficient gene knockdown with negligible cytotoxicity. *Biomaterials, 32*(11), 3106–3114.

Pooja, D., Kulhari, H., Singh, M. K., Mukherjee, S., Rachamalla, S. S., & Sistla, R. (2014). Dendrimer-TPGS mixed micelles for enhanced solubility and cellular toxicity of taxanes. *Colloids and Surfaces. B, Biointerfaces, 121*, 461–468.

Prausnitz, M. R., Mitragotri, S., & Langer, R. (2004). Current status and future potential of transdermal drug delivery. *Nature Reviews Drug Discovery, 3*, 115–124.

Qian, J., & Berkland, C. (2015). PH-sensitive triblock copolymers for efficient SiRNA encapsulation and delivery. *Polymer Chemistry, 6*(18), 3472–3479.

Qiu, Y., Lam, J. K. W., Leung, S. W. S., & Liang, W. (2016a). Delivery of RNAi therapeutics to the airways – From bench to bedside. *Molecules, 21*, E1249.

Qiu, C., Wei, W., Sun, J., Zhang, H.-T., Ding, J.-S., Wang, J.-C., & Zhang, Q. (2016b). Systemic delivery of SiRNA by hyaluronan-functionalized calcium phosphate nanoparticles for tumor-targeted therapy. *Nanoscale, 8*(26), 13033–13044.

Ralph, G. S., Radcliffe, P. A., Day, D. M., Carthy, J. M., Leroux, M. A., Lee, D. C. P., Wong, L.-F., Bilsland, L. G., Greensmith, L., Kingsman, S. M., et al. (2005). Silencing mutant SOD1 using RNAi protects against neurodegeneration and extends survival in an ALS model. *Nature Medicine, 11*, 429.

Rappaport, J., Hanss, B., Kopp, J. B., Copeland, T. D., Bruggeman, L. A., Coffman, T. M., & Klotman, P. E. (1995). Transport of phosphorothioate oligonucleotides in kidney: Implications for molecular therapy. *Kidney International, 47*(5), 1462–1469.

Ren, L., Zhang, Y., Cui, C., Bi, Y., & Ge, X. (2017). Functionalized graphene oxide for anti-VEGF SiRNA delivery: Preparation, characterization and evaluation in vitro and in vivo. *RSC Advances, 7*(33), 20553–20566.

Rettig, G. R., & Behlke, M. A. (2012). Progress toward in vivo use of SiRNAs-II. *Molecular Therapy, 20*(3), 483–512.

Rose, S. D., Kim, D. H., Amarzguioui, M., Heidel, J. D., Collingwood, M. A., Davis, M. E., Rossi, J. J., & Behlke, M. A. (2005). Functional polarity is introduced by dicer processing of short substrate RNAs. *Nucleic Acids Research, 33*(13), 4140–4156.

Rossi, J. J. (2009). Cholesterol paves the way for topically applied viricides. *Cell Host & Microbe, 5*(1), 6–7.
Sabbioni, S., Callegari, E., Manservigi, M., Argnani, R., Corallini, A., Negrini, M., & Manservigi, R. (2006). Use of herpes simplex virus type 1-based amplicon vector for delivery of small interfering RNA. *Gene Therapy, 14*(5), 459–464.
Sailor, M. J., & Ji-Ho, P. (2012). Hybrid nanoparticles for detection and treatment of cancer. *Advanced Materials, 24*(28), 3779–3802.
Santel, A., Aleku, M., Keil, O., Endruschat, J., Esche, V., Fisch, G., Dames, S., Löffler, K., Fechtner, M., Arnold, W., et al. (2006). A novel SiRNA-lipoplex technology for RNA interference in the mouse vascular endothelium. *Gene Therapy, 13*(16), 1222–1234.
Sawant, R. R., & Torchilin, V. P. (2012). Multifunctional nanocarriers and intracellular drug delivery. *Current Opinion in Solid State & Materials Science, 16*(6), 269–275.
Schedin, F., Geim, A. K., Morozov, S. V., Hill, E. W., Blake, P., Katsnelson, M. I., & Novoselov, K. S. (2007). Detection of individual gas molecules adsorbed on graphene. *Nature Materials, 6*(9), 652–655.
Schwarz, D. S., Hutvágner, G., Du, T., Xu, Z., Aronin, N., & Zamore, P. D. (2003). Asymmetry in the assembly of the RNAi enzyme complex. *Cell, 115*(2), 199–208.
Sinden, R. R., Pearson, C. E., Potaman, V. N., Ussery, D. W. (1998). DNA: Structure and function. In Verma, R. S. (Ed.). *Advances in genome biology* (Vol. 5, pp. 1–141).
Singer, O., Marr, R. A., Rockenstein, E., Crews, L., Coufal, N. G., Gage, F. H., Verma, I. M., & Masliah, E. (2005). Targeting BACE1 with SiRNAs ameliorates Alzheimer disease neuropathology in a transgenic model. *Nature Neuroscience, 8*(10), 1343–1349.
Singhania, A., Y Wu, S., & McMillan, N. (2011). Effective delivery of PEGylated SiRNA-containing lipoplexes to extraperitoneal tumours following intraperitoneal administration. *Journal of Drug Delivery, 2011*, 192562.
Smith, S. B., Cui, Y., & Bustamante, C. (1996). Overstretching B-DNA: The elastic response of individual double-stranded and single-stranded DNA molecules. *Science (80), 271*(5250), 795–799.
Song, J.-J., Liu, J., Tolia, N. H., Schneiderman, J., Smith, S. K., Martienssen, R. A., Hannon, G. J., & Joshua-Tor, L. (2003). The crystal structure of the Argonaute2 PAZ domain reveals an RNA binding motif in RNAi effector complexes. *Nature Structural Biology, 10*, 1026.
Song, J.-J., Smith, S., Hannon, G. J., Joshua-Tor, L., Song, J. J., Smith, S. K., Hannon, G. J., & Joshua-Tor, L. (2004). Crystal structure of Argonaute and its implications for RISC slicer activity. *Science, 305*, 1434–1437. Vol. 305.
Song, E., Zhu, P., Lee, S.-K., Chowdhury, D., Kussman, S., Dykxhoorn, D. M., Feng, Y., Palliser, D., Weiner, D. B., Shankar, P., et al. (2005). Antibody mediated in vivo delivery of small interfering RNAs via cell-surface receptors. *Nature Biotechnology, 23*(6), 709–717.
Svenson, S., & Tomalia, D. A. (2012). Dendrimers in biomedical applications-reflections on the field. *Advanced Drug Delivery Reviews, 64*(SUPPL), 102–115.
Tang, G. (2005). SiRNA and MiRNA: An insight into RISCs. *Trends in Biochemical Sciences, 30*(2), 106–114.
Taratula, O., Garbuzenko, O. B., Kirkpatrick, P., Pandya, I., Savla, R., Pozharov, V. P., He, H., & Minko, T. (2009). Surface-engineered targeted PPI dendrimer for efficient intracellular and intratumoral SiRNA delivery. *Journal of Controlled Release, 140*(3), 284–293.
Tatiparti, K., Sau, S., Kashaw, S., & Iyer, A. (2017). SiRNA delivery strategies: A comprehensive review of recent developments. *Nanomaterials, 7*(4), 77.
Thomas, M., Lu, J. J., Chen, J., & Klibanov, A. M. (2007). Non-viral SiRNA delivery to the lung. *Advanced Drug Delivery Reviews, 59*(2–3), 124–133.
Torchilin, V. P. (2014). Multifunctional, stimuli-sensitive nanoparticulate systems for drug delivery. *Nature Reviews. Drug Discovery, 13*(11), 813–827.
Tseng, Y. C., Mozumdar, S., & Huang, L. (2009). Lipid-based systemic delivery of SiRNA. *Advanced Drug Delivery Reviews, 61*(9), 721–731.

Urbiola, K., Blanco-Fernández, L., Ogris, M., Rödl, W., Wagner, E., & de Ilarduya, C. T. (2018). Novel PAMAM-PEG-peptide conjugates for SiRNA delivery targeted to the transferrin and epidermal growth factor receptors. *Journal of Personalized Medicine, 8*(1), 4.

Varkouhi, A. K., Foillard, S., Lammers, T., Schiffelers, R. M., Doris, E., Hennink, W. E., & Storm, G. (2011). SiRNA delivery with functionalized carbon nanotubes. *International Journal of Pharmaceutics, 416*(2), 419–425.

Vauthier, C., Zandanel, C., & Ramon, A. L. (2013). Chitosan-based nanoparticles for in vivo delivery of interfering agents including SiRNA. *Current Opinion in Colloid & Interface Science, 18*(5), 406–418.

Verma, P., Prajapati, S. K., & Prajapati, R. N. (2012). Performance evaluation of Pamam dendrimer based clotrimazole formulations. *International Journal of Pharmaceutics, 2*. (October), 203–209.

Wang, J., Lu, Z., Wientjes, M. G., & Au, J. L.-S. (2010). Delivery of SiRNA therapeutics: Barriers and carriers. *The AAPS Journal, 12*(4), 492–503.

Wang, F., Gao, L., Meng, L. Y., Xie, J. M., Xiong, J. W., & Luo, Y. A. (2016). Neutralized noncharged polyethylenimine-based system for efficient delivery of SiRNA into heart without toxicity. *ACS Applied Materials & Interfaces, 8*(49), 33529–33538.

Wartiovaara, J., Öfverstedt, L. G., Khoshnoodi, J., Zhang, J., Mäkelä, E., Sandin, S., Ruotsalainen, V., Cheng, R. H., Jalanko, H., Skoglund, U., et al. (2004). Nephrin strands contribute to a porous slit diaphragm scaffold as revealed by electron tomography. *The Journal of Clinical Investigation, 114*(10), 1475–1483.

Watanabe, K., Harada-Shiba, M., Suzuki, A., Gokuden, R., Kurihara, R., Sugao, Y., Mori, T., Katayama, Y., & Niidome, T. (2009). In vivo SiRNA delivery with dendritic poly(l-lysine) for the treatment of hypercholesterolemia. *Molecular BioSystems, 5*(11), 1306.

Whitehead, K. A., Langer, R., & Anderson, D. G. (2009). Knocking down barriers: Advances in SiRNA delivery. *Nature Reviews Drug Discovery, 8*(2), 129–138.

Wilson, R. C., & Doudna, J. A. (2013). Molecular mechanisms of RNA interference. *Annual Review of Biophysics, 42*(1), 217–239.

Wolff, J. A., & Rozema, D. B. (2008). Breaking the bonds: Non-viral vectors become chemically dynamic. *Molecular Therapy, 16*(1), 8–15.

Wolfrum, C., Shi, S., Jayaprakash, K. N., Jayaraman, M., Wang, G., Pandey, R. K., Rajeev, K. G., Nakayama, T., Charrise, K., Ndungo, E. M., et al. (2007). Mechanisms and optimization of in vivo delivery of lipophilic SiRNAs. *Nature Biotechnology, 25*(10), 1149–1157.

Wu, S. Y., Chang, H. I., Burgess, M., & McMillan, N. A. J. (2011). Vaginal delivery of SiRNA using a novel PEGylated lipoplex-entrapped alginate scaffold system. *Journal of Controlled Release, 155*(3), 418–426.

Wu, C., Gong, F., Pang, P., Shen, M., Zhu, K., Cheng, D., Liu, Z., & Shan, H. (2013). An RGD-modified MRI-visible polymeric vector for targeted SiRNA delivery to hepatocellular carcinoma in nude mice. *PLoS One, 8*(6), e66416.

Wu, J., Qu, W., Williford, J. M., Ren, Y., Jiang, X., Jiang, X., Pan, D., Mao, H. Q., & Luijten, E. (2017). Improved SiRNA delivery efficiency via solvent-induced condensation of micellar nanoparticles. *Nanotechnology, 28*(20), 204002.

Xia, Y., Tian, J., & Chen, X. (2016). Biomaterials effect of surface properties on liposomal SiRNA delivery. *Biomaterials, 79*, 56–68.

Xia, Y., Lin, Z., Li, Y., Zhao, M., Wang, C., Guo, M., Zhang, B., & Zhu, B. (2017). Targeted delivery of SiRNA using RGDfC-conjugated functionalized selenium nanoparticles for anticancer therapy. *Journal of Materials Chemistry B, 5*(33), 6941–6952.

Xu, C. F., & Wang, J. (2015). Delivery systems for SiRNA drug development in cancer therapy. *Asian Journal of Pharmaceutical Sciences, 10*(1), 1–12.

Yang, Y., Li, J., Liu, F., & Huang, L. (2012). Systemic delivery of SiRNA via LCP nanoparticle efficiently inhibits lung metastasis. *Molecular Therapy, 20*(3), 609–615.

Yezhelyev, M. V., Qi, L., O'Regan, R. M., Nie, S., & Gao, X. (2008). Proton-sponge coated quantum dots for SiRNA delivery and intracellular imaging. *Journal of the American Chemical Society, 130*(28), 9006–9012.

Yin, F., Hu, K., Chen, Y., Yu, M., Wang, D., Wang, Q., Yong, K. T., Lu, F., Liang, Y., & Li, Z. (2017). SiRNA delivery with PEGylated graphene oxide nanosheets for combined photothermal and genetherapy for pancreatic cancer. *Theranostics, 7*(5), 1133–1148.

Young, S. W. S., Stenzel, M., & Jia-Lin, Y. (2016). Nanoparticle-SiRNA: A potential cancer therapy? *Critical Reviews in Oncology/Hematology, 98*, 159–169.

Yu, H., Zou, Y., Wang, Y., Huang, X., Huang, G., Sumer, B. D., Boothman, D. A., & Gao, J. (2011). Overcoming endosomal barrier by amphotericin B-loaded dual PH-responsive PDMA- b-PDPA micelleplexes for SiRNA delivery. *ACS Nano, 5*(11), 9246–9255.

Yu-Wai-Man, C., Tagalakis, A. D., Manunta, M. D., Hart, S. L., & Khaw, P. T. (2016). Receptor-targeted liposome-peptide-SiRNA nanoparticles represent an efficient delivery system for MRTF silencing in conjunctival fibrosis. *Scientific Reports, 6*, 21881.

Zeng, Y., Yang, Z., Li, H., Hao, Y., Liu, C., Zhu, L., Liu, J., Lu, B., & Li, R. (2017). Multifunctional nanographene oxide for targeted gene-mediated thermochemotherapy of drug-resistant tumour. *Scientific Reports, 7*, 43506.

Zhang, Y., Cristofaro, P., Silbermann, R., Pusch, O., Boden, D., Konkin, T., Hovanesian, V., Monfils, P. R., Resnick, M., Moss, S. F., et al. (2006). Engineering mucosal RNA interference in vivo. *Molecular Therapy, 14*(3), 336–342.

Zhang, Y.-M., Yang, Y., Zhang, Y., & Liu, Y. (2016). Polysaccharide nanoparticles for efficient SiRNA targeting in cancer cells by supramolecular PKa shift. *Scientific Reports, 6*(28848), 1–11.

Zhou, J., Patel, T. R., Fu, M., Bertram, J. P., & Saltzman, W. M. (2012). Octa-functional PLGA nanoparticles for targeted and efficient SiRNA delivery to tumors. *Biomaterials, 33*(2), 583–591.

Zhou, J., Wu, Y., Wang, C., Cheng, Q., Han, S., Wang, X., Zhang, J., Deng, L., Zhao, D., Du, L., et al. (2016). PH-sensitive nanomicelles for high-efficiency SiRNA delivery in vitro and in vivo: An insight into the design of polycations with robust cytosolic release. *Nano Letters, 16*(11), 6916–6923.

Zhu, L., Perche, F., Wang, T., & Torchilin, V. P. (2014). Matrix metalloproteinase 2-sensitive multifunctional polymeric micelles for tumor-specific co-delivery of SiRNA and hydrophobic drugs. *Biomaterials, 35*(13), 4213–4222.

Zhu, H., Liu, W., Cheng, Z., Yao, K., Yang, Y., Xu, B., & Su, G. (2017). Targeted delivery of SiRNA with PH-responsive hybrid gold nanostars for cancer treatment. *International Journal of Molecular Sciences, 18*(10), 2029.

Silver-Based Polymeric Nanocomposites as Antimicrobial Coatings for Biomedical Applications

Navneet K. Dhiman, Shekhar Agnihotri, and Ravi Shukla

Abstract Hospital-acquired infections (HAIs) pose one of the major challenges to therapeutic applications of biomedical devices under clinically relevant conditions. A paradigm shift in understanding and pathogenesis of biofilm formation has constantly been forcing professionals to adopt some novel and effective, yet affordable anti-adhesive/anti-biofilm technologies for successful long-term implantation of devices without infections. The intriguing physicochemical properties of a biomaterial's surface is crucial to develop novel coating technologies where the anti-fouling feature of the device must also be accompanied with its long-term antibacterial performance without introducing toxicity to mammalian cells and the drug resistance. One of the best strategies to minimize nosocomial infections is through using biocompatible polymers that exhibit either an innate biocidal characteristic or may be surface-modified to impart antimicrobial features to a biomaterial by introducing biocidal agents, such as antibiotics, antimicrobial peptides, and more recently silver nanoparticles (AgNPs). Nano-silver has been widely accepted as the most efficacious metal that is well-adorned with antimicrobial properties due to its oligodynamic action, multifaceted mechanisms of biocidal action and low cytotoxicity to humans. The present chapter thus provides an exhaustive information about various surface modifications strategies for biomaterial coatings, which can be employed to immobilize silver nanoparticles onto polymeric composites with a few common

N. K. Dhiman
Department of Biotechnology, Thapar Institute of Engineering and Technology, Patiala, Punjab, India

S. Agnihotri (✉)
Department of Biotechnology, Thapar Institute of Engineering and Technology, Patiala, Punjab, India

TIFAC-CORE in Agro & Industrial Biotechnology, Thapar Institute of Engineering and Technology, Patiala, Punjab, India
e-mail: shekhar.agnihotri@thapar.edu

R. Shukla (✉)
Nanobiotechnology Research Laboratory and Centre for Advanced Materials & Industrial Chemistry, School of Science, RMIT University, Melbourne, VIC, Australia
e-mail: ravi.shukla@rmit.edu.au

S. Singh, P. K. Maurya (eds.), *Nanotechnology in Modern Animal Biotechnology*, https://doi.org/10.1007/978-981-13-6004-6_4

goals, i.e. broad-spectrum antimicrobial nature, higher efficacy, stability and promoting reuse. Various nano-silver based polymeric composites of both natural and synthetic origin will be discussed as potential coating materials candidates for implants (vascular grafts, endotracheal tubes, and catheters), wound dressings, surgical mesh and other porous scaffolds. The application of AgNPs-polymeric nanocomposites into several forms such as thin films, fibers, hydrogels, and multilayered structures will be correlated with their clinical relevance. Lastly, potential toxicity and safety concerns using these nanocomposites will also be discussed.

Keywords Surface modification · Immobilization · Biomedical implants · Cytotoxicity · Silver release · Antibacterial · Biocidal

1 Introduction

The latest innovations in material technology have maximized their utilization in the healthcare sector as biomedical implants, vascular grafts, and catheters in order to improve life quality of critical patients (Knetsch and Koole 2011; Salwiczek et al. 2014). Despite significant achievements in employing these biomedical devices under in vivo conditions, the commercial use of such devices involves certain inevitable complications at the implant site where only a handful of biomaterials can confirm their safety under practically relevant conditions. This can be understood by the fact that the inherent surface characteristics of a biomaterial, regardless of their composition act as a 'battleground' for microbial communities to attach, colonize and spread thereby facilitating their survival as microbial biofilms. Other well-recognized issues associated with nosocomial infections are (i) poor hygiene practices among hospital co-workers, (ii) rapid expansion of invasive devices, (iii) increased demand among old-aged patients, and (iv) diminished efficacy of used antibiotics. Medical devices and implants can still be invaded by pathogens leading to severe infections viz. biomaterial-associated infections (BAIs) (Ocsoy et al. 2018; Von Eiff et al. 2005; Hetrick and Schoenfisch 2006; Huang et al. 2007), even if the advanced sterilization and aseptic procedures were followed during the surgery. The prevalence of BAIs thus can impair the intended performance of biomedical devices while adding risks to human life with high mortality and morbidity. It also adds up an economic burden to the patients by repetitive surgeries, medications and prolonged stay at the hospital (Coma 2013). As a result, one of the major challenges faced by biomaterials and devices in vivo is not merely establishing an appropriate integration with the surrounding tissue, but also to prevent/mitigate bacterial colonization without compromising its intended function.

The preliminary clinical evidence showing microbial biofilms as causative agents in device infections were reported in early 1980s. Since then, some of the most common agents associated with BAIs have been reported as *Staphylococcus aureus, Staphylococcus epidermidis*, *Staphylococcus haemolyticus*, *Staphylococcus capitis*, *Staphylococcus saprophyticus, Staphylococcus warneri*, *Staphylococcus cohnii*,

Staphylococcus xylosus, Staphylococcus chromogens, Staphylococcus schleiferi, Staphylococcus lugdunensis, *Escherichia coli*, *Pseudomonas aeruginosa*, *Proteus mirabilis*, *Proteus vulgaris*, *Candida albicans* and many more (Agnihotri and Dhiman 2017). The ubiquitous nature of pathogenic microbes for biofilm formation is now well documented and follows a discrete set of events. As the first step, an initial attachment of microbial cells starts over biomaterial surface within a few hours of implantation. The presence of macromolecules (carbohydrates, proteins and uronic acids) at implant's site also known as 'conditioning film' alter the physicochemical properties i.e., hydrophobicity and roughness of biomaterial's surface, which in turn facilitate microbes to adhere. In the second step, an enormous cell proliferation facilitates the accumulation and aggregation of attached cells to form multiple layers leading to biofilm formation. Finally, the maturation of microbial biofilm takes place and small parts of the biofilm rupture making planktonic bacteria to escape from the biofilm and to invade newer clean surfaces at distant sites (Donlan 2001). Virtually, all indwelling medical devices are prone to biofilm formation. As soon as a biomaterial is implanted within the body, its surface is entirely covered by the host's protein and glycoproteins as the first response to host-biomaterial interactions. However, the presence of these proteins at biomaterial's surface calls for the race between tissue integration and microbial attachment (Gristina 1987). Thus, biofilm formation can result from microorganisms entering the wound or adhere to the implant either during surgery via pre-operative or post-operative procedures (Busscher et al. 2012; Lynch and Robertson 2008). Such primary risks factors can be evaded by following strict hygiene routines in the hospitals. However, the infections can never be completely avoided. Therefore, new therapies which can prevent infection either by inhibiting an initial microbial attachment and colonization onto the device or by killing microbial communities are yet to be sought.

The spread of infections in hospital environments is a serious and persistent issue among healthcare professionals, which often get worse by the emergence of drug-resistant pathogenic strains over long exposure. Greater understanding of the mechanism of such infections should be given critical importance in order to develop novel strategies to combat against them. During and after biofilm formation, microbial machinery alters their gene expression pattern so as to facilitate the production of extracellular polymeric substances (EPS) (Li et al. 2010; McArthur et al. 2000). EPS formation enhances their chances of survival by forming a protective sheet around the sessile bacteria in order to cope with the shear stress and attack by the host's immune system and/or antibiotics. Being highly hydrated (up to 95% water) and slimy in nature with porous architecture, biofilm layers aid in escaping the bacteria during antimicrobial treatments (Feng et al. 2011; Smith et al. 2012). All these features make the post-biofilm formation treatments of infection very difficult due to low penetration ability of systemically administered antibiotics (Li et al. 2010). Additionally, altered gene expression (up-regulation and/or down-regulation) and a shift in the bacterial metabolism within the biofilm matrix leads to a decreased sensitivity toward a variety of antibiotics targeting metabolism (Hall-Stoodley and Stoodley 2009). Combination therapy is also not a feasible option since an overuse of antibiotics and disinfectants chemicals lead to the emergence of more resistant bacterial biofilms, which increase their tolerance towards the patient's defence system (Høiby et al. 2010). In due course of time, removal of an infected implant through surgery becomes the only possible solution to replenish the damage caused.

From the above discussion, it is easy to understand that physicochemical characteristic of biomaterial surface and the external stimuli (physical, chemical, and biological) from the tissue microenvironment would play a major role for both–the targeted applications of biomaterials and microbial colonization. The surface modification of a biomaterial seems to be a viable and effective approach through which implant-host interactions can be improved in concurrence with inhibiting microbial attachment. This will potentially avoid most implant-related infections and diminish the chance of biofilm formation. As a promising strategy, this has urged scientists to put research efforts on developing antimicrobial surfaces and coatings that can be applied to biomedical devices so as to confer resistance against bacterial colonization.

2 Strategies for Designing Novel Antimicrobial Surfaces

While designing an antimicrobial surface, the inherent features of biomaterial surface such as composition, surface area/energy, wettability, thickness, topography (roughness and stiffness) are the critical factors affecting protein adsorption that in turn dictate the success of implant-host interactions (Salwiczek et al. 2014; Bazaka et al. 2012). Concurrently, these characteristics can also be exploited to minimize microbial adhesion (Knetsch and Koole 2011; Mittal 2013). Previous efforts have shown that an increase in hydrophilicity of surface may lead to a decrease in microbial adhesion (Hermansson 1999; Chia et al. 2011; Zhao et al. 2005). For instance, polyurethane surfaces have demonstrated lowering in bacterial attachment after modification with a hydrophilic moiety, polyethene oxide (Zhou et al. 2014a). In another study, increasing hydrophilicity of polyvinyl chloride surface through oxygen plasma modification though did not fully suppressed the biofilm formation, it greatly reduced adhesion of *P. aeruginosa* cells thereafter (Triandafillu et al. 2003). Similarly, topographical features like surface roughness have been recently reviewed for its importance in reducing microbial adhesion. It was observed that an ultra-smooth surface will be less likely to be populated with microbes than an irregular rough surface. The availability of more surface area may contribute to the generation of more adhesive force by the microorganisms (Pavithra and Doble 2008; Sousa et al. 2009). Therefore, these factors lead us to prevent or limit bacterial colonization on the biomaterial surface by following three different strategies (Fig. 1).

The first strategy involves the development of an anti-adhesive surface by modifying the hydrophilicity of the surface. The passive nature of substrate would aim at preventing bacteria to achieve various possibilities of adhesion at the surface. Polymeric moieties held together by non-covalent interactions either in the form of a thin layer or brushes over biomaterial surfaces are used for such non-adhesive, microbial repellent coatings (Zhao and Brittain 2000). It has been reported that the

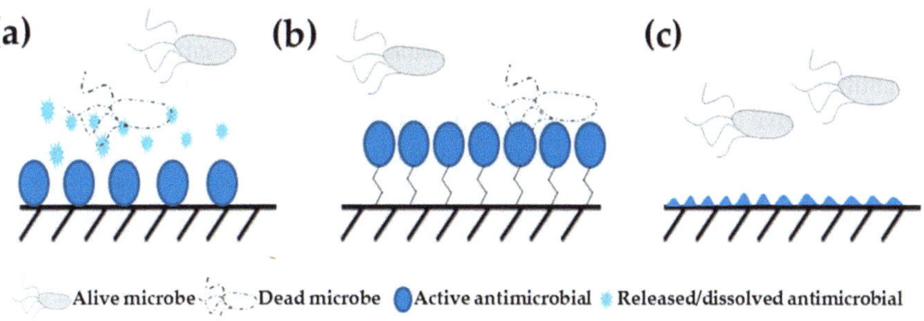

Fig. 1 Various strategies employed while designing biocidal coatings (**a**) anti-adhesiveness with modified surface topography to prevent microbial colonization, (**b**) killing of cells via direct-contact with antimicrobial agents, and (**c**) release of antibacterial agents

presence of charged high-density polymers/oligomers along with highly hydrophilic cationic groups limit the contact between bacteria and potential surface placement sites, thereby, minimizing bacterial adhesion (Bridges and García 2008; Roosjen et al. 2004). Some researchers believe that these polymers attract water molecules into the brush layer via hydrogen bonding, forming a repellent layer close to the surface where fewer proteins would adsorb over the surface of the biomaterial. This reduction in protein adsorption by several orders in magnitude offers the least possibility for microorganisms to bind through specific and/or non-specific interactions over the surface of biomedical devices (Knetsch and Koole 2011). Another challenge of designing anti-adhesive technologies relates to the current inability to design a universal surface treatment that can be applied to each bacterial species under all implant conditions.

The second strategy is based on killing the microorganisms as an outcome of 'direct-contact' with biomaterial's surface. Biocidal agents like quaternary ammonium compounds (QACs) (Nohr and Gavin 1994; Thorsteinsson et al. 2003), antimicrobial peptides (Coma 2013), enzymes (Thallinger et al. 2013), chitosan, polycations and phosphonium salts (Kanazawa et al. 1993) have been tethered chemically over implant's surface following this strategy. Being decorated with innate antimicrobial properties, these agent shows reduced cell viability and eventually kill pathogens if exposed over a long duration. These biocidal agents, however, may pose serious complications such as cytotoxicity, immunoreactivity, and genotoxicity to neighbouring healthier cells during biomaterial-tissue interactions under the account of their release in physiological conditions. Any such concerns may, however, be rectified through acute bonding with few other polymer chains like polyethylene glycol (Mittal 2013), the functionality of surface coating directing the biocidal action only against pathogenic cells yet facilitating the growth of mammalian cells is a technical challenge and has not been fully elucidated. A lot of tethering parameters (surface orientation and concentration) have to be optimized in order to fully exploit its antimicrobial potential.

The third strategy engages the 'release' of active biocidal agents such as antibiotics, metals (silver, copper, zinc oxide etc.), non-metal elements (e.g., carbon, selenium), organic substances (antibiotics, anti-infective peptides, chitosan, hyaluronic acid) and their combinations from the biomaterial surface. Systemic administration of antibiotics create undesired effects of resistance development, these can be minimized by active localized administration (Cloutier et al. 2015). Short-term release of antibiotics from "active" surface explains the diminution of biofilm formation and long-term release confines shielding fibrous capsule formation up to a certain extent (Zhang et al. 2014). However, apprehension over the evolution of antimicrobial resistance still remains an issue and toxicity risks hold high for these surfaces. Moreover, with the passage of time, the so-called reservoir surface runs out of biocidal agent and loses its bioactivity (Mittal 2013). All these strategies and their limitations hint towards a call for novel antimicrobial surface coatings in the biomedical field.

3 Antimicrobial Silver and Nano-silver for Coatings: A Historical Relevance

The antibacterial activity of the majority of metal-based coatings is attributed to its oligodynamic action, often linked to its ionic- or nano- form rather than the bulk regime. The use of coinage metals (silver, copper, zinc, gold) for various antimicrobial purposes viz. water disinfection, dental alloys, utensils have been documented since antiquity. Among all metals, silver shows the highest antimicrobial nature against broad varieties of pathogens such as bacteria, fungi, viruses, protozoa and is, therefore, an active ingredient in all antimicrobial based personal care products. The father of medicine, Hippocrates even believed the potential of using silver powder to heal ulcers and thus several practices have been documented using silver as a bactericide in hospitals (Alexander 2009). For instance, the administration of aqueous silver nitrate drops became a common therapy for newly born to treat ocular infections by *Neisseria Gonorrhoea* in the early 1800s (Silvestry-Rodriguez et al. 2007). At the end of this era, silver was established for its highest effectiveness and least toxicity to mammalian cells among all the elements in the periodic table in terms of antimicrobial properties (Alexander 2009). In the 1900s, people observed the use of silver dollars in prolonging the shelf life of daily perishable items like milk. Decades later, different types of silver compounds and products flooded the market including solutions (e.g., silver nitrate, oxide, bromide, chloride, and iodide), sutures, and colloids to cure bacterial infections (Alexander 2009). In 1920, The United States Food and Drug Administration (FDA) accepted the use of colloidal silver for wound management as silver demonstrated encouraging results in reducing wound surface inflammation. Over the next two decades, silver-based compounds were used throughout for alleviating infections, burns and wound healing purposes. One of the prominent examples of silver-based compounds is silver sulfadiazine which was used in medicine for the treatment of burns. Unfortunately, during the 1940s, the overhyped success of using penicillin and other antibiotics for treating infections diminished the brilliance of silver-based therapeutics.

However, with increasing incidences of multiple drug-resistant strains along with the latest advancements in nanoscience and nanotechnology, silver at its nanoscale i.e. silver nanoparticles again emerged as a potential antimicrobial agent with the highest extent of commercialization (Agnihotri et al. 2014, 2018; Bharti et al. 2015; Chakraborty et al. 2017). Today, nano-silver based consumer products can easily be identified in diverse areas of textiles, dietary supplements, food packaging, personal care and hygiene (cosmetics, sunscreen lotions, antibacterial sprays), biomedical devices (surgical coatings, medical implants, bandages, sutures) and water purification. Especially with clinical relevance, Acticoat™ (Gravante et al. 2009) and Actisorb® Plus 25 are commercially marketed as AgNPs-based wound dressings. AgNPs applied as coatings or additives to a variety of catheters such as polyurethane ventricular catheters are available as Silverline® and ON-Q SilverSoaker™ (Samuel and Guggenbichler 2004) and are also used in hand gels (Jain et al. 2009) and paints (Kumar et al. 2008) as a prolonged antibacterial disinfectant.

At the nanoscale, physicochemical properties of a material depend upon its atomic configuration and dimension which makes them popular and desirable for a vast array of applications. Silver nanoparticles (AgNPs) have a high surface area to volume ratio and active facets which imparts them a unique set of characteristics such as increased catalytic and antimicrobial activities even at low dosage (Franci et al. 2015). Particularly, AgNPs become more reactive and detrimental to microbes than their bulk counterpart (Domènech Garcia et al. 2014). Metallic silver has been shown to possess only weak antimicrobial activity which deteriorates fast and is strongly inhibited by protein adsorption to the silver (Atiyeh et al. 2007; Rai et al. 2009). The biocidal action of AgNPs is strongly dependent upon its size, shape, surface charge, the aggregation status and state of an application. As a rule of thumb, smaller sized (<10 nm) silver nanoparticles demonstrate stronger antibacterial action than larger particles with several order of magnitude (Agnihotri et al. 2014; Rai et al. 2009). On the other hand, truncated AgNPs demonstrated higher bacterial killing than spherical and rod-shaped nanoparticles with similar silver content (Pal et al. 2007; Panáček et al. 2006; Sotiriou and Pratsinis 2010). However considering several other factors like the ease in synthesis, reproducibility on size control, handling and recovery spherical AgNPs are still preferred over other shapes for use under practical relevant conditions (Agnihotri and Dhiman 2017; Agnihotri et al. 2013, 2014).

In addition to this, the biocidal action of silver nanoparticles is mediated by a multitude set of mechanisms, which render microbes to modulate their machinery developing resistance against them (Huh and Kwon 2011; Hajipour et al. 2012). For example, AgNPs may either bind directly to bacterial cell wall causing membrane damage via forming holes and pits or can alter the membrane potential resulting in membrane depolarization and loss of membrane integrity (Daima et al. 2014). AgNPs may also inhibit membrane transport by facilitating ionic imbalance, impaired respiration via binding with mitochondrial DNA, interruption of energy transduction and/or cell lysis, and eventually cell death (Rai et al. 2009; Pelgrift and Friedman 2013; Chen and Schluesener 2008; Gupta and Silver 1998).

Similarly, a burst release of reactive oxygen species (ROS) may cause severe oxidative stress damaging cell constituents, leading to lipid peroxidation, alteration of proteins, inhibition of enzymes, RNA and DNA damage (Chen and Schluesener 2008). At high concentrations, the ROS lead to cell death and at low doses cause severe DNA damage and mutations (Pan et al. 2010; Wang et al. 2011). Several other effects include direct inhibition of specific essential enzymes, induction of nitrogen reactive species (NRS) (Blecher et al. 2011), and induction of programmed cell death (Beyth et al. 2010). Recently elucidated mechanism of killing microbes via direct contact without internalization of nanoparticles seems promising and efficient than antibiotics (Agnihotri et al. 2013; Sondi and Salopek-Sondi 2004; Choi and Hu 2009).

4 Biogenic Silver Nanoparticles for Biomedical Applications

Silver nanoparticles can be synthesized through physical, chemical and biological methods where the route of synthesis somehow dictates their antimicrobial activity. However, biological methods are particularly attracting importance to be used in biomedical applications owing to their greener, eco-friendly and non-toxic features (Shukla et al. 2008, 2012; Nune et al. 2009; Abraham et al. 2018). In a few cases, biogenically synthesized nanoparticles also appeared to be more effective against clinically relevant pathogens than the nanoparticles synthesized via other routes (Agnihotri and Dhiman 2017). For instance, Ajitha et al. (2018) demonstrated superior antibacterial activity of flower-like AgNPs (average size, 36 nm) synthesized from *Phyllanthus amarus* leaf extract against clinically relevant strains *E. coli*, *Pseudomonas* spp., *Bacillus* spp., and *Staphylococcus* spp. while it also showed good antifungal property against fungal test pathogens like *Aspergillus niger*, *Aspergillus flavus* and *Penicillium* spp. Bindhu et al. (Bindhu and Umadevi 2013) reported the synthesis of AgNPs (average size, 9 nm) using *Hibiscus cannabinus* leaf extract and its antimicrobial activity against *E. coli*, *Proteus mirabilis* and *Shigella flexneri*. In another study, AgNPs (average size, range 9–11 nm)synthesized from leaf extracts of *Moringa oleifera* displayed antibacterial and antifungal characteristics against *S. aureus*, *E. faecalis*, *E. coli*, *P. aeruginosa*, *K. pneumoniae* and *Candida spp.*, respectively (Moodley et al. 2018). Few studies are exploiting several other sources such as fruit extracts for preparation of silver nanoparticles. In one such study, spherical AgNPs (size range 8–16 nm) were produced through *Manilkara zapota* (sapota pomace) fruit extract (Vishwasrao et al. 2018). These biogenic nanoparticles demonstrated good antibacterial activity against *E. coli*, *P. aeruginosa*, *K. pneumoniae* and moderate antimicrobial activity was observed against *S. aureus* and *B. subtilis*. Ibrahim & co-workers presented an eco-friendly, economical, rapid and facile method for AgNPs synthesis using banana peel extract (Ibrahim 2015). With an average particle size of 23.7 nm, these AgNPs showed excellent antibacterial activity against *E. coli*, *P. aeruginosa*, *B. subtilis*and *S. aureus* but very low antifungal property with the test pathogen *Candida albicans*. Also, in conjunction with antibiotic levofloxacin synthesized nanoparticles showed an increase in antimicrobial activity by 1.16 to 1.32-fold. Wypij et al. (2018) produced small, spherical, and polydispersed AgNPs from actinobacterial strain *Streptomyces xinghaiensis* OF1 and found their superior antimicrobial activity against *E. coli, P. aeruginosa, S. aureus, B. subtilis, C. albicans*and *Malassezia furfur.* In a recent study, AgNPs synthesized using marine red algae *Gelidium amansii* demonstrated a fairly good capacity to reduced biofilm growth of pathogenic strains like *S. aureus, B. pumilus, E. coli, P. aeruginosa, V. parahaemolyticus*, and *Aeromonas hydrophila* (Pugazhendhi et al., 2018). Further, authors also suggested the applicability of such AgNPs in the fabrication of anti-biofouling coatings in the biomedical sector.

Silver nanoparticles are also known to exhibit antiviral characteristics, both in combination with natural antimicrobial agents, such as curcumin and after surface modifications (Yang et al. 2016). In one such study, the antiviral efficacy of PVP coated AgNPs (69 mm) were tested against a test organism i.e., respiratory syncytial virus (Sun et al. 2008a).Similarly, a new AgNPs-based protocol for parallel evaluation of the antiviral and cytotoxic properties of AgNPs was tested against HIV-1 (Trefry and Wooley 2012). Orlowski et al. (2014) synthesized AgNPs of three different sizes (13, 33, and 46 nm) using tannic acid as a reducing and capping agent and claimed it as an effective antiviral agent against HSV-2 infection in a mouse model. Similarly, Mori et al. (2013) synthesized chitosan functionalized AgNPs of different sizes (3.5, 6.5 and 12.9 nm) and explored their antiviral activity against H1N1 influenza A virus. It was elucidated that while chitosan alone did not show any bioactivity, conjugation with AgNPs contributed for significantly increased antiviral efficacy. Despite all such studies reported to encourage their utilization for biocidal purposes, their practical applications are limited due to following reasons. Due to high surface reactivity, AgNPs have a tendency to aggregate which causes a non-homogeneous structure leading to a significant reduction in antimicrobial efficiency in the absence of any support material (Agnihotri et al. 2012, 2013, 2015; Gupta and Silver 1998; Li and Lenhart 2012; Morones et al. 2005). Furthermore, nanoparticles in the colloidal state cannot be used repeatedly over a long time and application becomes quite expensive under practically relevant conditions. To overcome these limitations research efforts have been on the rise for past few years with an objective to augment their antibacterial activities and encourage reusability. Most of the reported studies aimed at immobilizing or incorporating nanoparticles onto a solid support (Agnihotri et al. 2012, 2013, 2015; Zhou et al. 2014b; Bakare et al. 2016; Cao et al. 2010; Lin et al. 2013; Ifuku et al. 2015; Zheng et al. 2016; Chernousova and Epple 2013). Various immobilization approaches have been explored till date, however, their choice depends on factors such as size/shape, morphology, surface functionalization and stability of nanoparticles, and the type of solid support (Mukherji et al. 2012).

One of the important aspects while considering immobilization is the choice of a template as a support matrix. Amongst all the available support materials, polymers are of the prime choice because of their chemical and structural properties with the long chains allowing ease in incorporation and fine dispersion of nanoparticles. Additionally, the suitable functional groups of polymers can be used as targeted reactive sites for the controlled one-step synthesis of nanoparticles (Chen et al. 2005;

Dallas et al. 2011). Moreover, considering the limitations associated with using silver nanoparticles in colloidal phase, researchers are exploiting the antibacterial performance of silver in the form of nanocomposites as multiphase materials having at least one phase or dimension in the nanometer size range (Ocsoy et al. 2018; Camargo et al. 2009). These silver-polymer nanocomposites can be designed to form coatings, sheets, and films depending upon the area of application and suitability for antimicrobial activity that is required by direct contact with the material surface. Therefore, the following section provides an overview of the recent progress and state-of-art facts and methodologies on designing novel silver-polymer nanocomposites and their surface functionalization from a biomedical and healthcare perspective. Different chemistry based strategies for the generation of silver-polymeric nanocomposites are discussed here, that would mediate antimicrobial activities either an anti-reflective surface or contact killing and/or through a triggered release of silver ions from surfaces.

The focus here is given to silver-based nanocomposites where a polymeric matrix of biomedical importance is used for nanoparticles to be attached either through in-situ synthesis or using suitable cross-linkers through surface functionalization. The silver-polymer nanocomposites are discussed based on their fabrication, surface chemistry, coating methods, the location of the site of infection, stability and safety aspects. The significant achievements pertaining to antimicrobial nature of nano-silver coatings over biomedical devices and instruments would also be presented.

5 Polymers as Support Matrix

Previous studies have thoroughly exploited the use of anti-biofouling polymers for clinical purposes in the form of biomaterials because of their unique properties and versatile nature (Jagur-Grodzinski 1999; Jagur-Grodzinski 2006; Middleton and Tipton 2000). In addition to this, polymer molecules are also found to be very effective support for the stabilization of silver nanoparticles (Thirumurugan et al. 2009; Zezin et al. 2010) in composite microstructures. Various polymers of natural, synthetic or a combination of both have been utilized in healthcare. Natural or naturally occurring polymers are preferred over synthetic because of their abundance, low cost, biocompatibility, and biodegradability.

5.1 Natural Polymers

After cellulose, chitin is second in line for the most abundant cationic mucopolysaccharide on earth. Chemically, it is a long chain polymer of N-acetyl glucosamine and exists as a white, hard, inelastic material obtained from exoskeleton and internal supporting structure of crustaceans/invertebrates like crabs, shellfish, krill, and clams (Dutta et al. 2004). Chitosan is the deacetylated form of chitin consists of the copolymeric N-acetyl glucosamine and glucosamine moieties. Chitosan has a great biological importance thanks to its biocompatibility, biodegradability, intrinsic antimicrobial character, haemostatic, bone forming ability, wound healing knack. This makes chitosan a promising material to be used in various biomedical applications (Kumar 2000). It has been successfully applied as antibacterial bandages, hydrogel-based biomedical coatings for implants and designing artificial skin and membranes.

Alginate or alginic acid is an anionic polysaccharide distributed widely in cell walls of brown algae. It is a linear copolymer composed of β-D-mannuronic acid and α-L-guluronic acid joined by a 1–4 glycosidic bonds (Kumbar et al. 2014). Due to its inherent properties alginate is widely used in FDA approved wound dressings marketed as AlgiDERM® and Tegaderm®. Alginate-based hydrogels have also been implemented for the localized delivery of osteoblasts (Xu et al. 2008), myoblasts, (Orive et al. 2009), fibroblasts, keratinocytes (Hunt 2010) and adipose-derived stem cells (Pangas et al. 2003).

Hyaluronan or Hyaluronic acid is a linear anionic polysaccharide consisting of alternating units of N-acetyl-D-glucosamine and glucuronic acid. It has been traditionally isolated from rooster combs and bovine vitreous humour but recently microbial source i.e. *B. subtilis* has come forward for its extraction as well (Kumbar et al. 2014). It has several properties that make it unique such as it causes bacteriostasis (Carlson et al. 2004) and assists in tissue repair (Lloyd et al. 1998). HA has been formed into nanoparticles (Lee et al. 2007) and hydrogels (Gianolio et al. 2008; Hirakura et al. 2010) to work as a payload delivery vehicle. It has also been used in the synthesis of biological scaffolds for wound healing applications.

Collagen is the most abundant protein in the human body and is composed of polypeptide strands bearing tri-aminoacid blocks of glycine-X-Y, where X and Y most commonly are proline and hydroxyproline residues. Collagen has been extensively researched for various medical applications due to its biocompatibility, mechanical strength and enzymatic degradability by collagenases (Krane 2008). In addition, it can be easily processed with high solubility in acidic aqueous solutions allowing for the fabrication of collagen sponges (Matsuno et al. 2006), tubes (Bushnell et al. 2008), sheets (Komura et al. 2008), powders (Choi et al. 2009) and injectable microstructures (Huang et al. 2009). Collagen has been used for centuries as a suture material of which one form, catgut, is still sometimes utilized in surgery (Iamphongsai et al. 2009). Due to its structural integrity, it has been instrumental in load bearing applications such as tissue and skin engineering scaffolds.

Carrageenans are a family of linear sulfated polysaccharides extracted from red seaweeds. It has an ability to form a variety of different gels at room temperature; moreover, due to their gelling, thickening and stabilizing properties, it is mainly applied in the food industry. However, it is being inculcated in biomedical applications as well due to its prophylactic properties against viral infection causing agents.

5.2 *Synthetic Polymers*

Polycaprolactone (PCL) is semi-crystalline polyester and due to its ultra-low degradation under in vivo conditions with high drug permeability, it has been used as a long-term implant delivery device. It also makes a very good elastic biomaterial (Gunatillake et al. 2006) because of its very high elongation at breakage (4700%). In addition to that PCL and PCL, composites have been used as tissue engineering scaffolds for regeneration of bone (Zuo et al. 2010), ligament (Mountziaris et al. 2010), and skin (Chung et al. 2010). Polymethyl methacrylate (PMMA) is yet another commonly applied polymer in biomedical applications. It is a transparent thermoplastic having good biocompatibility with human tissue. It has been applied in the manufacturing of intraocular lenses used for the treatment of cataract, and in orthopaedic surgery and dentistry as bone cement (Meyers 1995). Similarly, polyurethanes are biocompatible, biostable, mouldable, strong polymers that possess ester bonds with terminal amide bonds that have a degradation rate similar to polyesters and polycarbonates. Polyurethanes have been used extensively in prostheses like cardiac assist devices (Asai et al. 2007), small vascular shunts (Uttayarat et al. 2010), and tracheal tubes (Poelaert et al. 2008). PEG is a synthetically derived polyether which has been of interest in polymeric drug delivery and tissue engineering for over 30 years (Kaetsu et al. 1980). Depending upon its molecular weight, it is known as polyethylene oxide or polyoxyethylene. PEG is used as an excipient in many pharmaceutical products depending upon its form such as solvents in oral liquids while solid variants are used in tablet binders, ointment bases, lubricants and film coatings (Smolinske 1992).

6 Silver-Polymer Nanocomposites

Antimicrobial silver nanocomposites have earned special consideration owing to their unique properties, which differentiate them from other antimicrobials. In recent times, many efforts have been made finding reproducible yet effective methods to produce polymer-based silver nanocomposites. One of the methods entails the incorporation and entrapment of preformed segregated nanoparticles inside a porous matrix via the ex-situ process, which often produces undesired nanostructures. The other method i.e. in-situ includes the simultaneous generation and immobilization of nanoparticles within a support matrix, which typically produces uniform, homogenous and monodispersed Ag-polymer nanocomposites. Therefore this approach is particularly suitable in the preparation of polymer-AgNPs nanocomposite coatings and has been extensively applied in biomedical applications.

AgNPs as antimicrobial filler in polymeric nanocomposites (Muñoz-Bonilla and Fernández-García 2012) has a diverse range of biomedical applications such as wound dressings, medical implants coatings, and tissue scaffolds. Depending upon such applications, fabrication of the polymeric support can be done into various structures and configurations such as solid support, nanofibers, thin films or coatings, and porous gel that act as a template for immobilization of silver nanoparticles (Mukherji et al. 2012). Medical coatings of such polymer nanocomposites have emerged as the most promising antimicrobial surface modifications technique for biomaterials and biomedical devices. On account of these coatings' nanoscale architecture along with microporous morphology, it facilitates the killing of microbial cells while promoting the adhesion, proliferation, and differentiation of surrounding tissue cells (Dahlin et al. 2011).

The nanocomposites which have been outlined in the conformations of coatings and films, the technique applied should ensure an even distribution of the nanoparticles within the matrix. Several techniques are being studied to manufacture silver-based polymer coatings and films. One of the widely used strategies in biomedical research is using plasma treatment. This treatment method involves three major aspects viz. plasma deposition, plasma etching, and plasma polymerization with a sole purpose as to modify the material's surfaces without shifting its bulk properties (Chu et al. 2002). For biomedical applications, a patterning technique called colloidal lithography can also be pooled with plasma processes to texture surfaces with plasma-assisted micro and nanofeatures (Sardella et al. 2006). Moreover, a unique combination of deposition of plasma polymer coatings with embedded Ag nanoparticles and the sputtering of Ag atoms from an Ag target while immobilizing AgNPs in the growing polymer matrix have been reported earlier (Körner et al. 2010). The deposition of plasma coatings using combined deposition/etching/sputtering processes enables the formation of precise, patterned surface with multifunctional features (Hegemann et al. 2007).

Among techniques for the deposition of silver coatings, the use of EPD i.e. electrophoretic deposition under the influence of an electric field has been demonstrated to be successful for coatings on conductive substrates (Mittal 2013). This particular method interests biomedical researchers because of high purity of the deposited material, greater control over thickness, morphology, and ability to form uniform deposits on intricately shaped substrates such as stents, prosthesis, surgical instruments. A group study demonstrated the practicality of this method by coating uniform and crack-free layers of silver- hydroxyapatite (Ag-HA) stabilized in a chitosan matrix on multiple substrates (Pang and Zhitomirsky 2008). Features such as release rate of silver ions when immersed in an aqueous medium can also be tuned by multilayer construction which is desirable for biocompatible antimicrobial coatings. On the other hand, coatings on thermally sensitive polymeric substrates like SiO_2, polypropylene (Dowling et al. 2001) can effectively be done by magnetron sputtering. Nonwoven fabrics to fix Ag nanoparticles have also been used by Mejia et al. (2010) to produce cotton–Ag composites. Another most commonly applied technique is the dip-coating process in which substrate to be coated is dipped into the desired solution to form a uniform layer over the surface. It is specifically employed in the creation of polyelectrolyte multilayers where oppositely charged electrolytes in alternative deposition manner lead to the development of coating with the desired number of layers and thickness. The incorporation of bioactive agents like silver in each layer has also been possible through this method which has made this technique quite versatile to be used in food packaging applications apart from their biomedical benefits.

6.1 *Silver Based Nanocomposites Derived from Natural Polymers*

Considering the issues related to implant rejection, inflammation at wound's site and toxicological consequences, the majority of biocidal coatings based on nano-silver are designed so as to impart good functionalities with high biocompatibility using non-toxic polymers of natural origin like chitosan, alginate, and hyaluronic acid. Among all, chitosan has been an ideal material for biomedical applications however its low mechanical strength restricted its applicability for many years. The usage of AgNPs as fillers not only improved its mechanical strength but it also contributed to improving its antimicrobial efficacy. For instance, Li and co-workers (2013) developed a chitosan/Ag complex coating for biomedical NiTi-based shape memory alloys which commonly suffer from post-surgery bacterial infection issues. The electrochemically deposited coatings with a thickness of 7.5 μm showed excellent bacterial inhibition against *E. coli* (Fig. 2) in the zone of inhibition studies (Li

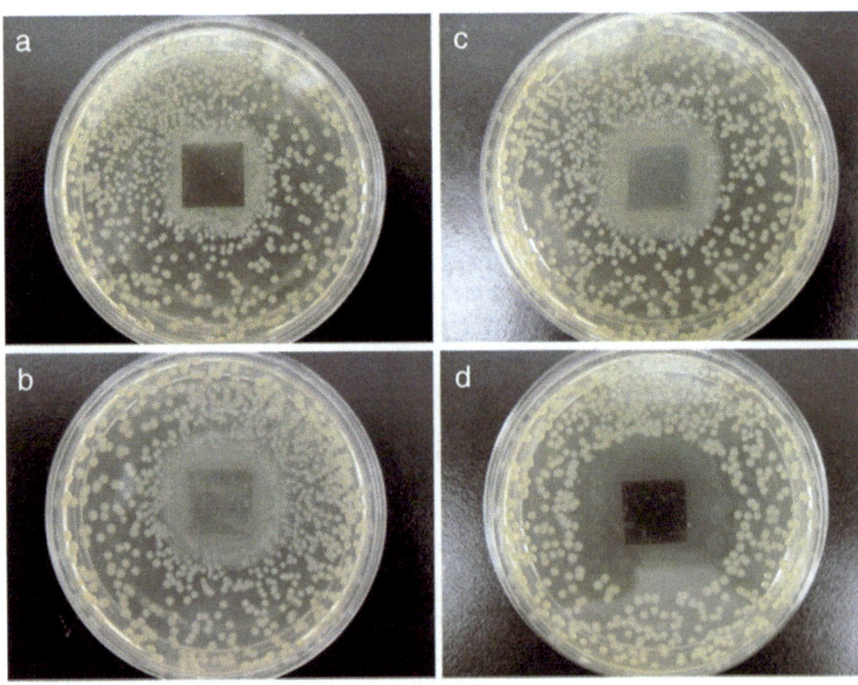

Fig. 2 The antibacterial results on the coated samples against *E. coli*: (**a**) bare NiTi substrate, (**b**) Ag/NiTi, (**c**) chitosan/NiTi, and (**d**) chitosan/Ag/NiTi. (Reproduced from Li et al. 2013)

et al. 2013). Cometa et al. (2017) combined chitosan and silver to form an antibacterial coating on poly (acrylic acid)-treated titanium substrates by an electro-deposition technique in order to combat prosthetic implant infection. Despite having a low concentration of silver release (<100 ppb up to 21 days), the resulting nanocomposite was found effective for alleviating the growth of *S. aureus* and *P. aeruginosa* pathogens. Moreover, the nanocomposite exhibited good cytocompatibility even after 7 days exposure to MG-63 osteoblasts cells. In another independent study, Bal et al. (2012) explored the use of the micro-spongy structure of *Luffa cylindrica* fibres (LCF) in wound healing applications to avoid infections owing to its ability of fluid absorption. The material was coated with chitosan/silver layer to endow the substrate with antimicrobial properties. The presence of AgNPs (average size, 20–150 nm) acted synergistically for enhanced antibacterial effect against *S. aureus* and *E. coli* as manifested by their zone of inhibition (ZOI) of 7–15 mm and 7–17 mm, respectively. Silk fibres (SF) from *Bombyx mori* (silkworm) have been applied in wound dressing, tendon reconstruction material, and silk sutures because of its slow degradation, biocompatibility, good tensile properties etc. In a study by Karhtikeyan et al. (2011), silk fibres were coated with chitosan impregnated silver nanoparticles, which displayed noticeable antibacterial activity with ZOI of 16 ± 3 mm compared to 6 ± 1 mm that of pristine silk fibres along with an improved thermal stability.

In today's scenario, where the prevention of biofilm on metallic implants and intravascular catheters is the main focus, developing an ideal 'active coating' could be a potential solution to circumvent nosocomial infections. The antifouling coating would be expected to deliver the antimicrobial agent at the implantation's site with high, persistent and controlled concentration during the critical short-term implantation period without reaching the systemic toxicity level of the agent. On the other hand, a good interface between a biomaterial and the bone would be of the primary target for implants such as a bone prosthesis. In such cases, synthetic hydroxyapatite, which is a natural constituent of bones can be employed as a major support matrix for the designing of biomedical coatings. Although hydroxyapatite forms strong bonding with bone, it lacks any intrinsic antimicrobial character, which limits its wide applicability. Therefore, in order to impart such character silver and chitosan have gained considerable interest to manufacture biocidal hydroxyapatite coatings. Chen et al. (2006) showed the antibacterial efficacy of magnetron co-sputtered silver -containing hydroxyapatite (Ag-HA) coating over Ti surfaces against *S. epidermidi-sand S. aureus*. However, Pang et al. (Pang and Zhitomirsky 2008) developed a chitosan-based silver-hydroxyapatite (CH-Ag-HAp) coating via electrochemical deposition (Fig. 3) with the tunable release of silver ions (Ag^{+}) from various layers. Yan et al. (2015) also prepared electrochemical coating on anodized titanium substrate consisting of chitosan, silver, and hydroxyapatite. Antibacterial assay of silver loaded coatings conducted against *S. aureus* and *E. coli* had an efficacy of 99.1% and 99.3% respectively compared to 81.2% and 73.4% of coatings that lack Ag. Also, good biocompatibility and viability with MC3T3-E1 cells were obtained on account of such coatings. More recently, Yu et al. (2018) prepared a hybrid coating where CS/Ag/Hap was fabricated on the surface of Ti by electrochemical deposition and

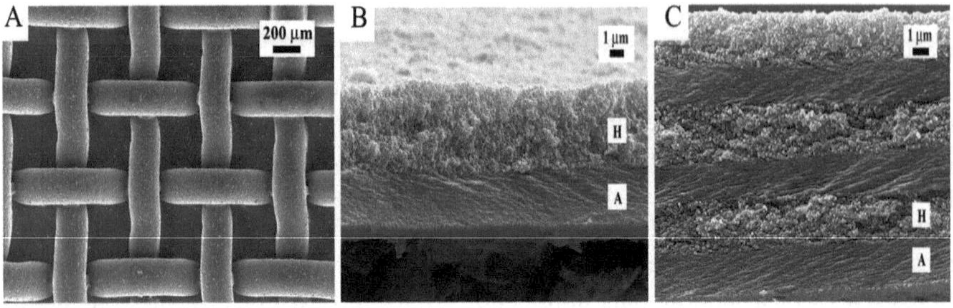

Fig. 3 SEM images (**a**) of the CH/Ag/HAp composite coating on a stainless steel gauze (**b**) cross-sections of the HA–chitosan (H)/Ag–chitosan (**a**) bi-layer coatings at deposition times of 15 min for layers A, and (**a**) 10 min for layers H (**c**) Multilayer coatings with different numbers of layers. (Reproduced with permission from Pang and Zhitomirsky 2008)

final film of lysozyme for amplified antibacterial effect was created by a spin coating process. This coating not only demonstrated an excellent antimicrobial efficiency of over 95.28% and 98.02% in 12 h against *E. coli* and *S. aureus*, respectively but also retained it up to 5 days with almost no cytotoxic effect.

In the same line of work, Pishbin et al. (2010) prepared composite orthopaedic coatings with antibacterial capability containing chitosan, bioglass particles, and AgNPs via single-step electrophoretic deposition (EPD) on stainless steel 316 model. A low concentration of silver ions (<2.5 ppm) detected in simulated body fluid was efficiently antibacterial against *S. aureus* up to 10 days. The coatings with lesser AgNPs concentration (<342 µg) were also biocompatible and osteoconductive supporting the proliferation of MG-63 osteoblast-like cells culture up to 7 days.

Chitosan has been modified and/or used in combination with other polymers, minerals and proteins to develop a highly biocompatible multifunctional coating with silver nanoparticles. For instance, Jennings et al. (2015) embedded porous calcium phosphate microspheres along with silver nanoparticles in chitosan coatings intended for the prevention of infection in orthopaedic and dental implants. These coatings reduced the bacterial viability by up to 90% of bone and dental pathogens such as *S. aureus*, *Prevotella denticola*, and *Porphyromonas gingivalis*. Another nano-Ag based coating using chitosan and gelatin as the biopolymeric matrix was fabricated through electrophoretic deposition where the overall killing capacity against *S. aureus* (50–100%) and *E. coli* (47–100%) was found to be linear with Ag concentration in coatings (Ma et al. 2017). Authors claimed a slight cytotoxicity against MC3T3-E1 cells for the coatings with a higher silver concentration.

Research activities have also been focused on designing antimicrobial coatings for bone implants using fiber-reinforced methacrylate composite. These composites have been applied in 15 cranial reconstructions, which showed promising results and have successfully replaced the metallic implants in orthopaedics owing to its tuneable mechanical strength. The porous structure of such composites allows easy integration with bone structure however it is also highly susceptible to bacterial infection due to the absence of any biocidal property within the material. To tackle this situation, non-cytotoxic silver-polysaccharide coating using lactose modified chitosan (chitlac) has been prepared (Nganga et al. 2013). The coatings were highly potent against *S. aureus* and *P. aeruginosa* when exposed to a low volume of concentrated bacterial suspension. Marsich et al. (2013) also prepared similar coatings biocompatible with human ADSC for thermoset biomaterials. Although the coatings showed good anti-bacterial and anti-biofilm activity against *S. aureus* and *P. aeruginosa* up to 3 weeks, bactericidal effect was obstructed by the presence of serum proteins. In a similar study, Chitlac-AgNPs coating (with AgNPs concentration of 5 Mm) on methacrylic surfaces intended for dental applications reduced the bacterial biomass of *Streptococcus mutans* by 80% (Ionescu et al. 2015). Further, modified chitosan in the form of chitosan acetate with AgNPs synergistically kill microbial cells of MRSA, *P. aeruginosa*, *Proteus mirabilis*, and *Acinetobacter baumannii* within 30 mins of incubation (Huang et al. 2011). In vivo studies on burn infected mice model showed an increase in survival rate from 64.3% to 21.4–0% when compared with untreated models. Therefore, results suggested the use of such nanocomposites as a topical antimicrobial on burn wounds.

Poly (ethylene terephthalate) (PET) polymers have been used in the medical field since ages. Its biostable form i.e. Dacron is applied for membranes, vascular grafts, surgical meshes, catheters, ligament and tendon repair etc. But like any other substrate, it is also prone to bacterial infections. Yuan et al. (2010) successfully constructed chitosan and heparin multilayers via layer by layer assembly loaded with silver ions on PET substrates. The polymeric template acted as in situ nanoreactors for AgNPs which participated in providing antibacterial effect against *E. coli* which lasted for more than a month. Similarly, AgNPs (10–50 nm) doped chitosan/polyvinylpyrrolidone (PVP) thin films were prepared by dip coating on PET-based implants which remained stable even after 35 days of immersion in phosphate buffer saline (PBS). An impulsive antibacterial efficacy with 100% elimination rate within 5 min against *S. aureus* and *E. coli* was demonstrated. However, cytotoxicity of these films with human umbilical vein endothelial cells was dependent upon silver concentration up to a threshold value of 0.25 mM (Wang et al. 2012). Other than PET, various synthetic polymers like polycaprolactone have also been used as wound dressing materials. For example, Nhi et al. (2016) made electrospun polycaprolactone plasma-treated membrane coated with chitosan-silver nanoparticles gel, which showed superior antibacterial activity against *E. coli, P. Aeruginosa, S. Sciuri*, and *S. Aureus* with ZOI ranging from 2.5 to 3.5 mm than the standard wound dressing material. In vitro and in vivo tests also suggested good biocompatibility with fibroblast cells.

Public places like in hospitals, armrests of the chair, beddings sometimes have leather covers which are susceptible to bacterial growth on account of absorption of sweat/sebum etc. This scenario can result in skin irritations, allergies when the subject comes in contact with it. For this purpose, antibacterial coatings based on deposition of silver nanoparticles could represent an interesting solution to reduce such unpleasant consequences. Liu et al. (2017) developed eco-friendly antibacterial coatings based upon PEGylated chitosan modified silver nanoparticles via electrostatic interaction where chitosan and Ag^+ ion in coatings displayed the synergistic antibacterial activity against both *E. coli* and *S. aureus* within 2 h (Fig. 4).

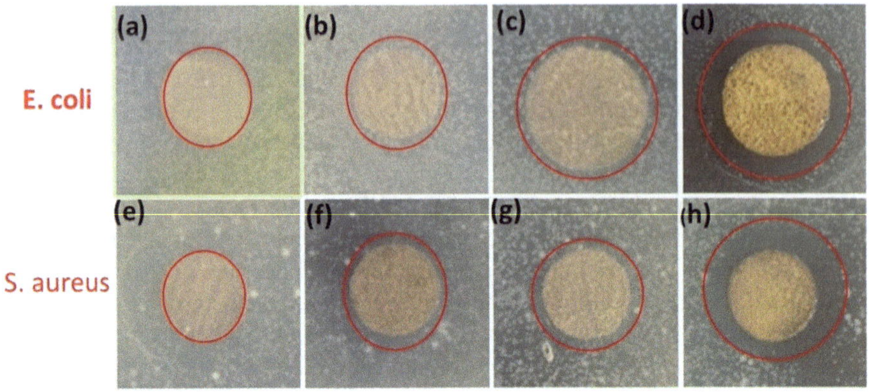

Fig. 4 The ZOI of leather samples coated with water (**a, e**) as a blank control, CS (**b, f**), PEG-g-CS (**c, g**) and PEG-g-CS@AgNPs, cultured with *E. coli* and *S. aureus*. (Reproduced with permission from Liu et al. 2017)

In the family of polysaccharide-silver coatings, several other biopolymers like lignin, alginate, carrageenan, and starch alone or in collaboration with chitosan, inorganic fillers are also being employed in designing antibacterial coatings. Specifically, the starch-based thin films with chitosan and AgNPs (20–25 nm) have improved tensile and oxygen gas barrier properties applicable to packaging and biomedical implants. In addition to this, films demonstrated enhanced antimicrobial activity against *E. coli*, *S. aureus* and *B. cereus* with ZOI ranging from 30 to 33 mm (Yoksan and Chirachanchai 2010). Similarly, Arockianathan et al. (2012) employed sago starch to prepare a coating/film for potential wound dressing along with chitosan and AgNPs. The composites with AgNPs provided optimal antibacterial activity with no significant variation in the gross healing pattern when compared to gentamycin loaded starch composites. Similarly, lignin has been used with bioceramics like hydroxyapatite on account of its brittleness to attain corrosion stability and surface porosity that enables osteogenesis. Eraković et al. (2012) prepared bone implant coatings on titanium using hydroxyapatite and lignin via electrophoretic deposition and also incorporated AgNPs to impart antibacterial character. The coatings showed optimal bioactivity in Kokubo's simulated body fluid.

Alginates are known to effectively promote wound healing by maintaining a moist environment over the wound (Augustine and Rajarathinam 2012; Nanno et al. 2009). Alginate gel polymers and its composites such as sodium alginate have been used in many biomedical applications including drug delivery, sutures and wound dressings. Alginate-based surgical sutures, although have been coated with antimicrobial agents like antibiotics, triclosan etc. they often experience a high risk of exposure owing to allergies, inflammation and microbial resistance. Alginate can also be effectively used for the immobilization of silver nanoparticles. Therefore, alginate-silver coatings have been studied to prevent bacterial infections on biomedical devices. For instance, Dubas et al. (2011) coated standard polyamide surgical sutures with AgNPs capped with sodium alginate via layer by layer assembly.

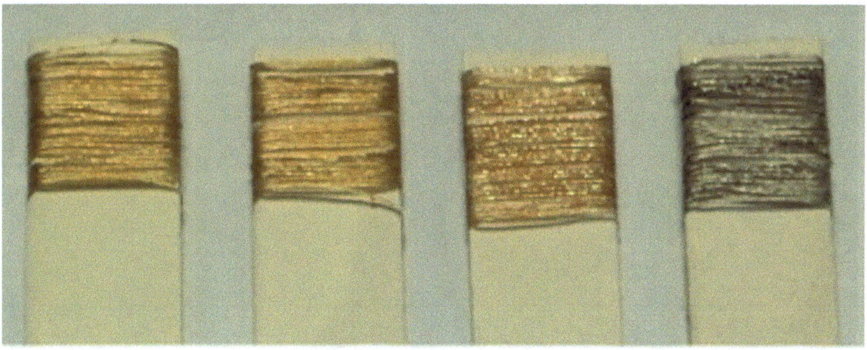

Fig. 5 Polyamide suture coated with 20 layers of silver nanoparticles capped with various sodium alginate ratios (Alginate: $AgNO_3$) in mM. (Reproduced with permission from Dubas et al. 2011)

The antibacterial property of coated sutures was however highly dependent upon capping agent concentration as 0.1 mM alginate provided 76% bacterial reduction against *S. aureus* compared to 13% in case of 5 mM (Fig. 5). In a similar trend, an antibacterial coating over an absorbable surgical gut suture was facilitated using immobilized AgNPs through slurry dipping technique was prepared (Augustine and Rajarathinam 2012). The antibacterial assay conducted against *E. coli* and *S. aureus* displayed significant inhibitory effects as compared to uncoated sutures. Such positive results depict the applicability of this novel suture in surgery for the prevention of surgical wound infection (Augustine and Rajarathinam 2012).

Peripherally inserted central catheters (PICCs) are widely used for nutrition and administration of medication in the neonatal intensive care unit (NICU). These catheters tend to disrupt skin's integrity of patients causing bacterial and fungal growth over these catheters. The patches or dressings loaded with antiseptic or antimicrobial agents thus have been employed to reduce central venous catheter infections. Hill et al. (2010) demonstrated the safety and bioactivity of antibacterial sterile polyurethane foam patches coated with nano-silver, alginate and maltodextrin matrix in NICU patients. The antibacterial nature of coated patches provided the required protection against microbial flora with no adverse skin reactions. In lieu of such applications, it has been formulated that flexible and transferrable layer by layer (LbL) nanosheets are convenient tools as coating materials. Ito et al. (2016) recently fabricated a novel antimicrobial coating material on PVA substrate by embedding silver nanoparticles (AgNPs) in an LbL nanosheet composed of layers of chitosan and sodium alginate by means of a photo reduction method. The coatings with a higher concentration of AgNPs showed a significant decrease (about 99%) in bacterial colonies of Methicillin-resistant *S. aureus* (MRSA). Cotton fabrics being used as a wound dressing material, the antibacterial coatings over cotton have been proposed recently (Zahran et al. 2014). In this study, researchers coated the nanocomposite suspension comprised of AgNPs–alginate over cotton fabric using a simple pad-dry-cure technique. The coatings were physically adhered to the

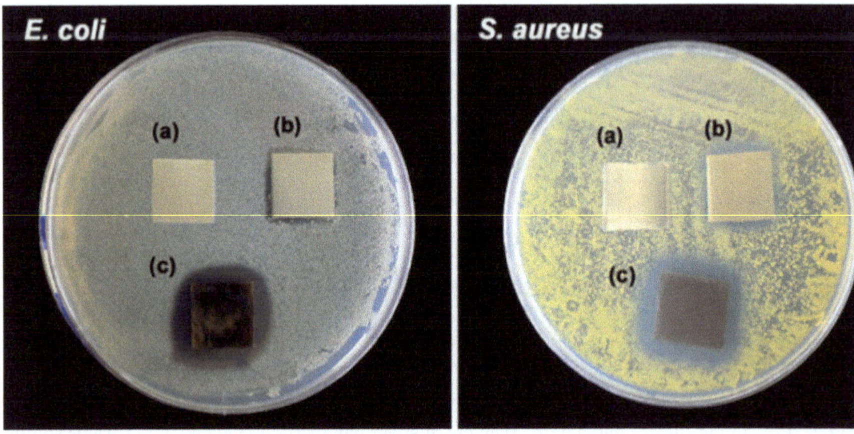

Fig. 6 Antibacterial activities of (**a**) original titanium, (**b**) DAL/CHI multilayer coated titanium, and (**c**) AgNPs-DAL/CHI multilayer coated titanium against *E. coli* and *S. aureus*. (Reproduced with permission from Zhang et al. 2013)

fabric as detected by SEM and displayed an excellent antibacterial activity against the tested bacteria, *E. coli*; *S. aureus* and *P. aeruginosa* with ZOI ranging from 1 to 3 mm. As per the broth antibacterial assays tested for *S. aureus*, the antibacterial cotton demonstrated a 95% decrease in bacterial viability. Although the antibacterial performance of cotton fabrics was slightly reduced after successive washings, an efficient antibacterial activity still remained there. Zhang et al. (2013) also modified the surface of titanium orthopaedic implant with an antibacterial coating made from dopamine-modified alginate/chitosan multilayers. AgNPs (50 nm) were immobilized in these layers, which reportedly inhibited the growth of *E. coli* and *S. aureus* with ZOI ranging from 2.85 ± 0.24 to 2.13 ± 0.15 mm, respectively (Fig. 6). The coatings however, slightly decreased the bioactivity of L929 fibroblast cells but were still found to be biocompatible.

Similar to the previous study, a few other studies have encouraged the use of polydopamine (PDA), for immobilizing silver nanoparticles. For instance, Wang et al. (2015a) prepared AgNPs-PDA bilayers over pre-treated silicone catheter surface and an outer antifouling layer of poly(sulfobetaine methacrylate-co-acrylamide) to ensure free diffusion of Ag from the surface. This complex coating reduced colonization of the urinary catheter by uropathogens i.e. *Proteus mirabilis* by approximately two orders of magnitude. Double layered catheter surfaces resisted encrustation for up to 45 days in comparison with 6 days of commercially available silver-coated catheters. Polyurethane (PU) substrates coated with AgNPs-Polydopamine with antifouling perfluoroalkyl layer were designed recently (Xu et al. 2017). The coated surfaces showed enhanced antibacterial activity against *E. coli* and *S. aureus* whilst reducing the platelet adhesion on the PU surface. In another study, AgNPs-PDA were coated on electrospun PVA nanofibers intended for broad range of practical medical applications where the presence of coating suppressed the bacterial growth of *E. coli* while bringing down the survival rates up to only 3.3% (Son et al. 2013).

Polymers like gelatin are most commonly used in food packaging applications but nowadays its applicability has widened into the biomedical sector as well. Recently, El Hajj et al. (2015) developed an AgNPs loaded methacrylated gelatin (GelMA) hydrogel for antimicrobial coating of biomedical implants. Results indicated inhibition in growth of *E.coli* and *S.aureus* at a very low concentration of AgNPs without any negative effect on the viability of rat aortic smooth muscle cells. Similarly, Li et al. (2014) with the help of spin-assisted LbL assembly technique prepared a coating of sulfhydrylated chitosan (Chi-SH) and gelatin gel embedded with AgNPs on a titanium implant. Kirby-Bauer test for antibacterial activity of such coated surfaces revealed significant ZOIs while the introduction of AgNPs into multilayer films did not inhibit the ALP activity of osteoblasts. Wang et al. (2015b) developed hydrogel coatings on conductive substrates with chitosan, gelatin, and nano-silver via electrodeposition. These diverse shaped films/coatings improved conductivity and in vitro antibacterial activity against *E. coli* and *S. aureus* with ZOI ranged between 9.2 and 13.8 mm. The research demonstrated a wide applicability of biopolymer-based biomedical coatings as skin biomaterials, artificial muscles and neuroprosthetic implants which have also been summarized in Table 1.

6.2 Silver Based Nanocomposites Derived from Synthetic Polymers

Man-made synthetic polymers are almost as manifold as the natural ones. Newly developed polymers have rapidly entered the medical application, such as the polyesters and polyamides are being used to synthesize synthetic suture materials (Maitz 2015). Synthetic polymers gained high attraction for various reasons, one of them is a wide range of tunable physical and chemical properties (Lendlein 2010). Such polymers fulfill structural and mechanical properties quite well in comparison to natural polymers. These polymers can be used for targeted interaction between the material and the body since they resemble closely to biological tissue than inorganic materials owing to their carbon-based chemistry. Also, for optimization of the biocompatibility properties, reactive groups in the polymers usually offer the possibility for biofunctionalization of the surface. Both biodegradable and non-biodegradable synthetic polymers are applied in biomedical applications according to the need. For instance, orthopaedic fixation and ligament augmentation (Hofmann 1992). Biodegradable synthetic polymers ideally stay in the body only as long as they serve their function and then they disappear without the need for a second surgical intervention (Nair and Laurencin 2007; Bat et al. 2014).

Table 1 Silver nanocomposite coatings based upon natural polymers and their biomedical applications

Silver based Nanocomposites (NCs)	Substrate	Size of AgNPs	Activity	Microbes tested	Evaluation parameters	Biomedical applications	References
Ag/chitosan	Ni-Ti	ND	AB	*E. coli*	ND	Antibacterial shape memory alloys	Li et al. (2013)
Ag/chitosan	*Luffa cylindrica*fibres	20–150 nm	AB	*S. aureus, E.coli*	ZoI: 7–17 mm	Antibacterial coatings for wound healing	Bal et al. (2012)
Ag/chitosan/hap	Titanium	ND	AB	*S. aureus, E.coli*	99.3% inhibition	Antibacterial orthopediccoatings	Yan et al. (2015)
Ag/chitlac	Methacrylic	ND	AB	*Streptococcus mutans*	80% inhibition	Antibacterial dental material	Ionescu et al. (2015)
Ag/starch/chitosan	ND	20–25 nm	AB	*E. coli*, *S. aureus* and *B. cereus*	ZoI: 30–33 mm	Biomedical implants and packaging	Yoksan and Chirachanchai (2010)
Ag/sodium alginate	Polyamide	ND	AB	*S.aureus*	76% inhibition	Antibacterial surgical sutures	Dubas et al. (2011)
Ag/dopamine modified alginate /chitosan	Titanium	50 nm	AB	*S. aureus, E.coli*	ZoI: 2.13–2.85 mm	Antibacterial orthopedic implant	Zhang et al. (2013)
Ag/chitosan/gelatin	Conductive	ND	AB	*E. coli, S.aureus*	ZoI: 9.2–13.8 mm	Antibacterial hydrogel coating for implants	Wang et al. (2015b)
Ag/bioactive glass/chitosan	Stainless steel 316	<50 nm	AB	*S. aureus*	ZoI:16 mm	Antimicrobial orthopediccoatings	Pishbin et al. (2013)
Ag/polydopamine (PDA)/ poly(sulfobetaine methacrylate-co-acrylamide)	Silicone	ND	AB	*Proteus mirabilis*	ND	Antimicrobial coatings for silicone urinary catheter	Wang et al. (2015a)
Ag/PEGylated chitosan	ND	ND	AB	*S. aureus, E. coli*	ND	Antimicrobial leather coatings for medical textiles	Liu et al. (2017)

AB Antibacterial, *ND* Not Determined

6.2.1 Poly(dimethylsiloxane) (PDMS)

In advanced applications, the synthetic polymers are typically not present as bulk materials; they are formed as coatings on biomedical devices. These coatings can be non-structured, homogeneous, cross-linked, polymer brushes and/or layer-by-layer deposited films. The following paragraphs will discuss such coatings based on synthetic polymers with nanosilver mainly for the provision of the antimicrobial surface for the implants and biomedical devices. Among synthetic polymers, silicones/siloxanes are highly inert biostable elastomers which are hydrophobic used in manufacturing catheters (medical grade silicone), intraocular lenses, coatings, adhesives, and breast augmentation implants (Yilgör and McGrath 1988). Among them, the most commonly applied silicone in the medical industry is poly(dimethylsiloxane) (PDMS) which has been used in association with silver and its nanoparticulate form in few applications. For example, Oktay and Kayaman-Apohan (2013) prepared PDMS based organic-inorganic hybrid formulations by sol-gel method embedded with AgNPs to achieve an antimicrobial high-performance coating. UV cured and nanosilver-containing hybrid coated polycarbonate panels showed 99% reduction in bacterial colonies with antibacterial activity against *E. coli* (Log CFU reduction >4.6) higher than that against *S. aureus* (Log CFU reduction >3.1). The sol-gel ratio in the nanocomposite coatings did not affect their antibacterial performance but affected the thermal and mechanical properties as modulus, ultimate strength values of hybrid coatings increased with increasing sol-gel content so did the decomposition temperature. Kim et al. (2017) fabricated PDMS films with AgNPs (~500 nm) via in-situ synthesis through citrate reduction of $AgNO_3$ precursor. With Ag content <0.05% of the total film weight, the PDMS film exhibited a reduction of *E. coli* and *S. aureus* with values of $\log_{10}4.8$ and $\log_{10}5.7$, respectively. Also, PDMS films that underwent MTT assay showed no cytotoxicity with fibroblast cells.

PDMS with HAp to improve latter's mechanical properties for its intended use in versatile biomedical applications has been quite interesting. PDMS provides the soft polymer backbone to the silver-doped HAp that in turn contributes biological and antimicrobial properties to the polymer nanocomposite. AgNPs/HAp/PDMS based composite films, layers, and coatings have been prepared. In a recent study, the silicon substrate surface was coated with nano-silver doped hydroxyapatite/PDMS layers via thermal evaporation technique (Ciobanu et al. 2015). The coated silicon surfaces showed significant antifungal biofilm activity against *Candida albicans* with inhibition at initial adherence phase and mature biofilm phase. Such coating should prevent or lag the fungal biofilm development on silicon-based implants. Similar nanocomposite layers were also prepared recently by Iconaru et al. (2017) where such coatings demonstrated not only antibacterial but antifungal activity against *E. coli, S. aureus,* and *C. albicans.* The antimicrobial activity of Ag loaded composites sustained even after 48 h. In a similar domain, Groza et al. (2016) applied different composite coatings containing either Ag or Zn as a functionalizing agent on titanium (Ti) substrate with PDMS as an interlayer. The PDMS improved the quality of coatings while embedded metals showed antifungal properties against *C. albicans.* Both silver and zinc loaded HAp-PDMS surfaces showed significantly higher biological activity in comparison to pristine Ti, Ti-PDMS, and HAp-PDMS surfaces. Not much surprise, fungal colonization on Ag/HAp-PDMS was reduced faster than Zn/HAp-PDMS showing the higher biocidal potential of the silver-based coating where fungicidal activity was lasted up to 72 h (Fig. 7).

Tran et al. (2015) synthesized Ag-doped coatings with different titanium dioxide–PDMS ratios. These coatings when applied on polyether ether ketone discs, commonly applied as spinal implant material inhibited bacterial growth of *S. aureus* and *S.epidermidis* in a dose-dependent manner. The higher concentration of dopant i.e. Ag in coatings resulted in larger ZOI than in control samples, which is an indication of release of Ag ions in such bactericidal concentrations. Zhou et al. (2007) attempted to improve the antibacterial capability of the urinary catheter by preparing a nanocomposite coating containing PDMS, clay, chitosan and AgNPs through intercalation. Results demonstrated that the nanocomposites were found to be bacteriostatic against urinary bacterial pathogens *E. coli*, *P. aeruginosa*, and *S. aureus* and showed a mildly fungistatic effect against *C. albicans.* Besides nanoparticles, silver nanowires (AgNWs) have also been employed as a potent bactericidal agent for coating applications (Polívková et al. 2017). With relatively higher aspect ratios AgNWs could replace AgNPs in polymer nanocomposites. For instance, Tang et al. (2014) prepared a uniform and pure AgNWs, which showed excellent and long-lasting antibacterial activity against *E. coli* and *S. aureus*, microorganisms that are frequently associated with nosocomial infections. Zhao et al. (2016) also developed a broad-spectrum and robust antimicrobial thin film coating based on graphene wrapped AgNWs. They found that this hybrid coating showed significant antimicrobial activity against *E. coli*, S. *aureus* and the fungus *Candida albicans*. Similarly, Jiang et al. (Jiang and Teng 2017) fabricated a thin polymeric film/coating based upon PDMS functionalized with AgNWs (diameter-100 nm; length-50 μm). Authors demonstrated that when the loading amount of AgNWs in PDMS films increased from 0.2 to 1 mg, the mortality of bacterial cells of *E. coli* and *S. aureus* was increased from 80% to 100%. PDMS/AgNWs films demonstrated a long-term antibacterial effect for at least 1 month and also demonstrated an excellent biocompatibility with human dermal fibroblasts.

6.2.2 Polyurethanes

Polyurethanes (PU) are one of the multifaceted polymeric materials that are synthesized with multiple chemistries and properties. Some modified polyurethanes with aromatic or aliphatic components such as polyester, polyether, and polycarbonate have been used for medical purposes (Maitz 2015). PU is applied in adhesives, medical synthetic materials, coatings, and construction materials etc. Coatings made from this polymer are widely used due to their durability and overall good balance of mechanical properties. For instance, Khwanmuang et al. (2017a) prepared self-disinfecting, effectively practical and optically clear coatings consisting of nanosilver and polyurethane upon glass substrates. The Ag/PU coatings resulted in>99.99% reduction in bacterial cells of problematic strains such as *E. coli, S.*

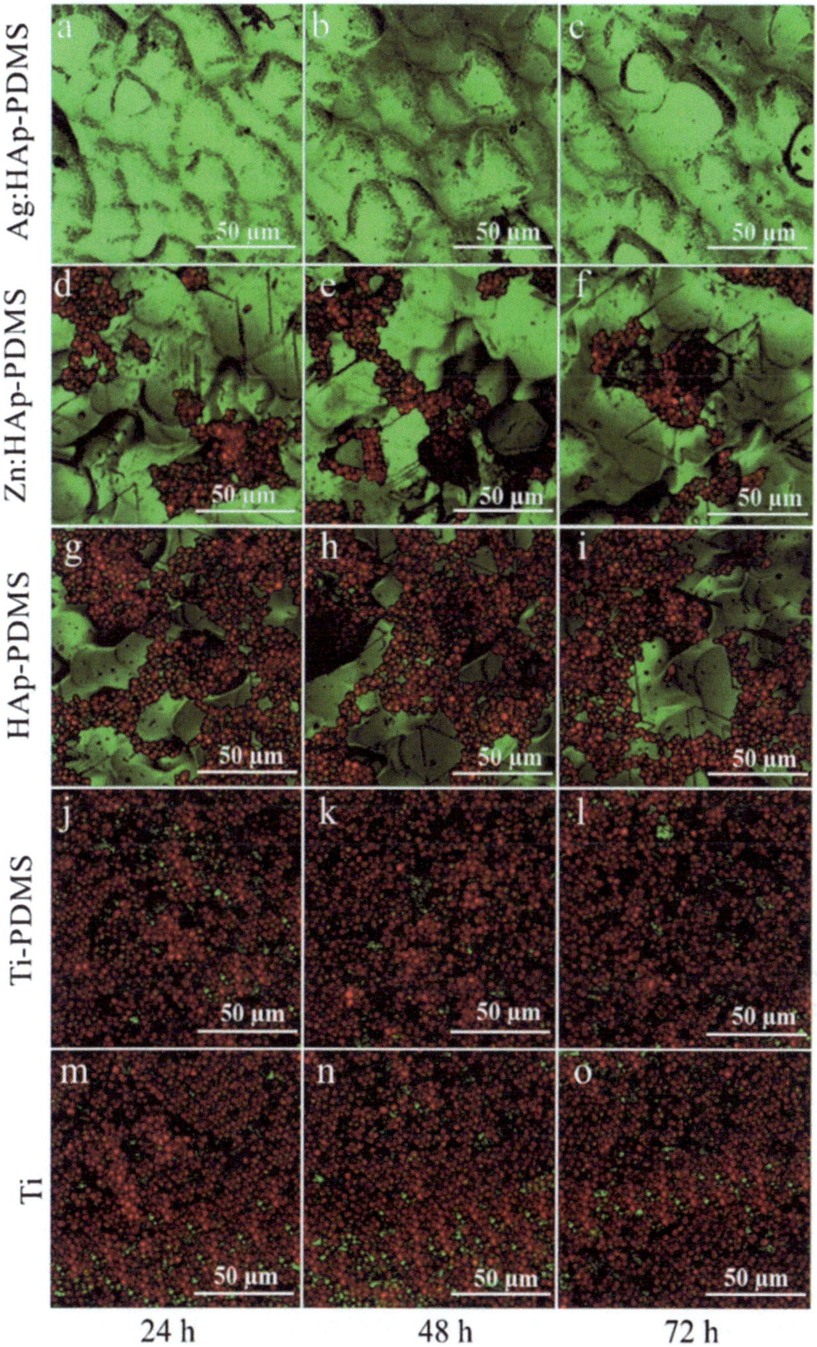

Fig. 7 CLSM of biofilm image of *C. albicans* biofilm stained with propidium iodide developed ondifferent substrates at various time intervals (**a**) Ag:HAp-PDMS after 24 h; (**b**) Ag:HAp-PDMS after 48 h; (**c**) Ag:HAp-PDMS after 72 h; (**d**) Zn:HAp-PDMS after 24 h; (**e**) Zn:HAp-PDMS after 48 h; (**f**) Zn:HAp-PDMS after 72 h; (**g**) HAp-PDMS after 24 h; (**h**) HAp-PDMS after 48 h; (**i**) HAp-PDMS after 72 h; (**j**) Ti-PDMS after 24 h; (**k**) Ti-PDMS after 48 h; (**l**) Ti-PDMS after 72 h; (**m**) Ti after 24 h; (**n**) Ti after 48 h; (**o**) Ti after 72 h where Red: live cells; green: dead cells in the *C. albicans* biofilm. (Reproduced with permission from Groza et al. 2016)

aureus, E. faecalis, K. pneumonia, P. aeruginosa, P. mirabilis and *A. baumannii*. Also, the cytotoxic studies depicted the coatings to be safe and non-toxic with 70% cell viability of mammalian L929 cells. Similar AgNPs/PU films were made by in situ light initiated synthesis of AgNPs within the matrix of medical grade PU (Saez et al. 2015). Incorporation of AgNPs in polymer material did not affect its physical properties and biocompatibility when tested for cell proliferation of primary skin fibroblasts and human erythrocytes. However, the presence of AgNPs conferred antibiofilm and antibacterial characteristics against *P. aeruginosa* with only approximately 50% bacterial reduction.

Another form of PU, waterborne polyurethane (WPU) is also known for its application in coatings and adhesives (Kantouch and El-Sayed 2011; El-Sayed et al. 2010). It can be prepared by adding hydrophilic groups to the polymer chains. There are various types of WPU such as non-ionic, anionic, and cationic. Anionic WPUs form coordinate bonds with silver ions; therefore can be used to incorporate AgNPs through in-situ nucleation (Wattanodorn et al. 2014). Authors reported anionic WPU/AgNPs films with various concentrations of silver (100–2000 ppm) with enhanced the tensile strength and Young's modulus. The nanocomposite films showed a bacterial reduction of 99.99% and 53.97% against *E. coli* and *S. aureus* strains maintaining a sustained Ag^{+} release over 21 days. With a similar concept, Hung et al. (2007) studied the antibacterial activity and biocompatibility of Ag/PU nanocomposites onto which 5 nm AgNPs were embedded at various concentrations ranging from 15 to 113 ppm. The optimization results through response surface methodology (RSM) indicated a different surface morphology of the nanocomposite at Ag concentration of 30.2 ppm. Also at this concentration, Ag/PU composites demonstrated maximum bovine carotid arterial endothelial cells (BECs) attachment and proliferation. Ag/PU showed drastic bacterial adhesion reduction in the case of *B. subtilis* and *E. coli* than the original PU but failed for Ag^{+}-resistant *E. coli*. From the immunological point of view, these nanocomposites at the same concentration of 30.2 ppm depicted the lowest number of human monocytes and activated macrophages.

In the healthcare sector, covered self-expandable metal stents (SEMSs) have been developed to reduce the incidences of microbial infections as these events are quite often in its pristine form. Covering materials for such commercially available devices include polymers like poly(tetrafluoroethylene), silicon and polyurethane (Bang et al. 2012). Following a nano approach, Lee et al. (2016) produced AgNPs incorporated silicone covered SEMs to check its ability to reduce biofilm formation and tissue inflammation. In vitro results showed a 99.9% antimicrobial activity against *E. coli* and *K. pneumoniae* over 24 weeks among the test group while in control group antimicrobial activity dramatically decreased after 2 weeks (Fig. 8a). An increased expression of IL-10 in coated SEMs compared with control groups was found which indicated the role of Ag particles in anti-inflammatory effect. Moreover, in vivo pathological analysis showed minimal signs of erythematous change (Fig. 8b, c) around the stent in the test group.

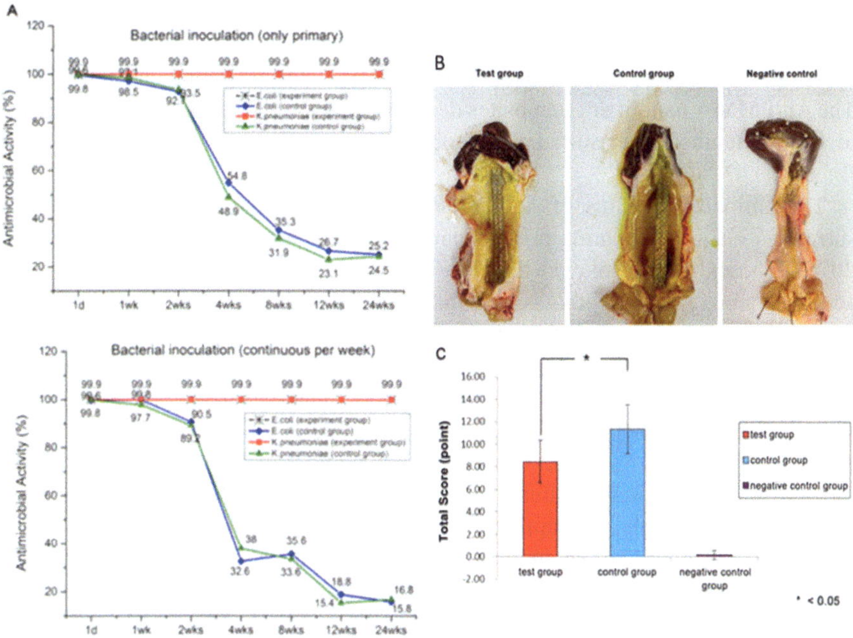

Fig. 8 (**a**) Antimicrobial activity over time.-Antimicrobial activity of stents in the test and control groups evaluated at 1 day, and 1, 2, 4, 8, 12, and 24 weeks, after placement. The test group showed 99.9% antimicrobial activity against *E. coli* and *K. pneumoniae* over 24 weeks. (**b**) Pathological analysis-Compared with the negative control, the test tissue sample showed mild erythematous mucosal change and a normal SEMS configuration. In the control group, more severe erythematous change and bile-clogged membrane stent were noted (**c**) Total pathology score was significantly lower in the test group. (Reproduced with permission from Lee et al. 2016, Open Access)

6.2.3 Poly (methyl methacrylate) (PMMA) and its Substitutes

Methyl methacrylate through radical polymerization can be transformed into rigid polymers such as PMMA, which can be utilized in orthopaedics and dentistry as bone and dental cement, respectively. Owing to its optical transparency, robustness and inert nature, they are equally popular in making intraocular lenses and as a packaging material. Additionally, the polymerization of hydroxyethyl methacrylate (HEMA) monomer can be achieved as a hydrogel which has good protein repellent and antifouling properties. Soft hydrogels of polymerised HEMA are also being used for various healthcare applications like haemocompatible (Maitz 2015; Tanaka and Mochizuki 2010), antimicrobial or lubricant coatings on contact lenses (Kim et al. 2002). Damm et al. (2005) prepared PMMA based nanocomposite randomly infused with spherical AgNPs of size range 5–50 nm. The slightly translucent nanocomposites with less than 0.12 wt.% of silver content released enough Ag^{+} ions (0.05 mg/l) in an aqueous medium to kill a significant count of bacteria. In another study, a novel PMMA nanofiber containing AgNPs were synthesized which showed promising antibacterial effects against *E. coli* and *S. aureus* bacteria for wound dressing, bioadhesive and coatings of biomedical materials (Kong and Jang 2008). The Ag/PMMA nanofiber composites demonstrated the ZOI of 45 mm as compared to 16 mm against silver sulfadiazine control samples through the modified Kirby-Bauer test.

The antibacterial films of AgNPs/PMMA which can be coated on to various substrates via a spin coating process have been designed recently (Lyutakov et al. 2015). These films displayed the dose and temperature dependent antibacterial activity against *E. coli*. Authors claim that the films with 10% AgNPs synthesized at 75 and 100 °C had more pronounced effects in bacterial reduction than the films with relatively lower (1%) AgNPs. Pulsed laser deposition (PLD) technique have also been adapted to deposit a thin coating of nano-silver/PMMA over silicon wafers (Petrochenko et al. 2017). Antibacterial studies using dynamic shake flask technique showed a significant reduction in the number of *E.coli* colonies while the biocidal coating elicited minimal cytotoxic response towards human bone marrow stromal cells. Recently, Siddiqui et al. (2018) investigated the antimicrobial behaviour of PMMA based nanocomposites with incorporated AgNPs at a concentration ranging from 0.75 to 2 wt.%. These PMMA/AgNPs were effectively bactericidal against *S.aureus* with ZOI varying from 24 to 27 mm as compared to 16–17 mm of control samples. The fungicidal behaviour of nanocomposites was also evaluated against *C. neoformans* with depicted encouraging results.

6.2.4 Poly (Lactic Acid), PLA

Poly(glycolic acid) (PGA), poly(lactic acid) (PLLA), and poly(D-lactic acid) (PDLA) are the polyesters of small aliphatic glycolic acid or lactic acid which have gained considerable clinical relevance because of their ease in degradation into monomeric form as physiological metabolites. These polymers can efficiently be used as immobilization matrix for drugs, metal ions and even silver nanoparticles for drug delivery and as biomedical coatings on vascular stents and other implants (Maitz 2015). For example, Xu et al. (2006) prepared biodegradable and antibacterial PLA ultrafine fibres containing AgNPs via electrospinning method. These nanofiber thin mats (Ag/PLA) had an accumulative release of Ag^+ ions up to 60 ppm, yet they displayed quite strongly effective antibacterial activity against *S. aureus* and *E. coli* with 98.5% & 94.2% reduction rates, respectively. Authors indicated that such mats could potentially be applied as a wound dressing material or anti-adhesion membranes. In another study, AgNPs/PLA nanocomposite films were synthesized with various AgNPs content i.e. 8–32% w/w (Shameli et al. 2010). Although all films displayed significant inhibition activity against *E. coli*, *S. aureus* and *V. parahaemolyticus* with ZOI ranging from 1.43 to 10.33 mm; 4–9.3 mm; 4–15 mm respectively, a significant ZOI was only observed with the highest AgNPs.

6.2.5 Poly (Lactide-co-Glycolide), PLGA

PLGA is yet another copolymericform which is widely utilized as biomaterials (Ma 2004; Park et al. 2005) for implanted medical devices viz. endotracheal tubes, urinary catheters (Sun et al. 2008b; Wang et al. 2004). However, like any other synthetic polymers, they are prone to be infected with bacterial pathogens which limit their in vivo applications. It has been recently demonstrated that loading PLGA with silver nanoparticles strongly reduces bacterial development, which is attributed to the modification in the surface properties of the polymer (Fortunati et al. 2011). Specifically, AgNPs in low concentration are able to induce surface morphological changes in the polymer film and affect nanocomposites' wettability and roughness. All these aspects marked great influence to bacterial adhesion onto the biomaterial surface (An and Friedman 1998). Liu et al. (2012) studied that the topographical changes occurred due to the loading of silver nanoparticles were responsible for eliciting anti-biofouling character to the Ag/PLGA nanocomposite, which were coated on stainless steel alloy under physiological conditions. As per data, the nanocomposite coatings with 1–2% nanosilver content reduced bacterial counts of *S. aureus* and *P. aeruginosa* within an hour of incubation from their initial bactericidal concentrations ranging from 10^3 to 10^5 CFU ml^{-1} (Fig. 9). Moreover adding AgNPs in PLGA decreased the adsorption of model protein and enhanced the osteoinductive growth factor BMP-2. Further, cytotoxicity tests depicted an active proliferation and differentiation of MT3T3-E1 preosteoblastic cells at a rate 1.88 times higher than that of pure PLGA. A similar effect was observed in a few other studies where incorporating AgNPs increased the biocidal activity of PLGA coatings *E. coli* (Paul et al. 2015) and exhibited anti-encrustation properties (Dayyoub et al. 2017).

6.2.6 Poly(vinyl alcohol) (PVA)

Being highly water-soluble, biocompatible, hemocompatible, easy to process, optically transparent and possessing low mammalian cytotoxicity, the usage of PVA has been quite popular for the stabilization of AgNPs in antimicrobial finishes (Bryaskova et al. 2010). Taking this into account, bi-functional coatings were developed using AgNPs/chitosan PVA hybrid nanocapsules onto which an anti-inflammatory drug naproxen was loaded (Mishra et al. 2017). The titanium-based metal implants coated with these core-shell nanohybrids demonstrated dual characteristics of biofilm inhibition and sustained drug release for anti-biofouling purposes. Precisely, the nano-coatings exhibited strong antibacterial action against *E. coli* and *S. aureus* where the ZOI was increased by increasing silver loading. Cell adhesion studies demonstrated enhanced adherence of cells to all the coated implants than their counterparts owing to their low wettability and nanoscaled structure of coating surface which mimics under *in vivo* condition. In another study, a novel hybrid thin films based on PVA/tetraethyl orthosilicate (TEOS) functionalized with

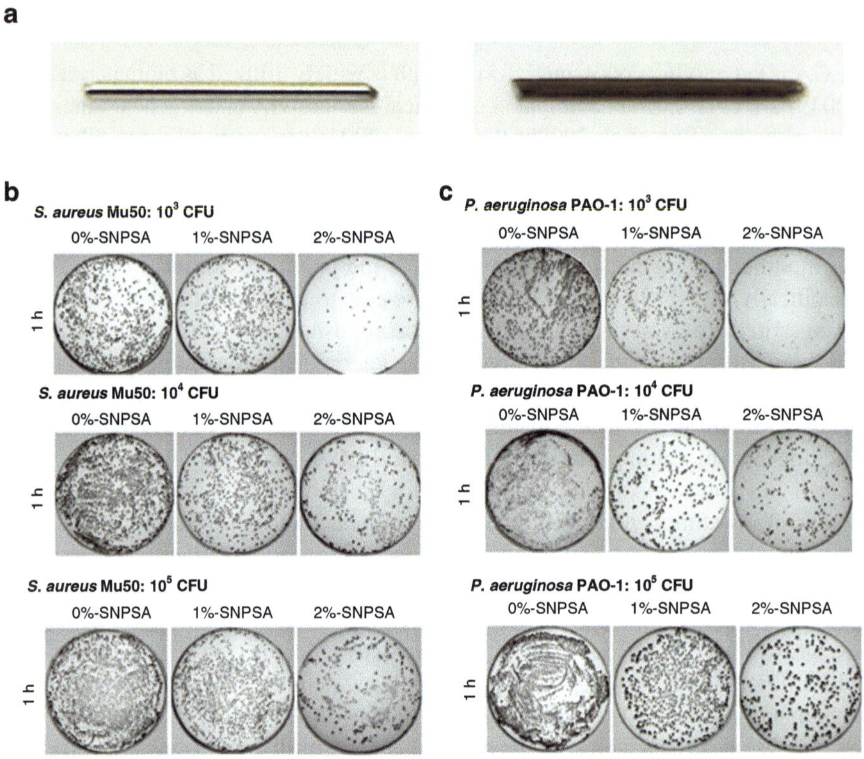

Fig. 9 (**a**) Silver nanoparticle/PLGA composite coated surface of a stainless steel alloy. Antimicrobial activity of SNPSAs against 10^3, 10^4, and 10^5 CFU (**b**) of *S. aureus* Mu50 (**c**) of *P. aeruginosa* PAO-1 at 1 h. (Reproduced with permission from Liu et al. 2012)

AgNPs (average size, 5–7 nm) were tested against hospital-acquired infection strains i.e., *E. coli, S. aureus* and *P. aeruginosa* with ZOIs 11.5 mm, 15.5 mm and 10 mm, respectively at 0.4% AgNPs concentration (Bryaskova et al. 2010). An optically transparent PVA based antibacterial coatings with AgNPs (average size, 10–20 nm) have also been reported which displayed an unprecedented bactericidal and fungicidal activities against clinically relevant strains *Bacillus atrophaeus, Bacillus megaterium, S. aureus, Enterococcus faecium, P. aeruginosa, Salmonella enterica, E. coli, Neurospora crassa*, and *S. cerevisiae* (Chan et al., 2011). Earlier, Galya et al. (2008) tested the antibacterial efficacy of PVA/AgNPs nanocomposites and claimed their intended use as a protective anti-biofouling coating for biomedical implants. As per the data, nanocoatings exhibited a strong antibacterial activity against *E. coli* and *S. aureus* while their mechanical strength was also reportedly increased even with the addition of only 0.5 wt.% AgNPs.

The occurrence of endotracheal intubation is commonly associated with hospital-acquired infections as the intubation device acts as a reservoir for bacterial colonization in the lungs. To reduce the incidence of bacterial colonization on tubes, hydrogel coatings based on PVA loaded with antimicrobial agents are gaining popularity

(Loo et al. 2014). The AgNPs (average size, 7–15 nm) loaded PVA hydrogels were found to be quite stable and non-toxic against human normal bronchial epithelial cells while they were found to be equally effective against *P. aeruginosa* and *S. auruginosa*. Within 18 h, *P. aeruginosa* colonies formed on pristine surface of hydrolysed PVA were calculated within a range 2–3.8 × 10^5 CFU cm^{-2}, which were reduced significantly to 1.2–9.4 × 10^4 CFU cm^{-2} in case with the AgNPs loaded PVA hydrogel.

PVA based hydrogels are also applied in wound care systems, where keeping a moist and disinfected environment at wound's site over a prolonged time is of the prime importance for antimicrobial dressing. The incorporation of AgNPs in wound dressing hydrogels thus enhances the efficacy of such systems and inhibits biofilm formation. Following this strategy, Anjum et al. (2015) prepared a nanocomposite wound dressing hydrogels based on AgNPs/PVA which were subsequently coated on cotton fabrics via dip coating. The colony formation assay demonstrated 100% bacterial reduction of *E. coli* and *S. aureus* in case of nanogel coated dressing fabrics and displayed significant ZOIs. With a cumulative release of 36 wt.% silver, the dressings exhibited high antibacterial efficacy and fast wound contraction over a period of 21 days in Swiss albino mice under in vivo conditions. On similar grounds, a transparent antibacterial patch of AgNPs loaded quaternized PVA (qPVA) hydrogel intended for wound dressing was fabricated recently (Bhowmick et al. 2016). AgNPs/QPVA hydrogels displayed noteworthy antibacterial efficacy (even after 96 h) against *P. aeruginosa, E. coli* and *S. aureus* with ZOIs ~20 mm compared to 11 mm of hydrogels devoid of nano-silver. In a different study, AgNPs were embedded onto PVA/melamine-formaldehyde coatings which were soaked onto polyvinyl foam and coated onto Whatman paper (Kakkar et al. 2015). The antimicrobial activity of prepared dressing materials was tested against foot ulcers causing microbes in diabetic patients, i.e. *S. aureus, P. vulgaris, P. aeruginosa* and *E. cloacae* (Kakkar et al., 2015).

6.2.7 Poly (ethylene glycol) (PEG)

Poly(ethylene glycol) is a hydrophilic linear polymer used as an antifouling coating on catheters, as hydrogel coatings on medical devices or as pore former in dialysis membranes. Since PEG has been known to confer protein resistance to a substrate, many researchers have attempted to fabricate substrates that resist bacterial adhesion using PEG (Banerjee et al. 2011). Park et al. (1998) reported the preparation of PEG-modified polyurethane substrates where various functional groups hydroxyl, amino, and sulfonate containing PEG materials were tested against *E. coli* and *S. epidermidis* under physiological conditions. PEG surfaces with terminal sulfonate groups were found to be the most effective in reducing bacterial attachment while the bacterial reduction was reportedly in proportion to the molecular weight of PEG. Additional functionalizing agents like AgNPs were also incorporated in polyelectrolyte multilayers of PEG polymers in order to enhance the bactericidal activity of the whole system. For instance, the efficacy of nano-silver embedded

PP-g-PEG polymer coated on ventricular catheters was tested for preventing catheter-related infections on a model shunt in rats (Hazer et al. 2016). Data indicated that the sterile and infected Ag-PP-g-PEG-coated ventricular catheters were more effective in reducing the bacterial count of *S. epidermis* than pristine polymeric coatings. In another study, Ag-PP-g-PEG polymer coatings were done on titanium bone screws and its antimicrobial effect was evaluated against MRSA in the lumbar spine of rabbit (Hazer et al. 2016). The bacterial colony count for modified titanium screws was much lower (17.2×10^3 CFU ml^{-1}) than the unmodified screws (200×10^3 CFUml^{-1}) with less biofilm formation. In another study, the biocidal and hemocompatible coatings on catheter surface consisting of AgNPs loaded star-PEG-heparin hydrogel layers were fabricated (Fischer et al. 2015). The resultant coatings (both single-layered and multilayered) displayed long-term antiseptic efficacy against *E. coli* and *S. epidermidis* with ZOIs ~6 mm and ~8 mm respectively, at a silver concentration of 4000 ppm. Recently, Baek et al. (2015) synthesized a hydrogel nanocomposite with poly (ethylene glycol) diacrylate (PEGDA), vinylpyrrolidone (VP) and AgNPs. The nanocomposite gel was further coated on nylon and aluminium substrates separately so as to demonstrate the usefulness of gel as a potent antifouling coating material. The coated nylon and aluminium substrates largely prevented adhesion of bacterial cells when they were exposed to *E. coli* at 37 °C over a period of 10 days (Fig. 10). In a different study, a novel cryogel composites based on chitosan and poly (ethylene glycol)diacrylate was developed after incorporating AgNPs (average size, 10–60 nm), which displayed up to 95% reduction in *E. coli* colonies upon exposure (Zou et al. 2017).

6.2.8 Poly (tetrafluoroethylene), PTFE

PTFE is known for its chemical stability, high order structural and mechanical strength, with high thermal resistance (Li et al. 2011; Shi et al. 2012). This chemically and biologically inert polymer can be used to apply coatings on biomedical devices accompanied with metallic nanoparticles to preclude even the slightest chance of infection. For example, Carvalho et al. (2016) deposited the PTFE/poly (amide) (PA) film coating incorporated with AgNPs over PTFE surface via magnetron sputtering method. The coated materials diminished the extent of colonization of hospital isolated bacteria *P. aeruginosa* compared to control materials. Their antibacterial activity was linked to high surface energy and negative zeta potential which enhances the difficulty of *P. aeruginosa* to interact with coatings due to charge repulsions (Gottenbos et al. 1999; Harkes et al. 1991). The polymer/metal nanocomposite coatings with two noble metallic nanoparticles silver and gold within the PTFE matrix have also been reported (Zaporojtchenko et al. 2006). The coated specimens portrayed significant antibacterial activity against *S. epidermidis, S. aureus, P. aeruginosa, E. coli, K. pneumoniae* with ZOIs of 6, 3, 4, 1 and 1 mm, respectively. Table 2 summarizes other significant achievements observed on antifouling coatings as biomaterials derived from synthetic polymers.

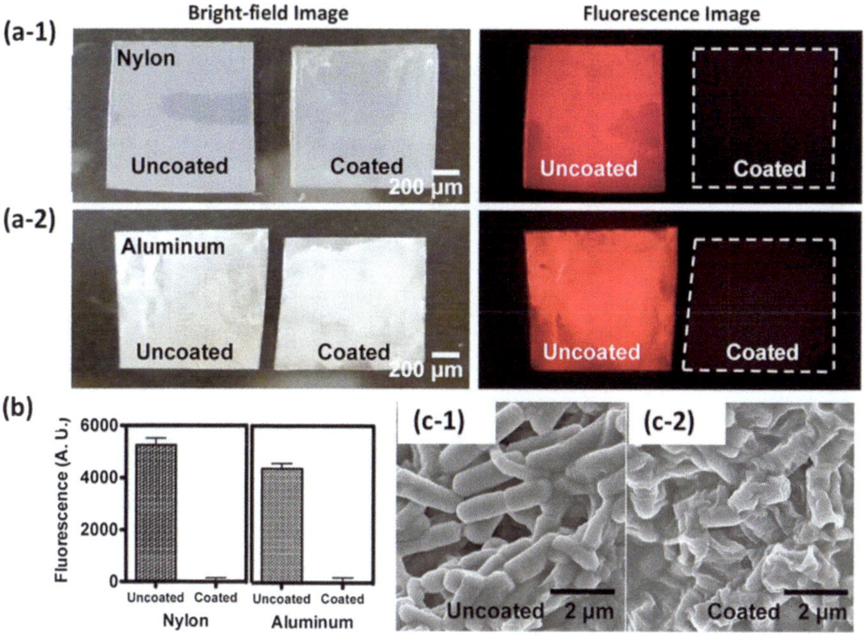

Fig. 10 Mitigation of biofilm formation on nylon and aluminium substrates using the AgNPs-PEGDA-PVP hydrogel composite as a coating material. (**a**) Optic (left) and fluorescence (right) images of tdTomato fluorescence from the uncoated and coated substrates. Images in (**a**-1) represent the nylon substrate and those in (**a**-2) the aluminium substrate. (**b**) Quantification of the fluorescence yield from the nylon and aluminium substrates either uncoated or coated by the AgNPs-PEGDA-PVP hydrogel composites. (**c**) SEM images of bacteria collected from the nylon surface either uncoated or coated by the gel composites. (Reproduced with permission from Baek et al. 2015)

6.2.9 Cellulose

Cotton-based fabrics are widely used due to its properties such as softness, breathability, moisture absorption and wear comfort (Pardini and Amalvy 2008). But cotton fabrics are also easily attacked by the microorganism and fungus owing to its large surface area and ability to retain moisture (Sundrarajan and Rukmani 2013). Therefore, necessary finishes like antibacterial finishing are required on the cotton fabrics. Such antibacterial cotton fabrics can be applied to medical linen, surgical clothing, medical curtains, and other medical textiles in order to prevent the hospital-acquired infections (HAIs). So far the research has been dedicated to textiles. Recently, Zhang et al. (2018) prepared an antibacterial finish of Ag/WPU-acrylate on pre-treated cotton fabric via silane click chemistry. The resultant cotton fabrics reduced bacterial adhesion of *E. coli* and *S. aureus* by 99.99% (Fig. 11). In addition to antibacterial activity and thermal stability, mechanical properties of the fabrics like breaking strength was also amplified from 355.8 to 378.2 N. El-Sayed et al. (2016) attempted the coating of fabric with antimicrobial AgNPs/WPU

Table 2 Silver nanocomposites based upon synthetic polymers and their biomedical applications

Silver based nanocomposites (NCs)	Substrate	Size of AgNPs	Activity	Microbes tested	Evaluation parameters	Biomedical applications	References
Ag/PDMS	Titanium	ND	AF	*C. albicans*	ND	Antifungal metal implant coatings	Groza et al. (2016)
Ag/chitosan/clay/PDMS	Silicone	ND	AB, AF	*E. coli, P. aeruginosa, S. aureus, C.albicans*	ND	Antimicrobial coatings for urinary catheter	Zhou et al. (2007)
Ag/PU	Glass	ND	AB	*E. coli, S. aureus, E. faecalis, K. pneumonia, P. aeruginosa, P. mirabilis, A. baumannii*	>99.99% inhibition	Antibacterial coatings	Khwanmuang et al. (2017a)
Ag/silicone	Metal stent	ND	AB	*E. coli, K. pneumoniae*	99% inhibition	Antibacterial coatings for self-expandable metal stents (SEMSs)	Lee et al. (2016)
Ag/PLA thin films	ND	3–4 nm	AB	*E. coli, S. aureus, V. parahaemolyticus*	ZoI: 9–15 mm	Antibacterial scaffold, biomedical coatings	Shameli et al. (2010)
Ag/PMMA	ND	ND	AB	*E. coli, S. aureus*	ZoI: 45 mm	Wound dressing, bioadhesive and coatings of biomedical materials	Kong and Jang (2008)
Ag/PLGA	Stainless steel	ND	AB	*S. aureus, P. aeruginosa*	ND	Antibacterial coatings for endotracheal tubes and catheters	Liu et al. (2012)
Ag/PU/PCL/PMMA		20–27 nm	AB	*E. coli*	10^6-fold reduction	Antibiofilm implants	Sawant et al. (2013)
Ag/PVA	Tetraaethylorthosilicate (TEOS)	5–7 nm	AB	*E. coli, S. aureus, P. aeruginosa*	ZoI:10–15.5 mm	Antimicrobial coatings	Bryaskova et al. (2010)

Ag/PTFE	ND	ND	AB	*S.epidermidis, S. aureus, P. aeruginosa, E. coli, K. pneumoniae*	ZoI:1-6 mm	Antimicrobial coatings, anti adhesive biomaterial	Zaporojtchenko et al. (2006)
Ag/WPU-acrylate	Pretreated cotton	ND	AB	*E. coli, S. aureus*	99.99% inhibition	Antimicrobial finish/ coatings for medical textiles	Zhang et al. (2018)
Ag/metallocene polyethylene (MPE)	ND	ND	AB	*E. coli, S. aureus*	99.98% inhibition	Antibacterial air spray coatings	Li et al. (2018)
Ag/polyaniline (PAni)	Medical polyurethane	ND	AB	*P. aeruginosa, B. subtilis*	90.6% inhibition	Implant coatings	Prabhakar et al. (2011)
Ag/PVA	Silk sericin (SS)	ND	AB	*E. coli, P. aeruginosa, S. aureus*	ZoI:21.7–28.2 mm	Coatings for wound healing materials	Cai et al. (2017)

AB Antibacterial, *AF* Antifungal, *ND* Not Determined

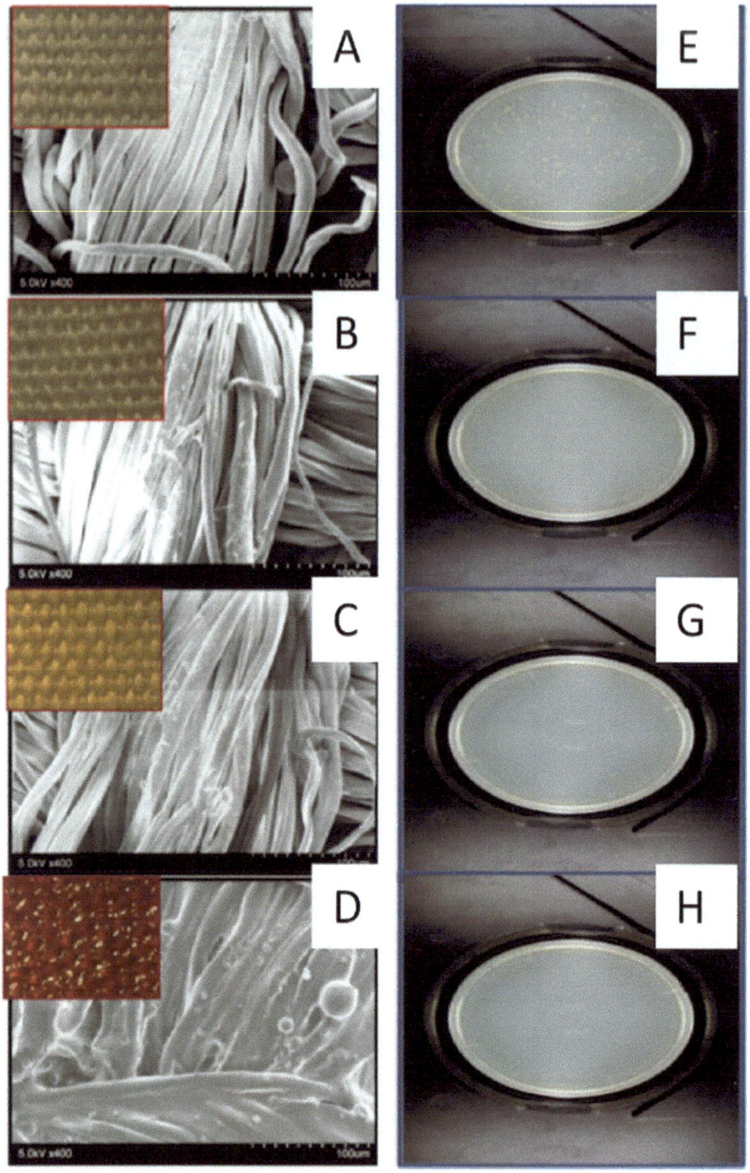

Fig. 11 FSEM images of (**a**) untreated cotton, (**b**) pre-treated cotton, (**c**) antibacterial cotton coated by Ag/WPUA-$C_{0.01}$, (**d**) antibacterial cotton coated by Ag/WPUA-$C_{0.1}$; The antibacterial activity of (**e**) untreated cotton, (**f**) cotton treated by Ag/WPUA-$C_{0.01}$, (**g**) cotton treated by Ag/WPUA-$C_{0.1}$ and (**f**) cotton treated by Ag/WPUA-C0.1 after 10 cycles of standard washing against *S. aureus*. (Reproduced with permission from Zhang et al. 2018)

nanocomposites and demonstrated a wide antibacterial and antifungal efficacy (13–100%) against *E. coli* and *C. albicans* of treated fabrics which depends upon Ag concentration.

Bactericidal coatings based upon Titania (TiO_2) and nanosilver are of considerable interest especially in antibacterial textile finishing (Sadu et al. 2014). A significant number of studies has reflected a positive response in decreasing the microbial adhesion. A study by Sadu et al. (2014) demonstrated the antibacterial activity of nanosilver doped titania coatings (nAg/TiO_2) based on a polyurethane matrix. The polyester fabric functionalized with nAg-TiO_2/polyurethane composites (particularly with 1% Ag content) using the dip-coating method has shown excellent antibacterial activity against *E. coli* and *S. epidermidis.* Effective bactericidal activity was observed under the black light illumination, which, in conjunction with Ag/TiO_2, completely inhibited any bacterial growth within 3 h of exposure. The coatings also maintained its durability and biological activity even after multiple cycles >30 of traditional textile washings. Chitichotpanya et al. (2017) prepared an antibacterial finishing of Ag/TiO_2/WPU on medical cotton fabrics without any additional organic reducing or stabilizing agent. The optimal formulation with contents of $AgNO_3$ (240 ppm) and TiO_2 NPs (980 ppm)exhibited excellent antibacterial activities against *K. pneumoniae* and *S. aureus* with over 99% reduction and improved mechanical properties as well as non-toxicity with cell viability >94% to mammalian L929 cells. Khwanmuang et al. (2017b) also synthesized similar nanocomposite coatings over glass sheets with a goal to employ them in real clinical applications. In order to examine the practicality of such coatings, these were put to test against *E. coli, S. aureus, E. faecalis, K. pneumoniae, P. aeruginosa, P. mirabilis,* and *A. baumanii* for any bactericidal activity. Within 24 h, the coatings showed a 99.99% bacterial reduction and cell viability of 90% in the cytotoxicity assay.

6.2.10 Other Polymers of Biomedical Importance

Polyolefins like poly (ethylene) (PE) and poly (propylene) (PP) are hydrophobic inert materials which do not degrade in vivo. These can be produced at different molecular weights and crystallinity, thus are categorized into low-density PE (LDPE) and high-density PE (HDPE). Each of them has different applications according to its properties such as LDPE is mostly soft therefore applied in packaging while HDPE is used for formation of stable devices for implantation such as sliding surfaces of artificial joints. Considering this, Li et al. (2018) prepared metallocene polyethylene (MPE)/nano-silver coatings by a facile air-spray method on polymer films. Results showed that the coatings with 50 ppm nano-Ag were enough for the antibacterial application, with high antibacterial efficiency against *E. coli* (>99.98%) and *S. aureus* (99.96%). Also, the amount of silver release was very low (1.2 ppb) even after 30 days, which is well below the permissible standards given by the WHO.

Polyaniline (PAni) has striking mechanical properties with processing advantages of various polymers which paves way for its potential utility in a wide variety of applications (Nanlin et al. 2009; Choudhury 2009). Studies have also shown that PAni has good antimicrobial activity (Stejskal et al. 2010; Kucekova et al. 2013) and efficacy against Gram-positive and Gram-negative bacteria and fungi (Shi et al. 2012; Seshadri and Bhat 2005). In addition, the incorporation of Ag into the polymer matrix of PAni enhances its conductivity as well as the antimicrobial properties of the Ag-PAni nanocomposite material (Prabhakar et al. 2011; Kucekova et al. 2013). Recently many researchers have demonstrated synergistic effects of Ag and PAni against *S. aureus* and *E. coli* (Jia et al. 2012). Prabhakar et al. (2011) modified the surface of medical grade polyurethane with a coating of PAni-AgNPs composite, where the attachment of *P. aeruginosa* and *B. subtilis* was reduced markedly on coated surfaces with inhibition rates of 90.6% and 50.5%, respectively. A substantial reduction in biofilm thickness (20%) was also observed in PAni-AgNPs coated PU surface. The biocompatibility of coated surface also improved with only 18% of 3T3 L1 cell death when compared to 41% with pristine PU after 48 h of incubation. Similarly, Agarwala et al. (2014) synthesized Ag-PPAni nanocomposite coatings over catheter via plasma deposition process. Antibacterial assays showed that coated catheters were capable of killing planktonic cells of most commonly encountered uropathogens and equally capable of eradicating biofilm formation. A ZOIs ranged from 13 to 22 mm was formed against MRSA, MRSE, VRE, *E.coli*, *K. pneumoniae*, *P. aeruginosa* and *P. mirabilis* as a function of increasing nano-Ag concentration in composites.

Silk sericin (SS), a globular protein synthesized by silkworm is biocompatible and biodegradable with additional properties like gelling ability, water-holding capacity and skin adhesion (Rajasekaran et al. 2011). Moreover, silk sericin could promote the adhesion and proliferation of human skin and accelerate burn or scald wound healing (Aramwit and Sangcakul 2007; Kundu et al. 2008). Therefore, it is regarded as a good candidate for wound dressing material but its application is limited due to its fragile nature. A recent few reports have improved its mechanical performance after blending with another polymer PVA where the antimicrobial properties were introduced by adding AgNPs (Cai et al. 2017). The SS/PVA/AgNPs film coated with polyelectrolyte multilayers (PEMs) displayed superior antibacterial efficiency against *E. coli, P. aeruginosa*, and *S. aureus* with ZOIs of 21.7 mm, 24.9 mm, and 28.2 mm respectively. Similarly, He et al. (2017) also created SS/PVA/AgNPs films which were effectively bactericidal with ZOIs increasing as a function of increasing UV irradiation time during in situ AgNPs synthesis in films. In case of inhibition of *E. coli*, ZOI increased from 19.2 to 21.2 mm while for *S. aureus* it was increased from 21.5 to 25.3 mm respectively with a surge in UV irradiation time from 10 to 60 min. Such films offer more choices to be potentially applied in antibacterial materials such as skin tissue engineering, coatings for wound healing materials.

7 Compatibility of Biomedical Coatings Containing Silver Nanoparticles

Regarding biocompatibility of silver coated medical devices/substrates, various studies have shown results with acute variability where some researchers have portrayed their significant biocompatibility while the others suggested a compromise. For medical device regulation, cytotoxicity testing is part of a biocompatibility evaluation, in which specific test methods are chosen based on the technological characteristics and intended use of a device. For example, In Sussman et al. (2015) prepared nanoAg-coated tissue culture polystyrene surfaces using a magnetron sputter coating which demonstrated more than 4 log reduction in *Escherichia coli* viability. However, when two different cytotoxicity assays (extract based and direct contact based) were compared, it was found that the extracts of nano-silver caused no cytotoxicity to L929 mouse fibroblasts, but cells cultured directly on nAg coatings showed a dose-dependent reduction in viability by up to 100%. ICP-MS results suggested that non-toxicity of extracts of nAg might be because the dissolved Ag becomes less cytotoxic over time owing to the reaction with cell culture media and serum. The findings of this study highlighted the potential value of direct-contact cytotoxicity testing for nAg in predicting biological interactions with cells or tissue in vivo. Shantiaee et al. (2011) also depicted the time-dependent cytotoxicity of nanoAg coated gutta-percha (root canal filling material) against mouse fibroblast cell line L929 where, maximum cytotoxicity was observed after 1 h and was decreased further to that in normal gutta-percha after 24 h and after 1 week, it reached the lowest level amongst the other test materials.

On the other hand, an in vitro study showing the biocompatibility of silver-coated orthopaedic external fixation pins was compared with stainless steel controls against human peripheral blood lymphocytes, fibroblasts NIH 3T3, and osteoblast-like cells (Bosetti et al. 2002). Compared to stainless steel implants, the coated implants did not show any genotoxicity or cytotoxicity. In fact, cells cultured over Ag-coated material demonstrated enhanced cell spreading and a higher cell count with respect to the uncoated material on the fourth day. Similarly, Liao et al. (2010) also concluded no cytotoxicity of nanoAg coated titanium plate against human gingival fibroblasts with excellent antibacterial activity against *S. aureus* and *E. coli.*

In a completely different study, the effect of structure and surface coatings on shape dependent environmental fate, cellular uptake, and the toxic effects of AgNPs was evaluated (Lu et al. 2010). Authors demonstrated that colloidal spherical AgNPs and silver nanoprisms of 30 nm sizes were nontoxic to human skin HaCaT keratinocyte cells even after 48 h of incubation. On the other hand, silver nitrate at the concentration of 10 μgml^{-1} was found to be highly toxic and also possessed phototoxicity in a dose-dependent manner. However, the environmental fate of such nanomaterials changed when this citrate coated AgNPs transformed into dried powder form. Toxicity data suggested that citrate coated silver nanopowder was toxic due to the chemical changes occurred during the drying process whereas PVP-coated silver nanoprism or nanoparticle powder was not toxic even after 3 weeks of sunlight exposure. The importance of capping agent on controlling AgNPs toxicity and cytotoxicity has been reported (Sanyasi et al. 2016). In this study, AgNPs capped with carboxymethyl tamarind polysaccharide showed prospective antibacterial activity against *E. coli* and *S. Typhimurium* without eliciting any cytotoxic response towards mouse macrophage RAW 264.7 cells (Fig. 12). Such studies shed a light on how biocompatible polymeric materials reduce the toxic effects of nanomaterials and their utilization to fabricate antibacterial surface coatings with good biocompatibility.

8 Toxicity and Safety Aspects of Nano Silver Based Nanocomposite Coatings

It is well documented that silver-based nanocomposites have been established as an ideal anti-biofouling material to be extensively used in biomedical and related healthcare applications. Out of nearly 800 commercial nano-based products available in the market, nearly 30% of them are claimed to contain AgNPs (Mittal 2013; Agnihotri et al. 2014). In the last few years, biomaterials coated with AgNPs have been witnessed to alleviate nosocomial infections in hospitals (Dastjerdi et al. 2010; Gao and Cranston 2008). However, the downside associated with such an extensive usage of nano-silver based products has also shown its toxicological consequences to the ecosystem (Mukherji et al. 2012). It is still a matter of debate whether long-term exposure to nanosilver causes serious toxic effects to eukaryotic cells and aquatic animals other than microbes. There are contradictory opinions and scientific evidence which address the fate and role of nanomaterials on health, surroundings by the scientific community and regulatory agencies. From last decades, the safety of silver-based products in the market is merely justified by the fact that silver in its ionic form is considered to have very low toxicity on humans except for its extreme dose, which may cause Argyria (Mittal 2013).

8.1 Silver Toxicity Towards Humans

In general, 'silver toxicity' refers to any deleterious effects on an organism upon exposure to silver. Obviously, if the practical intention is to disinfect or sterilize a specific microbial community, then the silver toxicity would be interpreted as its efficacy to reduce its growth e.g., antibacterial, antiviral, antifungal which would eventually benefit humans. However, if the same material exerts unintended or undesired impacts on higher organisms including humans, then such toxicity may be interpreted as a potential hazard. The evidence of silver toxicity to lower micro-organisms like bacteria, virus and fungi etc. has already been discussed in previous

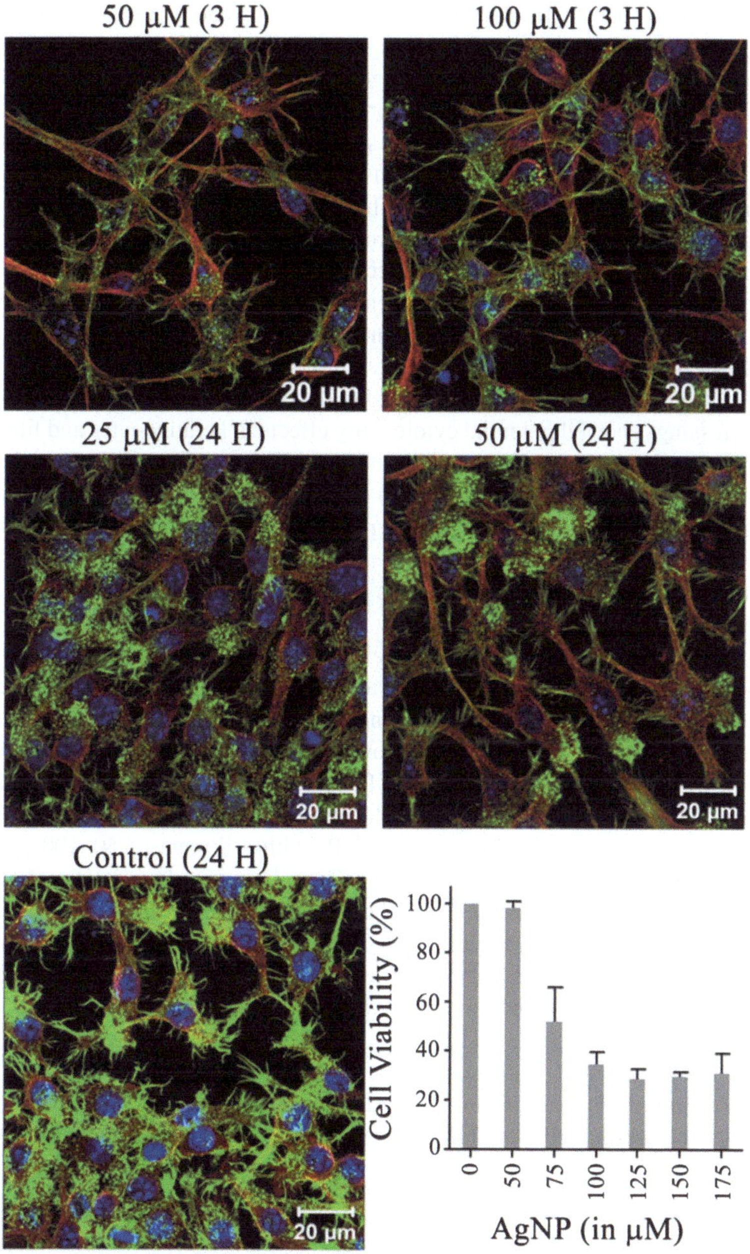

Fig. 12 Bio-compatibility of CMT-capped AgNPs to a mammalian cell. Confocal images of RAW 264.7 cells treated with CMT-capped AgNPs for different concentration and time duration followed by staining for actin cytoskeleton (green), microtubule (red) and DNA (blue) shown here. The percentage of cell viability assessed by MTT assay is presented as bar-graph at below (N = 4). (Reproduced with permission from Sanyasi et al. 2016, Open Access)

sections of this chapter. In this section, an insight on toxicity and biocompatibility of silver and silver nanocomposites on higher organisms will be provided.

Despite the fact that AgNPs have oligodynamic action, which indicates that it can kill microbes at a concentration where it is expected to be non-toxic to mammalian cells, a relatively large number of studies have also evidenced the biological activity of AgNPs detrimental to humans. Due to their small size, nanoparticles have access to skin, lungs, and brain (AshaRani et al. 2008). Previous studies suggested that the exposure of AgNPs to human lung epithelial cells generated reactive oxygen species, which can lead to oxidative stress and cellular damage. Therefore, it is important to study the cytotoxic potential for the application of silver nanoparticles that target specific cells or organs (Moghimi et al. 2001; Panyam and Labhasetwar 2003). Burd et al. (2007) studied the cytotoxicity of five different AgNPs-impregnated commercially available dressings and found that three of the silver dressings showed potential cytotoxicity effects in keratinocytes and fibroblast cultures. Similarly, toxic effects to mammalian cells have also been reported where a continuous exposure of AgNPs over 90 days substantially reduced lung function, produce inflammatory lesions in the lungs of rats (Sung et al. 2008), reduction of mitochondrial function and increment in membrane leakage of mouse spermatogonial stem cell and rat liver cells (Braydich-Stolle et al. 2005; Hussain et al. 2005). In a previous study, AgNPs have reportedly reduced the cellular glutathione levels in human fibrosarcoma and human skin/carcinoma cells as a consequence of oxidative stress, which resulted in cellular damage and lipid peroxidation (Arora et al. 2008). However, the dose necessary to induce necrosis (12.5 μgmL^{-1}) in both cell types was much higher than that required to provoke apoptosis (0.78–1.56 $\mu g\ mL^{-1}$). As a result, it was concluded that despite AgNPs toxicity towards mammalian cells, it is still possible to define a safe range for the application of silver nanoparticles as a topical antimicrobial agent. Similarly, AgNPs with a dosage of 50–100 $\mu g\ mL^{-1}$ induced apoptosis to 43.4% of fibroblast cells and produced necrosis to 40.2% of colon cancer cells, respectively. From the above studies, it can be concluded that longer-term studies in a variety of test systems and monitoring of humans exposed to silver nanoparticles are imperative to evaluate any potential toxicity of the nanoscale components of commercially available products, as well as for future products. Nevertheless, all silver-containing coatings may prove to be a real asset for the biomedical field in decreasing the number of medical device related infection. However, the nature of the silver particles as well as how these are incorporated into the coating will determine the efficacy of such modified medical devices.

9 Conclusions

Silver-based nanomaterials are becoming the new gold standard for medical application as these display outstanding biological and antimicrobial properties. On account of such superior characteristics, nano-silver based biomedical coatings are being preferred over others. Implantable medical devices, such as sutures, neurosurgical and venous catheters, have greatly benefited from the deposition of silver nanoparticles by reducing both patient infection and dependence on antibiotic use and the associated costs. Also, AgNPs are exploited to the maximum when accompanied with a suitable polymer matrix which boosts their antimicrobial performance along with stated benefits.

This chapter systematically summarizes various aspects of nano-silver based antimicrobial polymeric coatings where their application potential for biomedical purposes has been thoroughly reviewed. Several silver based polymer nanocomposite coatings derived from either natural or synthetic polymers are described in great detail with relevant supporting studies. Various coating deposition methods have been conferred and method of deposition of AgNPs were applied as per the specific application of substrate, chemical composition, level of antimicrobial activity and the extent of biocompatibility required. Specific application areas such as implantable short or long term medical devices, antimicrobial textiles, cellulosic sutures, bandages have also been discussed extensively. For each particular study, a method of fabricating nano-silver coatings, their topographic features and characterizations and lastly their strain-specific antimicrobial effects have been demonstrated. It was well documented that, even a small percentage of nano-silver used for the blend treatment is fair enough to ensure a good antibacterial capability. A repeated exposure of nano-silver at their higher concentrations may also pose some toxicological consequences to the systems and releasing environment. Therefore, some additional precautions would be needed to determine the fate and toxicity of silver when used for biomedical purposes. However, the experimental studies included in this chapter are derived from such studies where a rigorous long-term exposure of nano-silver based coatings warrants non-toxic effects to the ecosystem species. It is expected that the extended use of silver-based antimicrobial agents will be explored in many other fields for improving health and welfare.

References

Abraham, A. N., Sharma, T. K., Bansal, V., & Shukla, R. (2018). Phytochemicals as dynamic surface ligands to control nanoparticle–protein interactions. *ACS Omega, 3*, 2220–2229.

Agarwala, M., Barman, T., Gogoi, D., Choudhury, B., Pal, A. R., & Yadav, R. (2014). Highly effective antibiofilm coating of silver–polymer nanocomposite on polymeric medical devices deposited by one step plasma process. *Journal of Biomedical Materials Research Part B Applied Biomaterials, 102*, 1223–1235.

Agnihotri, S., & Dhiman, N. K. (2017). Development of nano-antimicrobial biomaterials for biomedical applications. In A. Tripathi & J. S. Melo (Eds.), *Advances in biomaterials for biomedical applications* (pp. 479–545). Singapore: Springer.

Agnihotri, S., Mukherji, S., & Mukherji, S. (2012). Antimicrobial chitosan–PVA hydrogel as a nanoreactor and immobilizing matrix for silver nanoparticles. *Applied Nanoscience, 2*, 179–188.

Agnihotri, S., Mukherji, S., & Mukherji, S. (2013). Immobilized silver nanoparticles enhance contact killing and show highest efficacy: Elucidation of the mechanism of bactericidal action of silver. *Nanoscale, 5*, 7328–7340.

Agnihotri, S., Mukherji, S., & Mukherji, S. (2014). Size-controlled silver nanoparticles synthesized over the range 5–100 nm using the same protocol and their antibacterial efficacy. *RSC Advances, 4*, 3974–3983.

Agnihotri, S., Bajaj, G., Mukherji, S., & Mukherji, S. (2015). Arginine-assisted immobilization of silver nanoparticles on ZnO nanorods: An enhanced and reusable antibacterial substrate without human cell cytotoxicity. *Nanoscale, 7*, 7415–7429.

Agnihotri, S., Dhiman, N. K., & Tripathi, A. (2018). Antimicrobial surface modification of polymeric biomaterials. In A. Tiwari (Ed.), *Handbook of antimicrobial coatings* (pp. 435–486). Amsterdam: Elsevier.

Ajitha, B., Reddy, Y. A. K., Jeon, H.-J., & Ahn, C. W. (2018). Synthesis of silver nanoparticles in an eco-friendly way using *Phyllanthus amarus* leaf extract: Antimicrobial and catalytic activity. *Advanced Powder Technology, 29*, 86–93.

Alexander, J. W. (2009). History of the medical use of silver. *Surgical Infections, 10*, 289–292.

An, Y. H., & Friedman, R. J. (1998). Concise review of mechanisms of bacterial adhesion to biomaterial surfaces. *Journal of Biomedical Materials Research, 43*, 338–348.

Anjum, S., Sharma, A., Tummalapalli, M., Joy, J., Bhan, S., & Gupta, B. (2015). A novel route for the preparation of silver loaded polyvinyl alcohol nanogels for wound care systems. *International Journal of Polymeric Materials and Polymeric Biomaterials, 64*, 894–905.

Aramwit, P., & Sangcakul, A. (2007). The effects of sericin cream on wound healing in rats. *Bioscience, Biotechnology, and Biochemistry, 71*, 2473–2477.

Arockianathan, P. M., Sekar, S., Kumaran, B., & Sastry, T. (2012). Preparation, characterization and evaluation of biocomposite films containing chitosan and sago starch impregnated with silver nanoparticles. *International Journal of Biological Macromolecules, 50*, 939–946.

Arora, S., Jain, J., Rajwade, J., & Paknikar, K. (2008). Cellular responses induced by silver nanoparticles: In vitro studies. *Toxicology Letters, 179*, 93–100.

Asai, T., Lee, M.-H., Arrecubieta, C., von Bayern, M. P., Cespedes, C. A., Baron, H. M., et al. (2007). Cellular coating of the left ventricular assist device textured polyurethane membrane reduces adhesion of *Staphylococcus aureus*. *The Journal of Thoracic and Cardiovascular Surgery, 133*, 1147–1153.

AshaRani, P., Low Kah Mun, G., Hande, M. P., & Valiyaveettil, S. (2008). Cytotoxicity and genotoxicity of silver nanoparticles in human cells. *ACS Nano, 3*, 279–290.

Atiyeh, B. S., Costagliola, M., Hayek, S. N., & Dibo, S. A. (2007). Effect of silver on burn wound infection control and healing: Review of the literature. *Burns, 33*, 139–148.

Augustine, R., & Rajarathinam, K. (2012). Synthesis and characterization of silver nanoparticles and its immobilization on alginate coated sutures for the prevention of surgical wound infections and the in vitro release studies. *International Journal of Nano Dimension, 2*, 205.

Baek, K., Liang, J., Lim, W. T., Zhao, H., Kim, D. H., & Kong, H. (2015). In situ assembly of antifouling/bacterial silver nanoparticle-hydrogel composites with controlled particle release and matrix softening. *ACS Applied Materials and Interfaces, 7*, 15359–15367.

Bakare, R., Hawthrone, S., Vails, C., Gugssa, A., Karim, A., Stubbs, J., 3rd, et al. (2016). Antimicrobial and cell viability measurement of bovine serum albumin capped silver nanoparticles (Ag/BSA) loaded collagen immobilized poly(3-hydroxybutyrate-co-3-hydroxyvalerate) (PHBV) film. *Journal of Colloid and Interface Science, 465*, 140–148.

Bal, K., Bal, Y., Cote, G., & Chagnes, A. (2012). Morphology and antimicrobial properties of Luffa cylindrica fibers/chitosan biomaterial as micro-reservoirs for silver delivery. *Materials Letters, 79*, 238–241.

Banerjee, I., Pangule, R. C., & Kane, R. S. (2011). Antifouling coatings: Recent developments in the design of surfaces that prevent fouling by proteins, bacteria, and marine organisms. *Advanced Materials, 23*, 690–718.

Bang, B. W., Jeong, S., Lee, D. H., Lee, J. I., Lee, S. C., & Kang, S.-G. (2012). The biodurability of covering materials for metallic stents in a bile flow phantom. *Digestive Diseases and Sciences, 57*, 1056–1063.

Bat, E., Zhang, Z., Feijen, J., Grijpma, D. W., & Poot, A. A. (2014). Biodegradable elastomers for biomedical applications and regenerative medicine. *Regenerative Medicine, 9*, 385–398.

Bazaka, K., Jacob, M. V., Crawford, R. J., & Ivanova, E. P. (2012). Efficient surface modification of biomaterial to prevent biofilm formation and the attachment of microorganisms. *Applied Microbiology and Biotechnology, 95*, 299–311.

Beyth, N., Yudovin-Farber, I., Perez-Davidi, M., Domb, A. J., & Weiss, E. I. (2010). Polyethyleneimine nanoparticles incorporated into resin composite cause cell death and trigger biofilm stress in vivo. *Proceedings of the National Academy of Sciences of the United States of America, 107*, 22038–22043.

Bharti, S., Agnihotri, S., Mukherji, S., & Mukherji, S. (2015). Effectiveness of immobilized silver nanoparticles in inactivation of pathogenic bacteria. *Journal of Environmental Research And Development, 9*, 849–856.

Bhowmick, S., Mohanty, S., & Koul, V. (2016). Fabrication of transparent quaternized PVA/silver nanocomposite hydrogel and its evaluation as an antimicrobial patch for wound care systems. *Journal of Materials Science. Materials in Medicine, 27*, 160.

Bindhu, M., & Umadevi, M. (2013). Synthesis of monodispersed silver nanoparticles using *Hibiscus cannabinus* leaf extract and its antimicrobial activity. *Spectrochimica Acta. Part A, Molecular and Biomolecular Spectroscopy, 101*, 184–190.

Blecher, K., Nasir, A., & Friedman, A. (2011). The growing role of nanotechnology in combating infectious disease. *Virulence, 2*, 395–401.

Bosetti, M., Masse, A., Tobin, E., & Cannas, M. (2002). Silver coated materials for external fixation devices: In vitro biocompatibility and genotoxicity. *Biomaterials, 23*, 887–892.

Braydich-Stolle, L., Hussain, S., Schlager, J. J., & Hofmann, M.-C. (2005). In vitro cytotoxicity of nanoparticles in mammalian germline stem cells. *Toxicological Sciences, 88*, 412–419.

Bridges, A. W., & García, A. J. (2008). Anti-inflammatory polymeric coatings for implantable biomaterials and devices. *Journal of Diabetes Science and Technology, 2*, 984–994.

Bryaskova, R., Pencheva, D., Kale, G. M., Lad, U., & Kantardjiev, T. (2010). Synthesis, characterisation and antibacterial activity of PVA/TEOS/Ag-Np hybrid thin films. *Journal of Colloid and Interface Science, 349*, 77–85.

Burd, A., Kwok, C. H., Hung, S. C., Chan, H. S., Gu, H., Lam, W. K., et al. (2007). A comparative study of the cytotoxicity of silver-based dressings in monolayer cell, tissue explant, and animal models. *Wound Repair and Regeneration, 15*, 94–104.

Bushnell, B. D., McWilliams, A. D., Whitener, G. B., & Messer, T. M. (2008). Early clinical experience with collagen nerve tubes in digital nerve repair. *Journal of Hand Surgery-American, 33*, 1081–1087.

Busscher, H. J., Van Der Mei, H. C., Subbiahdoss, G., Jutte, P. C., Van Den Dungen, J. J., Zaat, S. A., et al. (2012). Biomaterial-associated infection: Locating the finish line in the race for the surface. *Science Translational Medicine, 4*, 153rv110.

Cai, R., Tao, G., He, H., Guo, P., Yang, M., Ding, C., et al. (2017). In situ synthesis of silver nanoparticles on the polyelectrolyte-coated sericin/PVA film for enhanced antibacterial application. *Materials, 10*, 967.

Camargo, P. H. C., Satyanarayana, K. G., & Wypych, F. (2009). Nanocomposites: Synthesis, structure, properties and new application opportunities. *Materials Research, 12*, 1–39.

Cao, X., Tang, M., Liu, F., Nie, Y., & Zhao, C. (2010). Immobilization of silver nanoparticles onto sulfonated polyethersulfone membranes as antibacterial materials. *Colloids and Surfaces. B, Biointerfaces, 81*, 555–562.

Carlson, G. A., Dragoo, J. L., Samimi, B., Bruckner, D. A., Bernard, G. W., Hedrick, M., et al. (2004). Bacteriostatic properties of biomatrices against common orthopaedic pathogens. *Biochemical and Biophysical Research Communications, 321*, 472–478.

Carvalho, D., Sousa, T., Morais, P., & Piedade, A. (2016). Polymer/metal nanocomposite coating with antimicrobial activity against hospital isolated pathogen. *Applied Surface Science, 379*, 489–496.

Chakraborty, D., Sharma, V., Agnihotri, S., Mukherji, S., & Mukherji, S. (2017). Disinfection of water in a batch reactor using chloridized silver surfaces. *Journal of Water Process Engineering, 16*, 41–49.

Chan, C., Cheng, H., Djurišić, A. B., Ng, A., Leung, F. C., & Chan, W. K. (2011). Multicomponent antimicrobial transparent polymer coatings. *Journal of Applied Polymer Science, 122*, 1572–1578.

Chen, X., & Schluesener, H. (2008). Nanosilver: A nanoproduct in medical application. *Toxicology Letters, 176*, 1–12.

Chen, A., Wang, H., & Li, X. (2005). One-step process to fabricate Ag–polypyrrole coaxial nanocables. *Chemical Communications, 14*, 1863–1864.

Chen, W., Liu, Y., Courtney, H., Bettenga, M., Agrawal, C., Bumgardner, J., et al. (2006). In vitro anti-bacterial and biological properties of magnetron co-sputtered silver-containing hydroxyapatite coating. *Biomaterials, 27*, 5512–5517.

Chernousova, S., & Epple, M. (2013). Silver as antibacterial agent: Ion, nanoparticle, and metal. *Angewandte Chemie International Edition, 52*, 1636–1653.

Chia, T. W. R., Nguyen, V. T., McMeekin, T., Fegan, N., & Dykes, G. A. (2011). Stochasticity of bacterial attachment and its predictability by the extended Derjaguin-Landau-Verwey-Overbeek theory. *Applied and Environmental Microbiology, 77*, 3757–3764.

Chitichotpanya, P., Inprasit, T., & Chitichotpanya, C. (2017). In vitro assessment of antibacterial potential and mechanical properties of Ag-TiO2/WPU on medical cotton optimized with response surface methodology. *Journal of Natural Fibers*, 1–12.

Choi, O. K., & Hu, Z. Q. (2009). Nitrification inhibition by silver nanoparticles. *Water Science and Technology, 59*, 1699–1702.

Choi, J. S., Yang, H.-J., Kim, B. S., Kim, J. D., Kim, J. Y., Yoo, B., et al. (2009). Human extracellular matrix (ECM) powders for injectable cell delivery and adipose tissue engineering. *Journal of Controlled Release, 139*, 2–7.

Choudhury, A. (2009). Polyaniline/silver nanocomposites: Dielectric properties and ethanol vapour sensitivity. *Sensors and Actuators B: Chemical, 138*, 318–325.

Chu, P. K., Chen, J., Wang, L., & Huang, N. (2002). Plasma-surface modification of biomaterials. *Materials Science and Engineering R: Reports, 36*, 143–206.

Chung, S., Ingle, N. P., Montero, G. A., Kim, S. H., & King, M. W. (2010). Bioresorbable elastomeric vascular tissue engineering scaffolds via melt spinning and electrospinning. *Acta Biomaterialia, 6*, 1958–1967.

Ciobanu, C., Groza, A., Iconaru, S., Popa, C., Chapon, P., Chifiriuc, M., et al. (2015). Antimicrobial activity evaluation on silver doped hydroxyapatite/polydimethylsiloxane composite layer. *BioMed Research International, 2015*, 926513.

Cloutier, M., Mantovani, D., & Rosei, F. (2015). Antibacterial coatings: Challenges, perspectives, and opportunities. *Trends in Biotechnology, 33*, 637–652.

Coma, V. (2013). Polysaccharide-based biomaterials with antimicrobial and antioxidant properties. *Polímeros, 23*, 287–297.

Cometa, S., Bonifacio, M. A., Baruzzi, F., de Candia, S., Giangregorio, M. M., Giannossa, L. C., et al. (2017). Silver-loaded chitosan coating as an integrated approach to face titanium implant-associated infections: Analytical characterization and biological activity. *Analytical and Bioanalytical Chemistry, 409*, 7211–7221.

Dahlin, R. L., Kasper, F. K., & Mikos, A. G. (2011). Polymeric nanofibers in tissue engineering. *Tissue Engineering Part B, Reviews, 17*, 349–364.

Daima, H. K., Selvakannan, P. R., Kandjani, A. E., Shukla, R., Bhargava, S. K., & Bansal, V. (2014). Synergistic influence of polyoxometalate surface corona towards enhancing the antibacterial performance of tyrosine-capped Ag nanoparticles. *Nanoscale, 6*, 758–765.

Dallas, P., Sharma, V. K., & Zboril, R. (2011). Silver polymeric nanocomposites as advanced antimicrobial agents: Classification, synthetic paths, applications, and perspectives. *Advances in Colloid and Interface, 166*, 119–135.

Damm, C. (2005). Silver ion release from polymethyl methacrylate silver nanocomposites. *Polymers and Polymer Composites, 13*, 649.
Dastjerdi, R., Mojtahedi, M., Shoshtari, A., & Khosroshahi, A. (2010). Investigating the production and properties of Ag/TiO_2/PP antibacterial nanocomposite filament yarns. *The Journal of the Textile Institute, 101*, 204–213.
Dayyoub, E., Frant, M., Pinnapireddy, S. R., Liefeith, K., & Bakowsky, U. (2017). Antibacterial and anti-encrustation biodegradable polymer coating for urinary catheter. *International Journal of Pharmaceutics, 531*, 205–214.
Domènech Garcia, B., Muñoz Tapia, M., Muraviev, D., & Macanás de Benito, J. (2014). *Fabrication of polymer-metal nanocomposites with complex polymeric matrices for bactericidal and catalytic applications*. Barcelona: Universitat Autònoma de Barcelona.
Donlan, R. M. (2001). Biofilms and device-associated infections. *Emerging Infectious Diseases, 7*, 277.
Dowling, D., Donnelly, K., McConnell, M., Eloy, R., & Arnaud, M. (2001). Deposition of antibacterial silver coatings on polymeric substrates. *Thin Solid Films, 398*, 602–606.
Dubas, S. T., Wacharanad, S., & Potiyaraj, P. (2011). Tunning of the antimicrobial activity of surgical sutures coated with silver nanoparticles. *Colloids and Surfaces A: Physicochemical and Engineering Aspects, 380*, 25–28.
Dutta, P. K., Dutta, J., & Tripathi, V. (2004). Chitin and chitosan: Chemistry, properties and applications. *Journal of Scientific and Industrial Research, 63*, 20–31.
El Hajj, F., Hasan, A., Nakhleh, J., Osta, M., Darwish, G., Karam, P., et al. (2015). Nanosilver loaded GelMA hydrogel for antimicrobial coating of biomedical implants. In *IEEE: 2015 International Conference on Advances in Biomedical Engineering (ICABME)* (pp. 189–192).
El-Sayed, A. A., El Gabry, L., & Allam, O. (2010). Application of prepared waterborne polyurethane extended with chitosan to impart antibacterial properties to acrylic fabrics. *Journal of Materials Science. Materials in Medicine, 21*, 507–514.
El-Sayed, A. A., Salama, M., Salem, T., & Rehan, M. (2016). Synergistic combination of reduction and polymerization reactions to prepare silver/waterborne polyurethane nanocomposite for coating applications. *Indian Journal of Science and Technology, 9*, 1–10.
Eraković, S., Janković, A., Veljović, D., Palcevskis, E., Mitrić, M., Stevanović, T., et al. (2012). Corrosion stability and bioactivity in simulated body fluid of silver/hydroxyapatite and silver/hydroxyapatite/lignin coatings on titanium obtained by electrophoretic deposition. *The Journal of Physical Chemistry. B, 117*, 1633–1643.
Feng, W., Gao, X., McClung, G., Zhu, S., Ishihara, K., & Brash, J. L. (2011). Methacrylate polymer layers bearing poly (ethylene oxide) and phosphorylcholine side chains as non-fouling surfaces: In vitro interactions with plasma proteins and platelets. *Acta Biomaterialia, 7*, 3692–3699.
Fischer, M., Vahdatzadeh, M., Konradi, R., Friedrichs, J., Maitz, M. F., Freudenberg, U., et al. (2015). Multilayer hydrogel coatings to combine hemocompatibility and antimicrobial activity. *Biomaterials, 56*, 198–205.
Fortunati, E., Latterini, L., Rinaldi, S., Kenny, J., & Armentano, I. P. L. G. A. (2011). Ag nanocomposites: In vitro degradation study and silver ion release. *Journal of Materials Science. Materials in Medicine, 22*, 2735–2744.
Franci, G., Falanga, A., Galdiero, S., Palomba, L., Rai, M., Morelli, G., et al. (2015). Silver nanoparticles as potential antibacterial agents. *Molecules, 20*, 8856–8874.
Galya, T., Sedlařík, V., Kuřitka, I., Novotný, R., Sedlaříková, J., & Sáha, P. (2008). Antibacterial poly (vinyl alcohol) film containing silver nanoparticles: Preparation and characterization. *Journal of Applied Polymer Science, 110*, 3178–3185.
Gao, Y., & Cranston, R. (2008). Recent advances in antimicrobial treatments of textiles. *Textile Research Journal, 78*, 60–72.
Gianolio, D. A., Philbrook, M., Avila, L. Z., Young, L. E., Plate, L., Santos, M. R., et al. (2008). Hyaluronan-tethered opioid depots: Synthetic strategies and release kinetics in vitro and in vivo. *Bioconjugate Chemistry, 19*, 1767–1774.

Gottenbos, B., Van Der Mei, H., Busscher, H., Grijpma, D., & Feijen, J. (1999). Initial adhesion and surface growth of Pseudomonas aeruginosa on negatively and positively charged poly (methacrylates). *Journal of Materials Science. Materials in Medicine, 10*, 853–855.

Gravante, G., Caruso, R., Sorge, R., Nicoli, F., Gentile, P., & Cervelli, V. (2009). Nanocrystalline silver: A systematic review of randomized trials conducted on burned patients and an evidence-based assessment of potential advantages over older silver formulations. *Annals of Plastic Surgery, 63*, 201–205.

Gristina, A. G. (1987). Biomaterial-centered infection: Microbial adhesion versus tissue integration. *Science, 237*, 1588–1595.

Groza, A., Ciobanu, C. S., Popa, C. L., Iconaru, S. L., Chapon, P., Luculescu, C., et al. (2016). Structural properties and antifungal activity against *Candida albicans* biofilm of different composite layers based on Ag/Zn doped hydroxyapatite-polydimethylsiloxanes. *Polymers, 8*, 131.

Gunatillake, P., Mayadunne, R., & Adhikari, R. (2006). Recent developments in biodegradable synthetic polymers. *Biotechnology Annual Review, 12*, 301–347.

Gupta, A., & Silver, S. (1998). Molecular genetics: Silver as a biocide: Will resistance become a problem? *Nature Biotechnology, 16*, 888–888.

Hajipour, M. J., Fromm, K. M., Ashkarran, A. A., de Aberasturi, D. J., de Larramendi, I. R., Rojo, T., et al. (2012). Antibacterial properties of nanoparticles. *Trends in Biotechnology, 30*, 499–511.

Hall-Stoodley, L., & Stoodley, P. (2009). Evolving concepts in biofilm infections. *Cellular Microbiology, 11*, 1034–1043.

Harkes, G., Feijen, J., & Dankert, J. (1991). Adhesion of *Escherichia coli* on to a series of poly (methacrylates) differing in charge and hydrophobicity. *Biomaterials, 12*, 853–860.

Hazer, D. B., Sakar, M., Dere, Y., Altinkanat, G., Ziyal, M. I., & Hazer, B. (2016). Antimicrobial effect of polymer-based silver nanoparticle coated pedicle screws: Experimental research on biofilm inhibition in rabbits. *Spine, 41*, E323–E329.

He, H., Cai, R., Wang, Y., Tao, G., Guo, P., Zuo, H., et al. (2017). Preparation and characterization of silk sericin/pva blend film with silver nanoparticles for potential antimicrobial application. *International Journal of Biological Macromolecules, 104*, 457–464.

Hegemann, D., Hossain, M. M., & Balazs, D. J. (2007). Nanostructured plasma coatings to obtain multifunctional textile surfaces. *Progress in Organic Coating, 58*, 237–240.

Hermansson, M. (1999). The DLVO theory in microbial adhesion. *Colloids and Surfaces. B, Biointerfaces, 14*, 105–119.

Hetrick, E. M., & Schoenfisch, M. H. (2006). Reducing implant-related infections: Active release strategies. *Chemical Society Reviews, 35*, 780–789.

Hill, M., Baldwin, L., Slaughter, J., Walsh, W., & Weitkamp, J. (2010). A silver–alginate-coated dressing to reduce peripherally inserted central catheter (PICC) infections in NICU patients: A pilot randomized controlled trial. *Journal of Perinatology, 30*, 469.

Hirakura, T., Yasugi, K., Nemoto, T., Sato, M., Shimoboji, T., Aso, Y., et al. (2010). Hybrid hyaluronan hydrogel encapsulating nanogel as a protein nanocarrier: New system for sustained delivery of protein with a chaperone-like function. *Journal of Controlled Release, 142*, 483–489.

Hofmann, G. (1992). Biodegradable implants in orthopaedic surgery—A review on the state-of-the-art. *Clinical Materials, 10*, 75–80.

Høiby, N., Bjarnsholt, T., Givskov, M., Molin, S., & Ciofu, O. (2010). Antibiotic resistance of bacterial biofilms. *International Journal of Antimicrobial Agents, 35*, 322–332.

Huang, J., Murata, H., Koepsel, R. R., Russell, A. J., & Matyjaszewski, K. (2007). Antibacterial polypropylene via surface-initiated atom transfer radical polymerization. *Biomacromolecules, 8*, 1396–1399.

Huang, Z., Tian, J., Yu, B., Xu, Y., & Feng, Q. (2009). A bone-like nano-hydroxyapatite/collagen loaded injectable scaffold. *Biomedical Materials, 4*, 055005.

Huang, L., Dai, T., Xuan, Y., Tegos, G. P., & Hamblin, M. R. (2011). Synergistic combination of chitosan acetate with nanoparticle silver as a topical antimicrobial: Efficacy against bacterial burn infections. *Antimicrobial Agents and Chemotherapy, 55*, 3432–3438.

Huh, A. J., & Kwon, Y. J. (2011). "Nanoantibiotics": A new paradigm for treating infectious diseases using nanomaterials in the antibiotics resistant era. *Journal of Controlled Release, 156*, 128–145.
Hung, H.-S., & Hsu, S.-H. (2007). Biological performances of poly (ether) urethane–silver nanocomposites. *Nanotechnology, 18*, 475101.
Hunt, N. C. (2010). *An alginate hydrogel matrix for the localised delivery of a fibroblast/keratinocyte co-culture to expedite wound healing*. Birmingham: University of Birmingham.
Hussain, S., Hess, K., Gearhart, J., Geiss, K., & Schlager, J. (2005). In vitro toxicity of nanoparticles in BRL 3A rat liver cells. *Toxicology In Vitro, 19*, 975–983.
Iamphongsai, S., Eshraghi, Y., Totonchi, A., Midler, J., Abdul-Karim, F. W., & Guyuron, B. (2009). Effect of different suture materials on cartilage reshaping. *Aesthetic Surgery Journal, 29*, 93–97.
Ibrahim, H. M. (2015). Green synthesis and characterization of silver nanoparticles using banana peel extract and their antimicrobial activity against representative microorganisms. *Journal of Radiation Research and Applied Science, 8*, 265–275.
Iconaru, S., Chifiriuc, M., & Groza, A. (2017). Structural and antimicrobial evaluation of silver doped hydroxyapatite-polydimethylsiloxane thin layers. *Journal of Nanomaterials, 2017*, 7492515.
Ifuku, S., Tsukiyama, Y., Yukawa, T., Egusa, M., Kaminaka, H., Izawa, H., et al. (2015). Facile preparation of silver nanoparticles immobilized on chitin nanofiber surfaces to endow antifungal activities. *Carbohydrate Polymers, 117*, 813–817.
Ionescu, A., Brambilla, E., Travan, A., Marsich, E., Donati, I., Gobbi, P., et al. (2015). Silver–polysaccharide antimicrobial nanocomposite coating for methacrylic surfaces reduces *Streptococcus mutans* biofilm formation in vitro. *Journal of Dentistry, 43*, 1483–1490.
Ito, K., Saito, A., Fujie, T., Miyazaki, H., Kinoshita, M., Saitoh, D., et al. (2016). Development of a ubiquitously transferrable silver-nanoparticle-loaded polymer nanosheet as an antimicrobial coating. *Journal of Biomedical Materials Research Part B Applied Biomaterials, 104*, 585–593.
Jagur-Grodzinski, J. (1999). Biomedical application of functional polymers. *Reactive and Functional Polymers, 39*, 99–138.
Jagur-Grodzinski, J. (2006). Polymers for tissue engineering, medical devices, and regenerative medicine. Concise general review of recent studies. *Polymers for Advanced Technologies, 17*, 395–418.
Jain, J., Arora, S., Rajwade, J. M., Omray, P., Khandelwal, S., & Paknikar, K. M. (2009). Silver nanoparticles in therapeutics: Development of an antimicrobial gel formulation for topical use. *Molecular Pharmaceutics, 6*, 1388–1401.
Jennings, J. A., Pulgarin, D. A. V., Kunwar, D. L., Babu, J., Mishra, S., & Bumgardner, J. (2015). Bacterial inhibition by chitosan coatings loaded with silver-decorated calcium phosphate microspheres. *Thin Solid Films, 596*, 83–86.
Jia, Q., Shan, S., Jiang, L., Wang, Y., & Li, D. (2012). Synergistic antimicrobial effects of polyaniline combined with silver nanoparticles. *Journal of Applied Polymer Science, 125*, 3560–3566.
Jiang, S., & Teng, C. P. (2017). Fabrication of silver nanowires-loaded polydimethylsiloxane film with antimicrobial activities and cell compatibility. *Materials Science and Engineering C, Materials for Biological Applications, 70*, 1011–1017.
Kaetsu, I., Yoshida, M., & Yamada, A. (1980). Controlled slow release of chemotherapeutic drugs for cancer from matrices prepared by radiation polymerization at low temperatures. *Journal of Biomedical Materials Research. Part A, 14*, 185–197.
Kakkar, R., Madgula, K., Nehru, Y. S., & Kakkar, J. (2015). Polyvinyl alcohol-melamine formaldehyde films and coatings with silver nano particles as wound dressings in diabetic foot disease. *European Chemical Bulletin, 4*, 98–105.
Kanazawa, A., Ikeda, T., & Endo, T. (1993). Novel polycationic biocides: Synthesis and antibacterial activity of polymeric phosphonium salts. *Journal of Polymer Science: Part A, 31*, 335–343.
Kantouch, F., & El-Sayed, A. A. (2011). Acid dyeable and printable acrylic fabrics treated with cationic aqueous polyurethane. *Journal of Applied Polymer Science, 119*, 2595–2601.

Karthikeyan, K., Sekar, S., Devi, M. P., Inbasekaran, S., Lakshminarasaiah, C., & Sastry, T. (2011). Fabrication of novel biofibers by coating silk fibroin with chitosan impregnated with silver nanoparticles. *Journal of Materials Science. Materials in Medicine, 22*, 2721–2726.

Khwanmuang, P., Naparswad, C., Archakunakorn, S., Waicharoen, C., & Chitichotpanya, C. (2017a). Optimization of in situ synthesis of Ag/PU nanocomposites using response surface methodology for self-disinfecting coatings. *Progress in Organic Coating, 110*, 104–113.

Khwanmuang, P., Rotjanapan, P., Phuphuakrat, A., Srichatrapimuk, S., & Chitichotpanya, C. (2017b). In vitro assessment of Ag-TiO2/polyurethane nanocomposites for infection control using response surface methodology. *Reactive and Functional Polymers, 117*, 120–130.

Kim, S. H., Opdahl, A., Marmo, C., & Somorjai, G. A. (2002). AFM and SFG studies of pHEMA-based hydrogel contact lens surfaces in saline solution: Adhesion, friction, and the presence of non-crosslinked polymer chains at the surface. *Biomaterials, 23*, 1657–1666.

Kim, J. H., Park, H., & Seo, S. W. (2017). In situ synthesis of silver nanoparticles on the surface of PDMS with high antibacterial activity and biosafety toward an implantable medical device. *Nano Convergence, 4*, 33.

Knetsch, M. L., & Koole, L. H. (2011). New strategies in the development of antimicrobial coatings: The example of increasing usage of silver and silver nanoparticles. *Polymers, 3*, 340–366.

Komura, M., Komura, H., Kanamori, Y., Tanaka, Y., Suzuki, K., Sugiyama, M., et al. (2008). An animal model study for tissue-engineered trachea fabricated from a biodegradable scaffold using chondrocytes to augment repair of tracheal stenosis. *Journal of Pediatric Surgery, 43*, 2141–2146.

Kong, H., & Jang, J. (2008). Antibacterial properties of novel poly (methyl methacrylate) nanofiber containing silver nanoparticles. *Langmuir, 24*, 2051–2056.

Körner, E., Aguirre, M. H., Fortunato, G., Ritter, A., Rühe, J., & Hegemann, D. (2010). Formation and distribution of silver nanoparticles in a functional plasma polymer matrix and related Ag+ release properties. *Plasma Processes and Polymers, 7*, 619–625.

Krane, S. M. (2008). The importance of proline residues in the structure, stability and susceptibility to proteolytic degradation of collagens. *Amino Acids, 35*, 703.

Kucekova, Z., Kasparkova, V., Humpolicek, P., Sevcikova, P., & Stejskal, J. (2013). Antibacterial properties of polyaniline-silver films. *Chemical Papers, 67*, 1103–1108.

Kumar, M. N. R. (2000). A review of chitin and chitosan applications. *Reactive and Functional Polymers, 46*, 1–27.

Kumar, A., Vemula, P. K., Ajayan, P. M., & John, G. (2008). Silver-nanoparticle-embedded antimicrobial paints based on vegetable oil. *Nature Materials, 7*, 236–241.

Kumbar, S., Laurencin, C., & Deng, M. (2014). *Natural and synthetic biomedical polymers*. Amsterdam: Newnes.

Kundu, S. C., Dash, B. C., Dash, R., & Kaplan, D. L. (2008). Natural protective glue protein, sericin bioengineered by silkworms: Potential for biomedical and biotechnological applications. *Progress in Polymer Science, 33*, 998–1012.

Lee, H., Mok, H., Lee, S., Oh, Y.-K., & Park, T. G. (2007). Target-specific intracellular delivery of siRNA using degradable hyaluronic acid nanogels. *Journal of Controlled Release, 119*, 245–252.

Lee, T. H., Jang, B. S., Jung, M. K., Pack, C. G., Choi, J.-H., & Park, D. H. (2016). Fabrication of a silver particle-integrated silicone polymer-covered metal stent against sludge and biofilm formation and stent-induced tissue inflammation. *Scientific Reports, 6*, 35446.

Lendlein, A. (2010). Polymers in biomedicine. *Macromolecular Bioscience, 10*, 993–997.

Li, X., & Lenhart, J. J. (2012). Aggregation and dissolution of silver nanoparticles in natural surface water. *Environmental Science and Technology, 46*, 5378–5386.

Li, W., Zhou, J., Gu, J. S., & Yu, H. Y. (2010). Fouling control in a submerged membrane-bioreactor by the membrane surface modification. *Journal of Applied Polymer Science, 115*, 2302–2309.

Li, L., Jones, P. M., & Hsia, Y.-T. (2011). Characterization of a nanometer-thick sputtered polytetrafluoroethylene film. *Applied Surface Science, 257*, 4478–4485.

Li, P., Zhang, X., Xu, R., Wang, W., Liu, X., Yeung, K. W., et al. (2013). Electrochemically deposited chitosan/Ag complex coatings on biomedical NiTi alloy for antibacterial application. *Surface and Coating Technology, 232*, 370–375.

Li, W., Xu, D., Hu, Y., Cai, K., & Lin, Y. (2014). Surface modification of titanium substrates with silver nanoparticles embedded sulfhydrylated chitosan/gelatin polyelectrolyte multilayer films for antibacterial application. *Journal of Materials Science. Materials in Medicine, 25*, 1435–1448.

Li, L., Wang, Y., & Zhu, Y. (2018). Facile preparation and good performance of nano-Ag/metallocene polyethylene antibacterial coatings. *Journal of Coating Technology and Research, 15*, 593–602.

Liao, J., Anchun, M., Zhu, Z., & Quan, Y. (2010). Antibacterial titanium plate deposited by silver nanoparticles exhibits cell compatibility. *International Journal of Nanomedicine, 5*, 337.

Lin, J. J., Lin, W. C., Li, S. D., Lin, C. Y., & Hsu, S. H. (2013). Evaluation of the antibacterial activity and biocompatibility for silver nanoparticles immobilized on nano silicate platelets. *ACS Applied Materials and Interfaces, 5*, 433–443.

Liu, Y., Zheng, Z., Zara, J. N., Hsu, C., Soofer, D. E., Lee, K. S., et al. (2012). The antimicrobial and osteoinductive properties of silver nanoparticle/poly (DL-lactic-co-glycolic acid)-coated stainless steel. *Biomaterials, 33*, 8745–8756.

Liu, G., Li, K., Luo, Q., Wang, H., & Zhang, Z. (2017). PEGylated chitosan protected silver nanoparticles as water-borne coating for leather with antibacterial property. *Journal of Colloid and Interface Science, 490*, 642–651.

Lloyd, L., Kennedy, J., Methacanon, P., Paterson, M., & Knill, C. (1998). Carbohydrate polymers as wound management aids. *Carbohydrate Polymers, 37*, 315–322.

Loo, C.-Y., Young, P. M., Lee, W.-H., Cavaliere, R., Whitchurch, C. B., & Rohanizadeh, R. (2014). Non-cytotoxic silver nanoparticle-polyvinyl alcohol hydrogels with anti-biofilm activity: Designed as coatings for endotracheal tube materials. *Biofouling, 30*, 773–788.

Lu, W., Senapati, D., Wang, S., Tovmachenko, O., Singh, A. K., Yu, H., et al. (2010). Effect of surface coating on the toxicity of silver nanomaterials on human skin keratinocytes. *Chemical Physics Letters, 487*, 92–96.

Lynch, A. S., & Robertson, G. T. (2008). Bacterial and fungal biofilm infections. *Annual Review of Medicine, 59*, 415–428.

Lyutakov, O., Kalachyova, Y., Solovyev, A., Vytykacova, S., Svanda, J., Siegel, J., et al. (2015). One-step preparation of antimicrobial silver nanoparticles in polymer matrix. *Journal of Nanoparticle Research, 17*, 120.

Ma, P. X. (2004). Scaffolds for tissue fabrication. *Materials Today, 7*, 30–40.

Ma, K., Gong, L., Cai, X., Huang, P., Cai, J., Huang, D., et al. (2017). A green single-step procedure to synthesize ag-containing nanocomposite coatings with low cytotoxicity and efficient antibacterial properties. *International Journal of Nanomedicine, 12*, 3665.

Maitz, M. F. (2015). Applications of synthetic polymers in clinical medicine. *Biosurface and Biotribology, 1*, 161–176.

Marsich, E., Travan, A., Donati, I., Turco, G., Kulkova, J., Moritz, N., et al. (2013). Biological responses of silver-coated thermosets: An in vitro and in vivo study. *Acta Biomaterialia, 9*, 5088–5099.

Matsuno, T., Nakamura, T., Kuremoto, K.-I., Notazawa, S., Nakahara, T., Hashimoto, Y., et al. (2006). Development of β-tricalcium phosphate/collagen sponge composite for bone regeneration. *Dental Materials Journal, 25*, 138–144.

McArthur, S. L., McLean, K. M., Kingshott, P., St John, H. A., Chatelier, R. C., & Griesser, H. J. (2000). Effect of polysaccharide structure on protein adsorption. *Colloids and Surfaces. B, Biointerfaces, 17*, 37–48.

Mejia, M., Restrepo, G., Marin, J., Sanjines, R., Pulgarín, C., Mielczarski, E., et al. (2010). Magnetron-sputtered Ag surfaces. New evidence for the nature of the Ag ions intervening in bacterial inactivation. *ACS Applied Materials and Interfaces, 2*, 230–235.

Meyers, R. A. (1995). *Molecular biology and biotechnology: A comprehensive desk reference*. New York: Wiley.
Middleton, J. C., & Tipton, A. J. (2000). Synthetic biodegradable polymers as orthopedic devices. *Biomaterials, 21*, 2335–2346.
Mishra, S. K., Teotia, A. K., Kumar, A., & Kannan, S. (2017). Mechanically tuned nanocomposite coating on titanium metal with integrated properties of biofilm inhibition, cell proliferation, and sustained drug delivery. *Nanomedicine: Nanotechnology, Biology and Medicine, 13*, 23–35.
Mittal, V. (2013). *Polymer nanocomposite coatings*. Boca Raton: CRC Press, Taylor and Francis.
Moghimi, S. M., Hunter, A. C., & Murray, J. C. (2001). Long-circulating and target-specific nanoparticles: Theory to practice. *Pharmacological Reviews, 53*, 283–318.
Moodley, J. S., Krishna, S. B. N., Pillay, K., & Govender, P. (2018). Green synthesis of silver nanoparticles from *Moringa oleifera* leaf extracts and its antimicrobial potential. *Advances in Natural Sciences: Nanoscience and Nanotechnology, 9*, 015011.
Mori, Y., Ono, T., Miyahira, Y., Nguyen, V. Q., Matsui, T., & Ishihara, M. (2013). Antiviral activity of silver nanoparticle/chitosan composites against H1N1 influenza A virus. *Nanoscale Research Letters, 8*, 93.
Morones, J. R., Elechiguerra, J. L., Camacho, A., Holt, K., Kouri, J. B., Ramírez, J. T., et al. (2005). The bactericidal effect of silver nanoparticles. *Nanotechnology, 16*, 2346.
Mountziaris, P. M., Tzouanas, S. N., & Mikos, A. G. (2010). Dose effect of tumor necrosis factor-α on in vitro osteogenic differentiation of mesenchymal stem cells on biodegradable polymeric microfiber scaffolds. *Biomaterials, 31*, 1666–1675.
Mukherji, S., Ruparelia, J., & Agnihotri, S. (2012). Antimicrobial activity of silver and copper nanoparticles: Variation in sensitivity across various strains of bacteria and fungi. In N. Cioffi & M. Rai (Eds.), *Nano-antimicrobials: Progress and prospects* (pp. 225–251). Berlin: Springer.
Muñoz-Bonilla, A., & Fernández-García, M. (2012). Polymeric materials with antimicrobial activity. *Progress in Polymer Science, 37*, 281–339.
Nair, L. S., & Laurencin, C. T. (2007). Biodegradable polymers as biomaterials. *Progress in Polymer Science, 32*, 762–798.
Nanlin, S., Xuemei, G., Hemin, J., Jun, G., Chao, S., & Ke, Y. (2009). Antibacterial effect of the conducting polyaniline. *Journal of Materials Science and Technology, 22*, 289–290.
Nanno, K., Sugiyasu, K., Daimon, T., Yoshikawa, H., & Myoui, A. (2009). Synthetic alginate is a carrier of OP-1 for bone induction. *Clinical Orthopaedics and Related Research, 467*, 3149.
Nganga, S., Travan, A., Marsich, E., Donati, I., Söderling, E., Moritz, N., et al. (2013). In vitro antimicrobial properties of silver–polysaccharide coatings on porous fiber-reinforced composites for bone implants. *Journal of Materials Science. Materials in Medicine, 24*, 2775–2785.
Nhi, T. T., Khon, H. C., Hoai, N. T. T., Bao, B. C., Quyen, T. N., Van Toi, V., et al. (2016). Fabrication of electrospun polycaprolactone coated withchitosan-silver nanoparticles membranes for wound dressing applications. *Journal of Materials Science. Materials in Medicine, 27*, 156.
Nohr, R. S., & Gavin Macdonald, J. (1994). New biomaterials through surface segregation phenomenon: New quaternary ammonium compounds as antibacterial agents. *Journal of Biomaterials Science Polymer Edition, 5*, 607–619.
Nune, S. K., Chanda, N., Shukla, R., Katti, K., Kulkarni, R. R., Thilakavathi, S., et al. (2009). Green nanotechnology from tea: Phytochemicals in tea as building blocks for production of biocompatible gold nanoparticles. *Journal of Materials Chemistry, 19*, 2912–2920.
Ocsoy, I., Didar, T., Sumeyye, M., Cagla, C., Ahmet, K., & Funda, U. (2018). Biomolecules incorporated metallic nanoparticles synthesis and their biomedical applications. *Materials Letters, 212*, 45–50.
Oktay, B., & Kayaman-Apohan, N. (2013). Polydimethylsiloxane (PDMS)-based antibacterial organic–inorganic hybrid coatings. *Journal of Coating Technology and Research, 10*, 785–798.
Orive, G., De Castro, M., Kong, H.-J., Hernández, R. M., Ponce, S., Mooney, D. J., et al. (2009). Bioactive cell-hydrogel microcapsules for cell-based drug delivery. *Journal of Controlled Release, 135*, 203–210.

Orlowski, P., Tomaszewska, E., Gniadek, M., Baska, P., Nowakowska, J., Sokolowska, J., et al. (2014). Tannic acid modified silver nanoparticles show antiviral activity in herpes simplex virus type 2 infection. *PLoS One, 9*, e104113.

Pal, S., Tak, Y. K., & Song, J. M. (2007). Does the antibacterial activity of silver nanoparticles depend on the shape of the nanoparticle? A study of the gram-negative bacterium *Escherichia coli. Applied and Environmental Microbiology, 73*, 1712–1720.

Pan, X., Redding, J. E., Wiley, P. A., Wen, L., McConnell, J. S., & Zhang, B. (2010). Mutagenicity evaluation of metal oxide nanoparticles by the bacterial reverse mutation assay. *Chemosphere, 79*, 113–116.

Panáček, A., Kvitek, L., Prucek, R., Kolář, M., Večeřová, R., Pizúrová, N., et al. (2006). Silver colloid nanoparticles: Synthesis, characterization, and their antibacterial activity. *The Journal of Physical Chemistry. B, 110*, 16248–16253.

Pang, X., & Zhitomirsky, I. (2008). Electrodeposition of hydroxyapatite–silver–chitosan nanocomposite coatings. *Surface and Coating Technology, 202*, 3815–3821.

Pangas, S. A., Saudye, H., Shea, L. D., & Woodruff, T. K. (2003). Novel approach for the three-dimensional culture of granulosa cell–oocyte complexes. *Tissue Engineering, 9*, 1013–1021.

Panyam, J., & Labhasetwar, V. (2003). Biodegradable nanoparticles for drug and gene delivery to cells and tissue. *Advanced Drug Delivery Reviews, 55*, 329–347.

Pardini, O., & Amalvy, J. (2008). FTIR, 1H-NMR spectra, and thermal characterization of water-based polyurethane/acrylic hybrids. *Journal of Applied Polymer Science, 107*, 1207–1214.

Park, K. D., Kim, Y. S., Han, D. K., Kim, Y. H., Lee, E. H. B., Suh, H., et al. (1998). Bacterial adhesion on PEG modified polyurethane surfaces. *Biomaterials, 19*, 851–859.

Park, G. E., Pattison, M. A., Park, K., & Webster, T. J. (2005). Accelerated chondrocyte functions on NaOH-treated PLGA scaffolds. *Biomaterials, 26*, 3075–3082.

Paul, A., Kaverina, E., & Vasiliev, A. (2015). Synthesis of silver/polymer nanocomposites by surface coating using carbodiimide method. *Colloids and Surfaces A: Physicochemical and Engineering Aspects, 482*, 44–49.

Pavithra, D., & Doble, M. (2008). Biofilm formation, bacterial adhesion and host response on polymeric implants—issues and prevention. *Biomedical Materials, 3*, 034003.

Pelgrift, R. Y., & Friedman, A. J. (2013). Nanotechnology as a therapeutic tool to combat microbial resistance. *Advanced Drug Delivery Reviews, 65*, 1803–1815.

Petrochenko, P. E., Zheng, J., Casey, B. J., Bayati, M. R., Narayan, R. J., & Goering, P. L. (2017). Nanosilver-PMMA composite coating optimized to provide robust antibacterial efficacy while minimizing human bone marrow stromal cell toxicity. *Toxicology In Vitro, 44*, 248–255.

Pishbin, F., Simchi, A., Ryan, M., & Boccaccini, A. (2010). A study of the electrophoretic deposition of bioglass® suspensions using the Taguchi experimental design approach. *Journal of the European Ceramic Society, 30*, 2963–2970.

Pishbin, F., Mourino, V., Gilchrist, J. B., McComb, D. W., Kreppel, S., Salih, V., et al. (2013). Single-step electrochemical deposition of antimicrobial orthopaedic coatings based on a bioactive glass/chitosan/nano-silver composite system. *Acta Biomaterialia, 9*, 7469–7479.

Poelaert, J., Depuydt, P., De Wolf, A., Van de Velde, S., Herck, I., & Blot, S. (2008). Polyurethane cuffed endotracheal tubes to prevent early postoperative pneumonia after cardiac surgery: A pilot study. *The Journal of Thoracic and Cardiovascular Surgery, 135*, 771–776.

Polívková, M., Hubáček, T., Staszek, M., Švorčík, V., & Siegel, J. (2017). Antimicrobial treatment of polymeric medical devices by silver nanomaterials and related technology. *International Journal of Molecular Sciences, 18*, 419.

Prabhakar, P. K., Raj, S., Anuradha, P., Sawant, S. N., & Doble, M. (2011). Biocompatibility studies on polyaniline and polyaniline–silver nanoparticle coated polyurethane composite. *Colloids and Surfaces. B, Biointerfaces, 86*, 146–153.

Pugazhendhi, A., Prabakar, D., Jacob, J. M., Karuppusamy, I., & Saratale, R. G. (2018). Synthesis and characterization of silver nanoparticles using Gelidium amansii and its antimicrobial property against various pathogenic bacteria. *Microbial Pathogenesis, 114*, 41–45.

Rai, M., Yadav, A., & Gade, A. (2009). Silver nanoparticles as a new generation of antimicrobials. *Biotechnology Advances, 27*, 76–83.

Rajasekaran, E., Jency, S., & Panneerselvam, K. (2011). Carbon profile of commercially important sericin proteins of silkworm, *bombyx mori*. *Journal of Advanced Bioinformatics Applications and Research ISSN, 2*, 173–176.

Roosjen, A., van der Mei, H. C., Busscher, H. J., & Norde, W. (2004). Microbial adhesion to poly (ethylene oxide) brushes: Influence of polymer chain length and temperature. *Langmuir, 20*, 10949–10955.

Sadu, R. B., Chen, D. H., Kucknoor, A. S., Guo, Z., & Gomes, A. J. (2014). Silver-doped TiO 2/ polyurethane nanocomposites for antibacterial textile coating. *BioNanoScience, 4*, 136–148.

Saez, S., Fasciani, C., Stamplecoskie, K. G., Gagnon, L. B.-P., Mah, T.-F., Marin, M. L., et al. (2015). Photochemical synthesis of biocompatible and antibacterial silver nanoparticles embedded within polyurethane polymers. *Photochemical and Photobiological Sciences, 14*, 661–664.

Salwiczek, M., Qu, Y., Gardiner, J., Strugnell, R. A., Lithgow, T., McLean, K. M., et al. (2014). Emerging rules for effective antimicrobial coatings. *Trends in Biotechnology, 32*, 82–90.

Samuel, U., & Guggenbichler, J. (2004). Prevention of catheter-related infections: The potential of a new nano-silver impregnated catheter. *International Journal of Antimicrobial Agents, 23*, 75–78.

Sanyasi, S., Majhi, R. K., Kumar, S., Mishra, M., Ghosh, A., Suar, M., et al. (2016). Polysaccharide-capped silver nanoparticles inhibit biofilm formation and eliminate multi-drug-resistant bacteria by disrupting bacterial cytoskeleton with reduced cytotoxicity towards mammalian cells. *Scientific Reports, 6*, 24929.

Sardella, E., Favia, P., Gristina, R., Nardulli, M., & d'Agostino, R. (2006). Plasma-aided micro- and nanopatterning processes for biomedical applications. *Plasma Processes and Polymers, 3*, 456–469.

Sawant, S. N., Selvaraj, V., Prabhawathi, V., & Doble, M. (2013). Antibiofilm properties of silver and gold incorporated PU, PCLm, PC and PMMA nanocomposites under two shear conditions. *PLoS One, 8*, e63311.

Seshadri, D. T., & Bhat, N. V. (2005). Use of polyaniline as an antimicrobial agent in textiles. *Indian Journal of Fibre and Textile Research, 30*, 204–206.

Shameli, K., Ahmad, M. B., Yunus, W. M. Z. W., Ibrahim, N. A., Rahman, R. A., Jokar, M., et al. (2010). Silver/poly (lactic acid) nanocomposites: Preparation, characterization, and antibacterial activity. *International Journal of Nanomedicine, 5*, 573.

Shantiaee, Y., Dianat, S. O., Mohammad Khani, H., & Akbarzadeh Baghban, A. (2011). Cytotoxicity comparison of nanosilver coated gutta-percha with Guttaflow and normal gutta-percha on L929 fibroblast with MTT assay. *Shahid Beheshti University Dental Journal, 29*, 62–68.

Shi, Z., Zhou, H., Qing, X., Dai, T., & Lu, Y. (2012). Facile fabrication and characterization of poly (tetrafluoroethylene)@ polypyrrole/nano-silver composite membranes with conducting and antibacterial property. *Applied Surface Science, 258*, 6359–6365.

Shukla, R., Nune, S. K., Chanda, N., Katti, K., Mekapothula, S., Kulkarni, R. R., et al. (2008). Soybeans as a phytochemical reservoir for the production and stabilization of biocompatible gold nanoparticles. *Small, 4*, 1425–1436.

Shukla, R., Chanda, N., Zambre, A., Upendran, A., Katti, K., Kulkarni, R. R., et al. (2012). Laminin receptor specific therapeutic gold nanoparticles (^{198}AuNP-EGCg) show efficacy in treating prostate cancer. *Proceedings of the National Academy of Sciences of the United States of America, 109*, 12426–12431.

Siddiqui, N., Bhardwaj, A., Hada, R., Yadav, V. S., & Goyal, D. (2018). Synthesis, characterization and antimicrobial study of poly (methyl methacrylate)/Ag nanocomposites. *Vacuum, 153*, 6–11.

Silvestry-Rodriguez, N., Sicairos-Ruelas, E. E., Gerba, C. P., & Bright, K. R. (2007). *Silver as a disinfectant. Reviews of environmental contamination and toxicology* (pp. 23–45). New York: Springer.

Smith, R. S., Zhang, Z., Bouchard, M., Li, J., Lapp, H. S., Brotske, G. R., et al. (2012). Vascular catheters with a nonleaching poly-sulfobetaine surface modification reduce thrombus formation and microbial attachment. *Science Translational Medicine, 4*, 153ra132–153ra132.

Smolinske, S. C. (1992). *CRC handbook of food, drug, and cosmetic excipients*. Boca Raton: CRC press.

Son, H. Y., Ryu, J. H., Lee, H., & Nam, Y. S. (2013). Silver-polydopamine hybrid coatings of electrospun poly (vinyl alcohol) nanofibers. *Macromolecular Materials and Engineering, 298*, 547–554.

Sondi, I., & Salopek-Sondi, B. (2004). Silver nanoparticles as antimicrobial agent: A case study on *E. coli* as a model for gram-negative bacteria. *Journal of Colloid and Interface Science, 275*, 177–182.

Sotiriou, G. A., & Pratsinis, S. E. (2010). Antibacterial activity of nanosilver ions and particles. *Environmental Science and Technology, 44*, 5649–5654.

Sousa, C., Teixeira, P., & Oliveira, R. (2009). Influence of surface properties on the adhesion of *Staphylococcus epidermidis* to acrylic and silicone. *International Journal of Biomaterials, 2009*, 718017.

Stejskal, J., Sapurina, I., & Trchová, M. (2010). Polyaniline nanostructures and the role of aniline oligomers in their formation. *Progress in Polymer Science, 35*, 1420–1481.

Sun, L., Singh, A. K., Vig, K., Pillai, S. R., & Singh, S. R. (2008a). Silver nanoparticles inhibit replication of respiratory syncytial virus. *Journal of Biomedical Nanotechnology, 4*, 149–158.

Sun, B., Ranganathan, B., & Feng, S.-S. (2008b). Multifunctional poly (D, L-lactide-co-glycolide)/montmorillonite (PLGA/MMT) nanoparticles decorated by Trastuzumab for targeted chemotherapy of breast cancer. *Biomaterials, 29*, 475–486.

Sundrarajan, M., & Rukmani, A. (2013). Durable antibacterial finishing on cotton by impregnation of limonene microcapsules. *Advance Chemistry Letters, 1*, 40–43.

Sung, J. H., Ji, J. H., Yoon, J. U., Kim, D. S., Song, M. Y., Jeong, J., et al. (2008). Lung function changes in Sprague-Dawley rats after prolonged inhalation exposure to silver nanoparticles. *Inhalation Toxicology, 20*, 567–574.

Sussman, E. M., Casey, B. J., Dutta, D., & Dair, B. J. (2015). Different cytotoxicity responses to antimicrobial nanosilver coatings when comparing extract-based and direct-contact assays. *Journal of Applied Toxicology, 35*, 631–639.

Tanaka, M., & Mochizuki, A. (2010). Clarification of the blood compatibility mechanism by controlling the water structure at the blood–poly (meth) acrylate interface. *Journal of Biomaterials Science. Polymer Edition, 21*, 1849–1863.

Tang, C., Sun, W., Lu, J., & Yan, W. (2014). Role of the anions in the hydrothermally formed silver nanowires and their antibacterial property. *Journal of Colloid and Interface Science, 416*, 86–94.

Thallinger, B., Prasetyo, E. N., Nyanhongo, G. S., & Guebitz, G. M. (2013). Antimicrobial enzymes: An emerging strategy to fight microbes and microbial biofilms. *Biotechnology Journal, 8*, 97–109.

Thirumurugan, G., Shaheedha, S., & Dhanaraju, M. (2009). In vitro evaluation of antibacterial activity of silver nanoparticles synthesised by using *Phytophthora infestans*. *International Journal of ChemTech Research, 1*, 714–716.

Thorsteinsson, T., Loftsson, T., & Masson, M. (2003). Soft antibacterial agents. *Current Medicinal Chemistry, 10*, 1129–1136.

Tran, N., Kelley, M. N., Tran, P. A., Garcia, D. R., Jarrell, J. D., Hayda, R. A., et al. (2015). Silver doped titanium oxide–PDMS hybrid coating inhibits staphylococcus aureus and *Staphylococcus epidermidis* growth on PEEK. *Materials Science and Engineering C, Materials for Biological Applications, 49*, 201–209.

Trefry, J. C., & Wooley, D. P. (2012). Rapid assessment of antiviral activity and cytotoxicity of silver nanoparticles using a novel application of the tetrazolium-based colorimetric assay. *Journal of Virological Methods, 183*, 19–24.

Triandafillu, K., Balazs, D., Aronsson, B.-O., Descouts, P., Quoc, P. T., Van Delden, C., et al. (2003). Adhesion of *Pseudomonas aeruginosa* strains to untreated and oxygen-plasma treated poly(vinyl chloride) (PVC) from endotracheal intubation devices. *Biomaterials, 24*, 1507–1518.

Uttayarat, P., Perets, A., Li, M., Pimton, P., Stachelek, S. J., Alferiev, I., et al. (2010). Micropatterning of three-dimensional electrospun polyurethane vascular grafts. *Acta Biomaterialia, 6*, 4229–4237.

Vishwasrao, C., Momin, B., & Ananthanarayan, L. (2018). Green synthesis of silver nanoparticles using sapota fruit waste and evaluation of their antimicrobial activity. *Waste and Biomass Valorization*, 1–11.

Von Eiff, C., Jansen, B., Kohnen, W., & Becker, K. (2005). Infections associated with medical devices. *Drugs, 65*, 179–214.

Wang, Y., Challa, P., Epstein, D. L., & Yuan, F. (2004). Controlled release of ethacrynic acid from poly (lactide-co-glycolide) films for glaucoma treatment. *Biomaterials, 25*, 4279–4285.

Wang, S., Lawson, R., Ray, P. C., & Yu, H. (2011). Toxic effects of gold nanoparticles on Salmonella typhimurium bacteria. *Toxicology and Industrial Health, 27*, 547–554.

Wang, B.-L., Liu, X.-S., Ji, Y., Ren, K.-F., & Ji, J. (2012). Fast and long-acting antibacterial properties of chitosan-Ag/polyvinylpyrrolidone nanocomposite films. *Carbohydrate Polymers, 90*, 8–15.

Wang, R., Neoh, K. G., Kang, E. T., Tambyah, P. A., & Chiong, E. (2015a). Antifouling coating with controllable and sustained silver release for long-term inhibition of infection and encrustation in urinary catheters. *Journal of Biomedical Materials Research Part B Applied Biomaterials, 103*, 519–528.

Wang, Y., Guo, X., Pan, R., Han, D., Chen, T., Geng, Z., et al. (2015b). Electrodeposition of chitosan/gelatin/nanosilver: A new method for constructing biopolymer/nanoparticle composite films with conductivity and antibacterial activity. *Materials Science and Engineering C, Materials for Biological Applications, 53*, 222–228.

Wattanodorn, Y., Jenkan, R., Atorngitjawat, P., & Wirasate, S. (2014). Antibacterial anionic waterborne polyurethanes/Ag nanocomposites with enhanced mechanical properties. *Polymer Testing, 40*, 163–169.

Wypij, M., Czarnecka, J., Świecimska, M., Dahm, H., Rai, M., & Golinska, P. (2018). Synthesis, characterization and evaluation of antimicrobial and cytotoxic activities of biogenic silver nanoparticles synthesized from *Streptomyces xinghaiensis* OF1 strain. *World Journal of Microbiology and Biotechnology, 34*, 23.

Xu, X., Yang, Q., Wang, Y., Yu, H., Chen, X., & Jing, X. (2006). Biodegradable electrospun poly (L-lactide) fibers containing antibacterial silver nanoparticles. *European Polymer Journal, 42*, 2081–2087.

Xu, H. H., Weir, M. D., & Simon, C. G. (2008). Injectable and strong nano-apatite scaffolds for cell/growth factor delivery and bone regeneration. *Dental Materials, 24*, 1212–1222.

Xu, D., Su, Y., Zhao, L., Meng, F., Liu, C., Guan, Y., et al. (2017). Antibacterial and antifouling properties of a polyurethane surface modified with perfluoroalkyl and silver nanoparticles. *Journal of Biomedical Materials Research. Part A, 105*, 531–538.

Yan, Y., Zhang, X., Li, C., Huang, Y., Ding, Q., & Pang, X. (2015). Preparation and characterization of chitosan-silver/hydroxyapatite composite coatings onTiO_2 nanotube for biomedical applications. *Applied Surface Science, 332*, 62–69.

Yang, X. X., Li, C. M., & Huang, C. Z. (2016). Curcumin modified silver nanoparticles for highly efficient inhibition of respiratory syncytial virus infection. *Nanoscale, 8*, 3040–3048.

Yilgör, İ., & McGrath, J. E. (1988). *Polysiloxane containing copolymers: A survey of recent developments. Polysiloxane copolymers/anionic polymerization* (pp. 1–86). Berlin: Springer.

Yoksan, R., & Chirachanchai, S. (2010). Silver nanoparticle-loaded chitosan–starch based films: Fabrication and evaluation of tensile, barrier and antimicrobial properties. *Materials Science and Engineering C, Materials for Biological Applications, 30*, 891–897.

Yu, W.-Z., Zhang, Y., Liu, X., Xiang, Y., Li, Z., & Wu, S. (2018). Synergistic antibacterial activity of multi components in lysozyme/chitosan/silver/hydroxyapatite hybrid coating. *Materials and Design, 139*, 351–362.
Yuan, W., Fu, J., Su, K., & Ji, J. (2010). Self-assembled chitosan/heparin multilayer film as a novel template for in situ synthesis of silver nanoparticles. *Colloids and Surfaces. B, Biointerfaces, 76*, 549–555.
Zahran, M., Ahmed, H. B., & El-Rafie, M. (2014). Surface modification of cotton fabrics for antibacterial application by coating with AgNPs–alginate composite. *Carbohydrate Polymers, 108*, 145–152.
Zaporojtchenko, V., Podschun, R., Schürmann, U., Kulkarni, A., & Faupel, F. (2006). Physico-chemical and antimicrobial properties of co-sputtered Ag–Au/PTFE nanocomposite coatings. *Nanotechnology, 17*, 4904.
Zezin, A. B., Rogacheva, V. B., Feldman, V. I., Afanasiev, P., & Zezin, A. A. (2010). From triple interpolyelectrolyte-metal complexes to polymer-metal nanocomposites. *Advances in Colloid and Interface, 158*, 84–93.
Zhang, X., Li, Z., Yuan, X., Cui, Z., Bao, H., Li, X., et al. (2013). Cytotoxicity and antibacterial property of titanium alloy coated with silver nanoparticle-containing polyelectrolyte multilayer. *Materials Science and Engineering C, Materials for Biological Applications, 33*, 2816–2820.
Zhang, B. G., Myers, D. E., Wallace, G. G., Brandt, M., & Choong, P. F. (2014). Bioactive coatings for orthopaedic implants – recent trends in development of implant coatings. *International Journal of Molecular Sciences, 15*, 11878–11921.
Zhang, X., Zhu, M., Wang, W., & Yu, D. (2018). Silver/waterborne polyurethane-acrylate's antibacterial coating on cotton fabric based on click reaction via ultraviolet radiation. *Progress in Organic Coating, 120*, 10–18.
Zhao, B., & Brittain, W. J. (2000). Polymer brushes: Surface-immobilized macromolecules. *Progress in Polymer Science, 25*, 677–710.
Zhao, Q., Liu, Y., & Wang, C. (2005). Development and evaluation of electroless Ag-PTFE composite coatings with anti-microbial and anti-corrosion properties. *Applied Surface Science, 252*, 1620–1627.
Zhao, C., Deng, B., Chen, G., Lei, B., Hua, H., Peng, H., et al. (2016). Large-area chemical vapor deposition-grown monolayer graphene-wrapped silver nanowires for broad-spectrum and robust antimicrobial coating. *Nano Research, 9*, 963–973.
Zheng, Y., Cai, C., Zhang, F., Monty, J., Linhardt, R. J., & Simmons, T. J. (2016). Can natural fibers be a silver bullet? Antibacterial cellulose fibers through the covalent bonding of silver nanoparticles to electrospun fibers. *Nanotechnology, 27*, 055102.
Zhou, N.-1., Liu, Y., Li, L., Meng, N., Huang, Y.-X., Zhang, J., et al. (2007). A new nanocomposite biomedical material of polymer/clay–Cts–Ag nanocomposites. *Current Applied Physics, 7*, e58–e62.
Zhou, X., Zhang, T., Guo, D., & Gu, N. (2014a). A facile preparation of poly (ethylene oxide)-modified medical polyurethane to improve hemocompatibility. *Colloids and Surfaces A: Physicochemical and Engineering Aspects, 441*, 34–42.
Zhou, B., Li, Y., Deng, H., Hu, Y., & Li, B. (2014b). Antibacterial multilayer films fabricated by layer-by-layer immobilizing lysozyme and gold nanoparticles on nanofibers. *Colloids and Surfaces. B, Biointerfaces, 116*, 432–438.
Zou, X., Deng, P., Zhou, C., Hou, Y., Chen, R., Liang, F., et al. (2017). Preparation of a novel antibacterial chitosan-poly (ethylene glycol) cryogel/silver nanoparticles composites. *Journal of Biomaterials Science, Polymer Edition, 28*, 1324–1337.
Zuo, Y., Yang, F., Wolke, J. G., Li, Y., & Jansen, J. A. (2010). Incorporation of biodegradable electrospun fibers into calcium phosphate cement for bone regeneration. *Acta Biomaterialia, 6*, 1238–1247.

Recent Trends of Nano-material as Antimicrobial Agents

Taru Aggarwal, Ridhima Wadhwa, Noopur Thapliyal, Riya Gupta, Philip Michael Hansbro, Kamal Dua, and Pawan Kumar Maurya

Abstract Nanomaterial has been employed as an alternative to antibiotics, diagnostic tools and delivery of therapeutics. In particular, nanomaterial has grabbed the attention of researchers due to their antimicrobial properties due to the emergence of multi-drug resistance of several micro-organisms. The present chapter highlights the antimicrobial nanomaterials with their mechanism of action along with their broad spectrum applications such as silver nanomaterial is antimicrobial in nature and is effective in drug delivery. Metallic, non-metallic and natural/ biodegradable

---

Authors Taru Aggarwal and Ridhima Wadhwa have been equally contributed to this chapter.

T. Aggarwal
ICAR-National Research Centre on Plant Biotechnology,
Lal-Bahadur Shastri Centre, IARI, New Delhi, India

R. Wadhwa
Faculty of Life Science and Biotechnology, South Asian University,
New Delhi, India

N. Thapliyal · R. Gupta
Amity Institute of Biotechnology, Amity University, Noida, Uttar Pradesh India

P. M. Hansbro
Priority Research Centre for Healthy Lungs, University of Newcastle & Hunter Medical Research Institute, Newcastle, NSW, Australia

Centre for Inflammation, Centenary Institute, Sydney, NSW, Australia

Faculty of Science, University of Technology Sydney, Ultimo, NSW, Australia

K. Dua
Priority Research Centre for Healthy Lungs, University of Newcastle & Hunter Medical Research Institute, Newcastle, NSW, Australia

Centre for Inflammation, Centenary Institute, Sydney, NSW, Australia

Discipline of Pharmacy, Graduate School of Health, University of Technology Sydney, Ultimo, NSW, Australia

P. K. Maurya (✉)
Department of Biochemistry, Central University of Haryana,
Mahendergarh, Haryana, India
e-mail: pkmaurya@cuh.ac.in

S. Singh, P. K. Maurya (eds.), *Nanotechnology in Modern Animal Biotechnology*, https://doi.org/10.1007/978-981-13-6004-6_5

nanomaterials have been discussed as potential antimicrobial and their mode of action. The mechanism of antimicrobial nanomaterial is poorly understood, but oxidative stress, non-oxidative action, inhibition of cell adhesion, decline in biofilm formation, obstructed quoram sensing and metal ion release are attributed to be as the major reasons. In addition, the limitation and toxicity with the clinical and environmental applications are also described.

Keywords Nanomaterial · Antimicrobial · Drug resistance · Toxicity · Oxidative stress · Biofilm

1 Introduction

Since ages, microbial contamination is amongst the major factor for morbidity and mortality across the world. As per reports, in developing countries, almost half of the population is affected by microbial contaminants and causes more than three million people die annually (Armentano et al. 2014). Instead of, great advances in diagnostics and therapeutics; microbes continue to affect biomedical and healthcare sectors due to the development of antibiotic resistance (Schwartz et al. 1997). According to the WHO 2018 release on the high-level risk of antimicrobial resistance states that, worldwide across 22 countries, 500,000 individuals are suffering from antimicrobial resistance revealing the increasing risk of serious health alignments due to microbial infection (Organization 2018). For instance, in patients with antiretroviral treatment, the resistance of malaria for artemisinin is at its pace which increases the resistance of anti-human immunodeficiency virus (HIV) drug (Organization 2016). A number of contributing factors for such increase include the change in human lifestyle, industrialization, wars, and microbial genome mutations. These pathogens are not only responsible for the deterioration in healthcare but are also responsible for damaging crops, food spoilage, deterioration of textiles etc. Therefore, preservation of potency of existing antibiotics through a wiser use of their properties and developing better alternatives calls for an urgent quest.

Super-bacteria is resistant to almost all antibiotics due to their abuse. It has been shown that the resistance is because of gene called NDM (Hsueh 2010). The major three bacterial targets antibiotics are: cell wall synthesis, DNA replication mechanism, and translational mechanism. Antimicrobial mechanisms with nanomaterial against antibiotic-resistant microbes work by direct contact with the bacterial cell wall, without penetrating the cell, enhances release of antimicrobial metal ions from nanoparticle surfaces as shown in Fig. 1. This gives the hope that nanomaterial is considered less prone to promoting resistance. In the last few decades, nanostructure-based antimicrobial agents have drawn considerable attention to combat antimicrobial resistance. Nanomaterial holds unique characteristic features including electrical, optical, chemical and thermal. These unique properties provide application of nanomaterial in multidisciplinary fields of medicine, technology and industries (Refer Table 1). The basic properties of nanomaterial should be an inexpensive, effective and broad-spectrum effect.

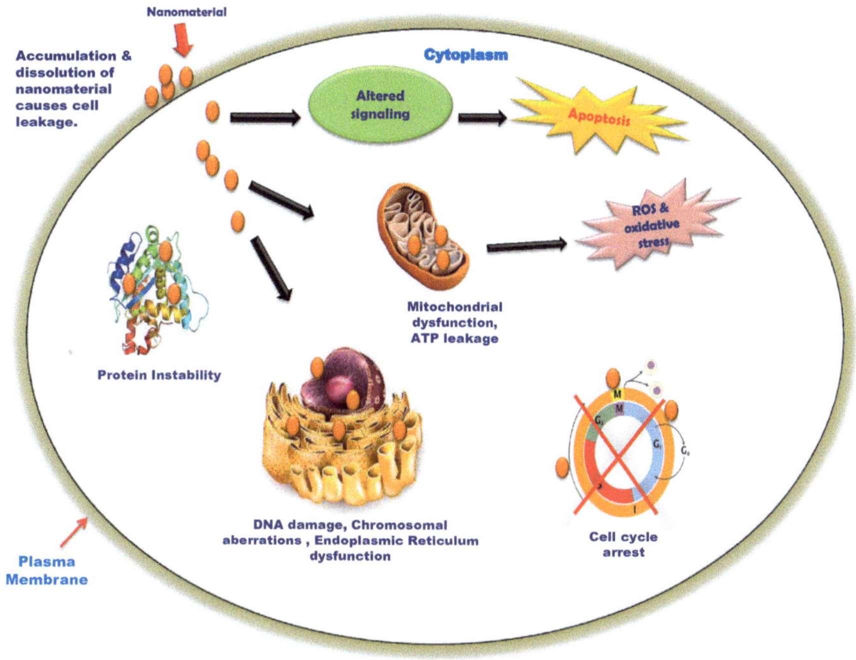

Fig. 1 Mechanism of nanomaterial against microbial cell: nanomaterial can cross the cell membrane barrier due to its accumulation, nano-size and shape. When nanomaterial enter the cytoplasm it can interfere with the cell organelles, proteins and signaling cascade as a result the cell could not survive due to apoptosis, cell cycle arrest, oxidative stress, protein instability, or damaged DNA

Table 1 Application and mechanism of nanomaterial

S.no.	Nanomaterial	Size (average)	Test micro-organism	Mechanism	Potential industrial application	References
1.	ZnO	12–60 nm	*E. coli, S. aureus*	Membrane disruption and ROS generation	Antimicrobial creams, lotions and ointments, sunscreen lotions, deodorants, ceramics, and self-cleaning glass	Gunalan et al. (2012)
2.	Ag	12–50 nm	*Pseudomonas aeruginosa, S. aureus, Candida albicans, E.coli, Vibrio cholerae, Salmonella typhi*	Membrane disruption, Ag ion interference with DNA replication,	Next generation antibiotics, medical, and health care products	Srisitthiratkul et al. (2011)

(continued)

Table 1 (continued)

S.no.	Nanomaterial	Size (average)	Test micro-organism	Mechanism	Potential industrial application	References
3.	Cu	100 nm	*E. coli, Bacillus subtilis, Candida albicans*	Protein inactivation via thiol interaction	Dental materials	Jadhav et al. (2011)
4.	Fe_3O_4	8–10 nm	*S. aureus, Staphylococcus epidermidis, Bacillus cereus, Klebsiellapneumonia*	Membrane disruption and ROS generation	Biomedical and antimicrobial applications	Ansari et al. (2017)
5.	Al_2O_3	11-60 nm	*E. coli, B. subtilis, Pseudomonas fluorescens*	Flocculation, dose dependent ROS and penetration of particle	Antibacterial applications	Jiang et al. (2009) and Simon-Deckers et al. (2009)
6.	TiO_2	17 nm	*E.coli, C. albicans*	Disruption of membrane	Next generation antibacterial and antifungal agent	Bahri-Laleh et al. (2011) and Simon-Deckers et al. (2009)
7.	SiO_2	20 nm	*E. coli, B. subtilis, P. fluorescens*	Disruption of membrane, flocculation	Biomedical and food applications	Jiang et al. (2009)
8.	Chitosan	40 nm	*E. coli, S. aureus, E. agglomerans*	Flocculation, membrane disruption	Biomedical devices, water filters, and instrument preparation	Kumar et al. (2017) and Qi et al. (2004)
9.	SWNT	0.83 nm and 5–50 nm	*E. coli, B. subtilis, P. aeruginosa, S. aureus*	Membrane disruption, interference with DNA replication	Medical devices, anti-biofouling membranes, and wastewater treatment	Liu et al. (2009)
10.	Dendrimers	NR	*P. aeruginosa, S. aureus,* strains of human cytomegavirus (HCMV), *C. albicans*	Kill biofilm cells, blocks virion attachment to target cells, membrane damage	Potential for drug delivery, anti-infective agents	Scorciapino et al. (2017)

2 Classification of Nanomaterial as Antimicrobial Agents

The increasing risk of antimicrobial resistance can be resolved with the help of upcoming approach to utilise the nanomaterial as antimicrobial. Nanomaterial possesses various physical, chemical and biological properties due to the nano-sized material. Different nano-material behaves differently against different microbes. Nanomaterial act by disrupting the bacterial membrane, hindering biofilm formation, acts as a carrier of antibiotics and acting against various mechanisms simultaneously. The nanomaterial causes antimicrobial action by either interacting directly with microorganisms or by oxidising the cell components or generates of reactive oxygen species which induces stress. Nanoparticles range from 1 to 100 nm in diameter. Depending upon composition and size, nanoparticles have unique properties in comparison to the bulk material. These are the surface area to volume ratio, surface Plasmon resonance, super-magnetization, surface-enhanced Raman scattering, photoluminescence, electric and heat conductivity and surface catalytic activity. As cell organelles and bio-molecules are in nano-size, nano-material can be combined with enzymes, antibodies, peptides, nucleic acids etc. Such modifications would provide specific functions to nanostructured material. Functionalization can be achieved by adsorption, linking with thiol groups, covalent bonding and electrostatic interactions (Sperling and Parak 2010). The nanomaterial used as antimicrobial can be classified on the basis of metallic, and non-metallic properties as mentioned in Fig. 2. Various reports focus on metals and metallic nanoparticles against micro-organisms (Chwalibog et al. 2010).

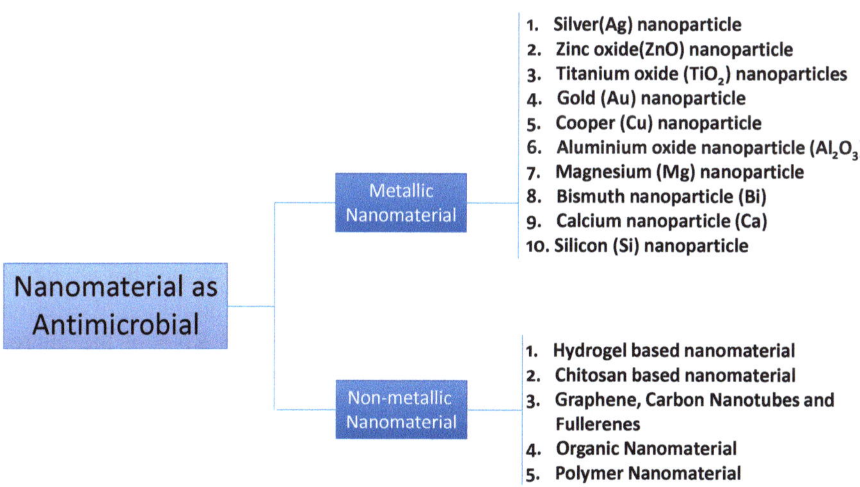

Fig. 2 Classification of nanomaterial as antimicrobial on the basic of metallic and non-metallic nature

2.1 Metallic Nanomaterial

Metals are potent antimicrobials and are vital for physiological activities in prokaryotic as well as eukaryotic cells such as iron acts as a cofactor for various enzymes, also essential for DNA replication, transcription and other metabolic processes (Andreini et al. 2008). Therefore, high levels of important metal ions are harmful to live organisms. Such nanostructured particles can be employed as they provide large surface area with increased reactivity. Several metal nanoparticles are known to possess antimicrobial properties such as Silver (Ag), Gold (Au), Copper (Cu), Zinc (Zn), Titanium (Ti), and Magnesium (Mg). Metal oxides have also been considered for their antimicrobial properties such as silver oxide (Ag_2O), titanium dioxide (TiO_2), silicon (Si), copper oxide (CuO), zinc oxide (ZnO), calcium oxide (CaO) and magnesium oxide (MgO). Metal oxide nanomaterial poses bactericidal due to the generation of reactive oxygen species (ROS), their physical structure and ion release (Fernando et al. 2018).

2.1.1 Silver (Ag) Nanoparticles

Silver salts and silver element are well known for their broad-spectrum antimicrobial properties. It has been used since ages to disinfect medical devices and for purification of water. It has been used as a pharmaceutical to recover from wounds, burns and other infections (Avalos et al. 2016). Therefore silver nanoparticle can act more efficiently as an antimicrobial agent. Ag nanoparticle is an antimicrobial which can act against gram positive and gram negative bacteria, as well as yeast (Luo et al. 2013). Ag nanoparticles possess various mechanisms for antimicrobial resistance. Ag+ interacts with sulphur and phosphorus groups of proteins present in the cell wall and cell membrane (Lara et al. 2010). Therefore, binds to negatively charged groups present resulting in holes in the membrane, leading to efflux of the cytoplasmic contents out of the cell along with the movement of H+ ions and this leads to cell death (Zhang et al. 2010). Nonetheless, Gram-positive bacteria are more susceptible than Gram-negative bacteria to the activity of Ag nanoparticles, as the ions get trapped in the lipopolysaccharide (LPS) of the Gram-negative bacteria which cannot penetrate in the cell (Lara et al. 2010). Within the microbial cell, Ag nanoparticles act by various mechanisms including inhibition of electron transport via cytochrome, binds to and damages DNA and RNA of the microorganism thereby, also inhibits DNA replication and cell cycle, prevents protein translocation by denaturing the 30S subunit of ribosomal, releases ROS which is toxic to the microbial cell (Huang et al. 2011). At nano-scale Ag possess anti-fungal, anti-bacterial and anti-viral properties. The antimicrobial properties of Ag contribute towards its wide application in medical devices, home appliance, some biosensors, etc.

Silver oxide (Ag_2O) nanoparticle was previously discovered, possessing the antimicrobial activity. These nanoparticles can be used as a substitute of antibiotics to a greater extent. The efficacy of Ag_2O was previously demonstrated on the basis of the effect these nanoparticle cause on E.coli. The DNA of the microbes losses the ability to DNA replication and arrests the cell cycle by causing DNA damage (Allahverdiyev et al. 2011). Hence with further research and advancement of various compounds, alloys can be generated for a better application.

2.1.2 Zinc Oxide (ZnO) Nanoparticles

ZnO containing nanomaterial has a potential antimicrobial effect especially against gram-positive and gram-negative bacteria; therefore, zinc oxide nanocomposites are being used in packing food (Espitia et al. 2013). The mode of antimicrobial activity of ZnO is the release of Zn2+ and ROS generation which damages the lipids and proteins of the cell membrane as well as that present inside the cell and interacts with essential metabolic pathways leading to cell death (Chupani et al. 2017). ZnO nanomaterial when coated with polyvinyl alcohol (PVA) increases the permeability of the membrane, easily enters into the cytoplasm and creates oxidative stress (Hajipour et al. 2012). ZnO nanomaterial, when combined with polymethylmethacrylate (PMMA), inhibits fungal biofilm formation that can treat denture stomatitis. Studies also suggest that ZnO also induces the production of p53 tumour suppressor protein that leads to apoptosis of cancer cells in human (Akhtar et al. 2012).

2.1.3 Titanium Oxide (TiO_2) Nanoparticles

The antimicrobial activity of TiO_2 is due to its structure as crystal and its specific size and shape. Titanium oxide (TiO_2) alone or conjugation with other antimicrobial agent is non-toxic and have antimicrobial activity. TiO_2 nanomaterial is used in varied products such as lotions, toothpaste, paints, coatings etc. due to whiteness properties and high refractive index. Due to its antimicrobial properties, it is used as a disinfectant in potable water.TiO_2 nanoparticles contain specific photocatalytic properties due to which it can act more effectively as an antimicrobial. This photocatalytic activity helps TiO_2 nanoparticles to generate ROS under UV-light. The mode of action of TiO_2 nanomaterial is by ROS generation especially –OH free radicals (Dizaj et al. 2014).

2.1.4 Gold (Au) Nanoparticles

Gold nanomaterial is a worth metallic nanomaterial due to their biocompatibility, low cytotoxicity compared to other nanomaterials, higher and ease of detection along with the capability of functionalization. It has been reported to damage cell membrane by changing membrane potential which leads to ATP loss and oxidative stress which further causes ROS generation resulting in microbial death (Abdel-Raouf et al. 2017). It is also used as a carrier in drug delivery by the ease of functionalization with thiol groups, low cytotoxicity and surface Plasmon resonance properties. Thus, biocompatibility, conjugation with functional groups, high absorption and optical properties help in targeted drug delivery and therapeutics (Chen et al. 2008). The size of Au nanoparticles is less than 2 mm; therefore, several studies speculate on the antimicrobial activity of Au nanoparticles. Au nanoparticlesuse as anticancer or antibacterial agents is due to irradiation with laser energy with the help of electrons which generate heat by excitation and oscillation (Riley and Day 2017).

2.1.5 Copper (Cu) Nanoparticles

Copper (Cu) nanoparticlesare amongst the best antimicrobial agents due to their chemical stability and resistance to heat. Cu nanoparticles are evaluated for the antibacterial and antifungal activities on various microorganisms which include *P. aeruginosa, S. aureus, Salmonella choleraesuis, C. albicans* and *B. subtilis* (Dizaj et al., 2014). Whereas, due to the rapid rate of oxidation Cu nanoparticles are not widely used. Therefore, Copper nanoparticles can be synthesised as copper oxides nanoparticles and copper nanoparticles loaded thin film which interacts with carboxyl and amine groups of the membrane of the microbial cell along with induces ROS with inhibits replication of DNA and protein synthesis (Blecher et al. 2011). Copper oxide (CuO) is a more cost efficient antimicrobial when compared with Ag and Au. It is more stable, both physically and chemically in relation to the others. It also possesses properties for easy miscible with the polymers (Huh and Kwon 2011).

2.1.6 Aluminium Oxide Nanoparticles (Al_2O_3)

The antimicrobial effect due to $Al_2O_3$is limited to mild inhibitory effect, it is also at high concentration, by disrupting cell wall. These nanoparticles are supposed to cause resistance in microbes (Qiu et al. 2012). In *E. coli*, Al_2O_3 nanoparticles travel through the cytoplasm and result in toxic effect (Hajipour et al. 2012). Their higher concentrations damages the cell wall but studies report that it only causes a low level of growth inhibition (Huh and Kwon 2011). Al_2O_3 nanoparticles increase the risk of horizontal gene transfer by 200-folds through conjugation especially in *E. coli* and *Salmonella* (Qiu et al. 2012). It damages the microbes through oxidative stress and promotes the expression of genes involved in conjugation along with suppression of genes that inhibit conjugation (Huh and Kwon 2011; Qiu et al. 2012).

2.1.7 Magnesium (Mg) Nanoparticles

Magnesium halogen conjugates and magnesium oxide (MgO) nanoparticles are the two types of magnesium-based nanoparticles used as the antimicrobial therapeutics. Magnesium halogen-containing nanoparticles act by inhibiting microbial enzymes while MgO containing nanoparticles work by ROS production leading to lipid peroxidation of the microbial cell membrane which leads to an outflow of cytoplasmic contents. For example, MgF_2nanoparticles work by lipid peroxidation of the microbial cell membrane leading to efflux of cytoplasmic contents along with a drop in cytoplasmic pH which thereby increases the membrane potential. MgF_2 has been successfully studied against *E. coli* and *S. aureus* for growth inhibition and prevent biofilm formation (Blecher et al. 2011).

2.1.8 Bismuth Nanoparticles (Bi)

Bi nanoparticles are effective against multi-drug resistant microbes when combined with X-rays thereby limiting the toxic effect on the host cells (Luo et al. 2013).When combined with X-rays Bi nanoparticles emits free radicals and electrons, these damages the bacterial DNA. These are effective against *P. aeruginosa* (Luo et al., 2013).

2.1.9 Calcium Nanoparticles (Ca)

CaO nanoparticles have strong antimicrobial activity, due to free and active oxygen species. According to the study by Jeong et al., antimicrobial CaO can be generated by heating $CaCO_3$ (Jeong et al. 2007). The mechanism of action of CaO is similar to MgO by acting on the cell wall. Due to increased oxidative stress and the generation of superoxide anions, the antimicrobial effect occurs. The other reason for antimicrobial activity is due to an increase in pH (Dizaj et al. 2014).

2.1.10 Silicon (Si) Nanoparticles

Antimicrobial action of SiO_2 nanoparticles would turn out to be more noteworthy due to more surface area. Si nanoparticles conjugated with the other biocidal metals, for example, Ag has been widely examined, Egger et al. announced the creation and examination of antimicrobial action of novel Ag–Si nanocomposite (Egger et al. 2009).The results suggest that Ag/SiO2 nanocomposites showed enhanced antimicrobial properties against *E. coli, S. aureus*, and *C. albicans.* The applications of nanocomposites are endless as it can be mixed and prepared with antimicrobial activity.

2.2 *Non-metallic Nanomaterial*

2.2.1 Hydrogel-Based Nanomaterial

These nitric oxide-releasing nanomaterials have antimicrobial potential against the broad spectrum of multi-drug resistant microbes. They are effective against multidrug resistant *S.aureus* (MRSA), *A. baumannii* (Friedman et al., 2008). They increase the synthesis of interferon-Ɣ, which inhibits angiogenesis in reducing the spread of microbes (Han et al. 2009). Later on, a study by Friedman et al., reports that when nitric oxide-releasing hydrogel when reacts with glutathione (GSH) produced S-nitrosoglutathione significantly decreases the microbial growth of MRSA, *E. coli*, *P. aeruginosa*, and *K. pneumonia* in comparison to independent inhibition by hydrogel or GSH (Friedman et al. 2011).

2.2.2 Chitosan-Based Nanomaterial

Chitosan is deacetylated monomeric units of chitin in a random manner derived from a polymeric chain of N- acetyl glucosamine and glucosamine (Huh and Kwon 2011). From the deacetylated units every C2 amino group of chitosan has pKa of 6.5 leading to protonation and pH lower than 6.5 which is associated with antimicrobial and anti-inflammatory properties of Chitosan (Friedman et al. 2013). The positive charge of Chitosan provides affinity towards negatively charged cell wall and cell membrane of microbes. This increases the influx in cell envelope causing osmotic damage, efflux of cytoplasmic contents (Friedman et al. 2013). It is unlikely to develop resistance against chitosan-based nanomaterial as the cell envelope of microbes is highly conserved to evolutionary changes so, it does not change with a single gene mutation. It also acts by inhibiting the mRNA during transcription, preventing growth and metabolic activities of the microbes especially in bacteria and fungus (Friedman et al. 2013). It reduces the activity of metalloproteins as chitosan chelates metals. By inhibiting secretion of inflammatory cytokines, it employees fibroblast cells and deposits collagen III, thereby promoting faster wound healing and prevents infection of wounds. Chitosan nanomaterial is effective against *S. aureus* in comparison to E. coli. It has been reported to have stronger activity against fungi and viruses compared to bacteria (Blecher et al. 2011). Nano-chitosan with low molecular weight has greater efficiency against gram-positive bacteria than gram-negative bacteria. Although, chitosan would be more effective against Gram-negative bacteria because of the presence of more negative charged cell envelope. Positively charged amino groups of chitosan have the ability to replace Ca2+ and Mg2+ ions involved in destabilizing the lipopolysaccharide of gram-negative bacteria which increases the permeability of the membrane (Friedman et al. 2013). Chitosannanoparticles are biodegradable antimicrobial nanoparticles which can be employed as an agent to combat antimicrobial resistance. The biodegradable nanoparticles are more advantageous as antimicrobial metal and metal oxide nanoparticle could not be used due to increased accumulation and toxicity.

2.2.3 Graphene, Carbon Nanotubes and Fullerenes

Graphene nanomaterial includes oxides, reduced oxides and nano-composites which are based on antimicrobial activity due to their surface properties, sheet effect leading to cell dysfunction and oxidative stress in the cell (Ocsoy et al. 2017). The layer-by-layer assembly of graphene oxide nanosheets attributes to: optical, dielectric and antibacterial aspects (Baranwal et al. 2018). The property to prevent microbial contamination, graphene-based nanomaterial can be employed in food packaging. Single-walled carbon nanotubes (SWCNTs) have been found efficient against both gram-negative and gram-positive bacteria as they are toxic to microbes which further disrupts membrane integrity along with induces oxidative stress (Dizaj et al. 2014). Therefore, carbon nanotubes (CNTs) have been used in filters to prevent bio-fouling and biofilm formation (Lee et al. 2010). The microbicidal property of fullerenes (C60) and its derivatives like fullerol has not yet been exploited much but is attributed to ROS generation and highly reactive singlet oxygen species formation respectively (Lyon et al. 2006).

2.2.4 Organic Nanomaterial

In the last few decades, a group of nanomaterial has attracted considerable interest including dendrimeric peptides, liposomes, polymer-based nanomaterial etc. A dendrimeric peptide containing multiple R (Arg) W (Trp) dipeptides synthesised against gram-negative and gram-positive bacteria which act via membranolytic method (Liu et al. 2007). G3KL, a novel antimicrobial dendrimeric peptide containing alternating branches of natural leucine and lysine amino acids effective against A. baumannii and P. aeruginosa as compared to standard antibiotics (Pires et al. 2015). A tetra-branched SB105 potentially inhibited replication of human cytomegalovirus (HCMV) strains in primary fibroblast and endothelial cells. Dendrimer SB105 prevents virions attachment to heparansulphate over the cell membrane (Luganini et al. 2010). The microbicidal properties of dendrimeric peptides are due to high surface area ratio, in vivo activity, affinity to carry both polar and non-polar drug molecules (Cheng et al. 2016).

Liposomes have been used since long as cargos of the drug due to their ability to mimic microbial cell membrane, which allows them to fuse with the infectious microbes. Thereby, allows unhindered delivery of the drug in the cell which causes oxidative stress and imbalanced ionic levels leading to cell death (Pushparaj Selvadoss et al. 2018). Similarly, polymer nanomaterial due to a stable structure, zeta potential, affinity to cargo drugs allows delivery of antimicrobial agents.

2.2.5 Polymer Nanomaterial

By imitating the general compound structure of antimicrobial peptides, polymers could be synthesised with antimicrobial characteristics by fusing cationic and hydrophobic moieties into the polymer chains. Interaction with the bacterial cell walls which possess negative chargetooccur due to the general cationic charge present on the polymer, while the hydrophobic partners enable the penetration inside the microbial membrane (Lam et al. 2017). The polymeric nanoparticles can be of various types on the basis of its architecture, such as self-assembly polymer nanoparticles and star nanoparticles. The type of antimicrobial activity is contributed by the type of polymeric nanomaterial and its specific characteristics. Polymer nanoparticlesare also useful for antimicrobial drug delivery due to its stable structure which enables the synthesis of nanoparticles with nano-size distribution, particle properties which can be specified by the selection of surfactant, organic solvent and the length of the polymer and presence of functional group on the polymer nanoparticles which can be chemically modified (Lakshminarayanan et al. 2018).

3 Application of Nanomaterial as Antimicrobial Agent

Nanoparticle obtained from either physical, chemical, or biological method as mentioned in (Table 2) consist of various applications. Nanoparticle possesses various application as antimicrobials such as water disinfectant, therapeutic, food packaging preserver, drug delivery agent, nano-fertilizer and nano-pesticides, antibacterial paper, antibacterial textile, biofortification and biodegradable nanoparticle for environment protection. For e.g., nanotechnology has provided alternative way for water disinfection. Nanomaterial result as an effective antimicrobial due to the high surface-to-volume ratio, crystallographic structure, and adaptability to various substrates. Several metal and metal oxide nanoparticles have been applied to the use of water disinfection. Silver nanoparticle (AgNP) are the most utilised nanoparticle for water disinfection (Liu et al. 2012). Another antimicrobial used as water

Table 2 Advantages and disadvantages of nanoparticle synthesis method

Method	Advantages	Disadvantages
Physical method	The solute system is not present	Not environment friendly
	Desirable size and shape of the nanoparticle can be obtained	Huge infrastructure required
	Interaction domains can be modified	More time consuming
	Utilize bulky nanoparticle	
Chemical method	Can be combined with the physical method	Hazardous chemicals involved
	Solution can be aqueous and non-aqueous	Accumulation of nanoparticle can occur
		Sometimes particle may not stabilize
		Not environment friendly
Biological method	Environment friendly	To be monitored
	Size and shape of the nanoparticle can be monitored	Media constituents
	No chemicals required	Environmental conditions,
	Cost-effective	Genetic makeup,
	Renewable synthesis	Cell growth conditions,
	Large scale synthesis	Enzyme activity

disinfectant is TiO_2 by causing ROS burden on microbial cells. Advantage of using TiO_2 for water disinfection include stability of TiO_2 in water and ingestion has low toxicity to human health (Liu et al. 2012).

4 Toxicity of Nanomaterial

Nanotechnology has increased critical advancement over the previous decades, which steer the revolution in the sphere of information, industry, medicine, aerospace aviation and food security. Nanotechnology has become a new research hot spot in the world. However, we cannot only focus to its benefits to the society and economy because its increased use has been creating potent environmental and health effects due to the toxicity of the nanoparticles. At high doses, anything to everything can be toxic but it is relevant to understand the ideal concentration of nanomaterial to be used. The toxicity of nanomaterials is determined by the base material, size, shape and coatings.

For toxicity studies, several research groups use distinct cell lines, culture environment and incubation periods. It is difficult to determine physiologically relevant cytotoxicity due to difference in toxicity parameters during the study by different groups. To understand the toxic effect several biological models includes cell lines, aquatic embryonic zebrafish (*Danio rerio*), and whole-animal tests such as rodents (mice/rat) (Girardi et al. 2017; Griffitt et al. 2007).

5 Metal Nanomaterial

Metallic nanoparticles are most extensively used engineered nanomaterials; however, there is limited understanding in context with environmental fate and effects. Comparatively bulk gold is safe, due to its remarkable characteristics; different researchers have evaluated cellular uptake and toxicity of gold nanoparticles. In the study by Goodman et al. reported non- cytotoxic effect of gold nanoparticles with immune system cells and reduction in harmful ROS in the cells. Their study in three different types of cells suggested the toxicity of 2 nm gold nanoparticles functionalized with both cationic and anionic surface groups which proved that cationic functionalization is less toxic than anionic particles, which might be attributed to the electrostatic interaction between the cationic group of nanomaterial and the negatively charged cell membrane (Goodman et al. 2004). Nanomaterial may show less or no cytotoxicity but may cause serious cellular damage.

Cytotoxicity is related to cell type; 33 nm citrate-capped gold nanospheres were non-cytotoxic in baby hamster kidney and human hepatocellular liver carcinoma cells, but cytotoxic to a human carcinoma lung cell line as reported by Patra et al. (2007).

Prolonged exposure to silver results in argyria marked by a blue-gray discoloration of the skin and other organs. Low-level exposure can lead silver deposition on skin and other parts of the body. Elevated levels of silver in air can cause breathing problems, lung and throat irritation, and stomach pain, mild allergic reactions over skin including rashes, swelling, and inflammation (Drake and Hazelwood 2005).

Griffitt et al. reported toxicity of metallic nanomaterial in aquatic organisms (zebrafish, daphnids and an algal species) Different organisms manifested with silver, copper, aluminium, nickel, and cobalt as both nanoparticles and soluble salts as well as to titanium dioxide nanoparticles resulted in nanosilver and nano-copper toxicity with 48-h lethal concentrations as low as 40 and 60 μg/L, respectively, in *Daphnia pulex* adults, whereas titanium dioxide was non-toxic (Griffith and Swartz 2006).

5.1 Metal Oxide Nanomaterial

These nanomaterial are vitally utilized as added substances in pharmaceuticals, beauty care products and colouring agents. TiO_2/ZnO based nanomaterial have water and strain resistant properties, so are used in sunscreens and stringy creams. A few studies have analyzed the detrimental effects of metal oxide nanomaterial.

A study by Grassian et al. reported 2–5 nm TiO_2 nanomaterial inhalation exposure and their aggregation for the formation of aerosols in the exposure chamber (Grassian et al. 2007). A significant inflammatory response was observed in mice, 3 weeks post subacute exposure to the aggregates (Fabian et al. 2008).

Much has not been reported regarding impact of nanomaterial on higher plants.

Nano-TiO_2 remarkably enhances photosynthesis and spinach development by boosting nitrogen fixation. Suspension of nano-alumina had no impact on California red kidney bean and ryegrass development (Wang et al. 2011). However, it prevents rooting in corn, cucumber, soybean, cabbage, and carrot. High concentration of nano-ferrophase obstructed popcorn development. Lin et al. analyzed nano-ZnO cell internalization and upward translocation in Loliumperenne (ryegrass) (Lin and Xing 2008). Scanning electron microscopy and transmission electron microscopy were used to demonstrate internalization of ZnO nanoparticles, ryegrass biomass loss, root tips shrinking and root epidermal and cortical cells vacuolation. During translocation of Zn from root to shoot does not attribute to risk in use of nano-ZnO (Lin and Xing 2008).

Franklin et al. stated relative toxicity of nano-ZnO, bulk ZnO, and $ZnCl_2$ in freshwater microalgae i.e. *Pseudokirchneriella subcapitata* (Franklin et al., 2007). This revealed the toxicity in 72 h with IC_{50} near 60 μg Zn/L (Franklin et al. 2007).

Karlsson et al. focused metal oxide nanoparticles (CuO, TiO_2, ZnO, $CuZnFe_2O_4$, Fe_3O_4, Fe_2O_3) and differentiated with carbon nanoparticles and multi-walled carbon nanotubes (MWCNT). This concluded that a various nanoparticles have different toxicities (Karlsson et al. 2008). CuO nanoparticles are prone to cytotoxicity and DNA damage (Chang et al. 2012). ZnO indicate effects viability of cells and caused DNA damage (Chang et al. 2012), while the TiO_2 particles (a blend of rutile and anatase) cause DNA damage (Zhu et al. 2010). In metal oxide nanomaterial (Fe_3O_4,

Fe_2O_3), no or low toxicity has been observed, however, $CuZnFe_2O_4$ particles leads to DNA injuries (Karlsson et al. 2008). Also, carbon nanotubes present cytotoxicity and cause DNA mutations. Xia et al. stated ROS and cytotoxicity of TiO_2, ZnO, and CeO_2, in RAW 264.7 and BEAS-2B cell lines which concluded that ZnO instigates lethality in both, leading to ROS, oxidant damage and cell death (Xia et al. 2006). Conversely, cellular uptake of nano-CeO_2 in low concentrations is genotoxic and produces ROS and oxidative stress (Zeyons et al. 2009).

5.2 *Carbon Nanomaterial*

Siliva et al. proved that ultrafine carbon particles effectively penetrate lungs compared to larger particles and have the ability to cross the blood-brain barrier leading to central nervous system (CNS) toxicity. This also proved that inhalation of CNTs can result in CNS toxicity rather releasing clotting agents from the lungs (Silva et al. 2011). Asbestoes fibre inhalation induces asbestosis, lung cancer, and malignant mesothelioma of pleura. Thus, asbestos is highly lethal than CNT due to their structural resemblances (Magrez et al. 2006).

Zhu et al. studied multi-walled carbon nanotubes (MWCNTs) in mouse embryonic stem (ES) cells for DNA damage. It was found that MWNTs accumulates and induces apoptosis by activating the tumor suppressor protein p53 in 2 h exposure. It also induces oxidative stress They also report elevated expression base excision repair protein 8-oxoguanine-DNA glycosylase 1 (OGG1), double-strand break repair protein Rad 51, phosphorylation of H2AX histone at serine 139, and SUMO modification of XRCC4 after treating with MWCNTs (Zhu et al. 2007).

5.3 *Quantum Dots*

Quantum Dots (QDs) are characterized by unique optical and electrical properties which furnish QDs as optimal fluorophores for biomedical imaging/diagnosis, e.g. fluorescent QDs conjugated with bioactive of DNA, protein and cell membrane receptors moieties to target specific process. Different QD have specific physicochemical properties, which determines potential toxicity.

Zhang et al. reported the impact of QDs via skin penetration. Their study accounted carboxylic acid coated QD655 and QD565 diffused in uppermost stratum corneum layer of skin, with constant flow of 8 and 24 h, proves to be cytotoxic (Zhang et al 2008).

Shiohara et al. also reported QD-induced cytotoxicity. 11-mercaptoundecanoic acid (MUA)-coated CdSe/ZnS QDs were cytotoxic to HeLa cell lines and primary human hepatocytes at 100 μg/mL (MTT assay) concentration. Using primary hepatocytes as a liver model, Deufus et al. found that CdSe-core QDs induced acute toxicity.

This study also proved that QDs cytotoxicity varies due to synthesis parameters, ultraviolet light exposure and surface coating. This suggests that cytotoxicity related to release of free Cd^{2+} ions because of CdSe lattice deterioration (Shiohara et al. 2004).

6 Summary, Outlook and Future Needs

In summary, we have discussed the types, behaviour, applications and toxicity of several nanomaterials in use. The toxic effect of nanomaterials has been studied but much is not known yet. Also, there is lack of knowledge, how nanomaterials interact with the environmental system. More research is required evaluate their stability in different test systems identify prospects for human use. Studies relating toxicity on diverse cell lines with varying incubation times are being reported, but divergent nanoparticle concentrations, cell lines as well as culture conditions results in poor lack knowledge of their mechanism, relevance of toxicity. Nevertheless, as per the above discussion, numerous research challenges within this field remain answered. Once the biomedical society acknowledges nanomaterials as tool for biomedical imaging, for future, only then their interaction with cells and organs will be understood.

Analytical methods can be employed for better understanding, the formation and activity mechanism of nanomaterials. Along with, the importance of nanomaterial purity is detected by analytical methods to identify impurities, appreciating the use of greener approaches. Indeed, they are too small to be detected by optical microscopes. The challenge is to reach global agreement on a battery of in vitro screening tests for human and environmental toxicity.

To control size and shape of the nanomaterials, surfactants used are toxic. Therefore, one of the aim is to identify alternatives to the use of surfactants or other substances for nanomaterial stability and shape during synthesis. New biomimetic approaches are pivotal to control shape are essential, which are promising biological derivatives for nanoparticle production. Intense work is required for the development of these methods. Green nanoscience guides design, production, and application of greener nanomaterials with broad spectrum compositions, sizes, shapes, and functionality. This will provide research opportunities and challenges for this community in the foreseeable future.

Imparting research on nanotechnology risks and advantages outside mainstream researchers is challenging, however, is fundamental for exchanges in light of sound science. This implies creating correspondence exercises that empower specialized data to be condensed, scrutinized and at last integrated for different invested individuals, including chiefs and customers. At last, a worldwide comprehension of nanotechnology-particular danger is fundamental assuming extensive and little businesses work on a level playing field, and creating economies are not to be denied basic data on planning safe nanotechnologies. If universally research community can take benefit of these circumstances then we can surely look towards the advent of safe nanotechnologies.

References

Abdel-Raouf, N., Al-Enazi, N. M., & Ibraheem, I. B. (2017). Green biosynthesis of gold nanoparticles using Galaxaura elongata and characterization of their antibacterial activity. *Arabian Journal of Chemistry, 10*, S3029–S3039.

Akhtar, M. J., Ahamed, M., Kumar, S., Khan, M. M., Ahmad, J., & Alrokayan, S. A. (2012). Zinc oxide nanoparticles selectively induce apoptosis in human cancer cells through reactive oxygen species. *International Journal of Nanomedicine, 7*, 845.

Allahverdiyev, A. M., Abamor, E. S., Bagirova, M., & Rafailovich, M. (2011). Antimicrobial effects of TiO2 and Ag2O nanoparticles against drug-resistant bacteria and leishmania parasites. *Future Microbiology, 6*(8), 933–940.

Andreini, C., Bertini, I., Cavallaro, G., Holliday, G. L., & Thornton, J. M. (2008). Metal ions in biological catalysis: From enzyme databases to general principles. *Journal of Biological Inorganic Chemistry, 13*(8), 1205–1218.

Armentano, I., Arciola, C. R., Fortunati, E., Ferrari, D., Mattioli, S., Amoroso, C. F., … Visai, L. (2014). The interaction of bacteria with engineered nanostructured polymeric materials: A review. *The Scientific World Journal, 2014*.

Ansari, H., Shabanian, M., & Khonakdar, H. A. (2017). Using a β-Cyclodextrin-functional Fe3O4 as a Reinforcement of PLA: Synthesis, Thermal, and Combustion Properties. *Polymer-Plastics Technology and Engineering, 56*(12), 1366–1373.

Avalos, A., Haza, A. I., Mateo, D., & Morales, P. (2016). Interactions of manufactured silver nanoparticles of different sizes with normal human dermal fibroblasts. *International Wound Journal, 13*(1), 101–109.

Bahri-Laleh, N., Correa, A., Mehdipour-Ataei, S., Arabi, H., Haghighi, M. N., Zohuri, G., … Cavallo, L. (2011). Moving up and down the titanium oxidation state in Ziegler– Natta catalysis. *Macromolecules, 44*(4), 778–783.

Baranwal, A., Srivastava, A., Kumar, P., Bajpai, V. K., Maurya, P. K., & Chandra, P. (2018). Prospects of nanostructure materials and their composites as antimicrobial agents. *Frontiers in Microbiology, 9*, 422.

Blecher, K., Nasir, A., & Friedman, A. (2011). The growing role of nanotechnology in combating infectious disease. *Virulence, 2*(5), 395–401.

Chang, Y.-N., Zhang, M., Xia, L., Zhang, J., & Xing, G. (2012). The toxic effects and mechanisms of CuO and ZnO nanoparticles. *Materials, 5*(12), 2850–2871.

Chen, P. C., Mwakwari, S. C., & Oyelere, A. K. (2008). Gold nanoparticles: From nanomedicine to nanosensing. *Nanotechnology, Science and Applications, 1*, 45.

Cheng, G., Dai, M., Ahmed, S., Hao, H., Wang, X., & Yuan, Z. (2016). Antimicrobial drugs in fighting against antimicrobial resistance. *Frontiers in Microbiology, 7*, 470.

Chupani, L., Zusková, E., Niksirat, H., Panáček, A., Lünsmann, V., Haange, S.-B., … Jehmlich, N. (2017). Effects of chronic dietary exposure of zinc oxide nanoparticles on the serum protein profile of juvenile common carp (*Cyprinus carpio* L.). *Science of the Total Environment, 579*, 1504–1511.

Chwalibog, A., Sawosz, E., Hotowy, A., Szeliga, J., Mitura, S., Mitura, K., … Sokolowska, A. (2010). Visualization of interaction between inorganic nanoparticles and bacteria or fungi. *International Journal of Nanomedicine, 5*, 1085.

Dizaj, S. M., Lotfipour, F., Barzegar-Jalali, M., Zarrintan, M. H., & Adibkia, K. (2014). Antimicrobial activity of the metals and metal oxide nanoparticles. *Materials Science and Engineering: C, 44*, 278–284.

Drake, P. L., & Hazelwood, K. J. (2005). Exposure-related health effects of silver and silver compounds: A review. *The Annals of Occupational Hygiene, 49*(7), 575–585.

Egger, S., Lehmann, R. P., Height, M. J., Loessner, M. J., & Schuppler, M. (2009). Antimicrobial properties of a novel silver-silica nanocomposite material. *Applied and Environmental Microbiology, 75*(9), 2973–2976.

Espitia, P. J. P., Soares, N. D. F. F., Teófilo, R. F., dos Reis Coimbra, J. S., Vitor, D. M., Batista, R. A., ... Medeiros, E. A. A. (2013). Physical–mechanical and antimicrobial properties of nanocomposite films with pediocin and ZnO nanoparticles. *Carbohydrate Polymers, 94*(1), 199–208.

Fabian, E., Landsiedel, R., Ma-Hock, L., Wiench, K., Wohlleben, W., & Van Ravenzwaay, B. (2008). Tissue distribution and toxicity of intravenously administered titanium dioxide nanoparticles in rats. *Archives of Toxicology, 82*(3), 151–157.

Fernando, S., Gunasekara, T., & Holton, J. (2018). Antimicrobial nanoparticles: Applications and mechanisms of action. *Sri Lankan Journal of Infectious Diseases, 8*(1), 2–11.

Franklin, N. M., Rogers, N. J., Apte, S. C., Batley, G. E., Gadd, G. E., & Casey, P. S. (2007). Comparative toxicity of nanoparticulate ZnO, bulk ZnO, and ZnCl2 to a freshwater microalga (Pseudokirchneriella subcapitata): The importance of particle solubility. *Environmental Science & Technology, 41*(24), 8484–8490.

Friedman, A. J., Han, G., Navati, M. S., Chacko, M., Gunther, L., Alfieri, A., & Friedman, J. M. (2008). Sustained release nitric oxide releasing nanoparticles: Characterization of a novel delivery platform based on nitrite containing hydrogel/glass composites. *Nitric Oxide, 19*(1), 12–20.

Friedman, A. J., Blecher, K., Schairer, D., Tuckman-Vernon, C., Nacharaju, P., Sanchez, D., ... Nosanchuk, J. D. (2011). Improved antimicrobial efficacy with nitric oxide releasing nanoparticle generated S-nitrosoglutathione. *Nitric Oxide, 25*(4), 381–386.

Friedman, A. J., Phan, J., Schairer, D. O., Champer, J., Qin, M., Pirouz, A., ... Modlin, R. L. (2013). Antimicrobial and anti-inflammatory activity of chitosan–alginate nanoparticles: A targeted therapy for cutaneous pathogens. *Journal of Investigative Dermatology, 133*(5), 1231–1239.

Girardi, F. A., Bruch, G. E., Peixoto, C. S., Dal Bosco, L., Sahoo, S. K., Gonçalves, C. O., ... Barros, D. M. (2017). Toxicity of single-wall carbon nanotubes functionalized with polyethylene glycol in zebrafish (*Danio rerio*) embryos. *Journal of Applied Toxicology, 37*(2), 214–221.

Goodman, C. M., McCusker, C. D., Yilmaz, T., & Rotello, V. M. (2004). Toxicity of gold nanoparticles functionalized with cationic and anionic side chains. *Bioconjugate Chemistry, 15*(4), 897–900.

Grassian, V. H., O'Shaughnessy, P. T., Adamcakova-Dodd, A., Pettibone, J. M., & Thorne, P. S. (2007). Inhalation exposure study of titanium dioxide nanoparticles with a primary particle size of 2 to 5 nm. *Environmental Health Perspectives, 115*(3), 397.

Griffith, L. G., & Swartz, M. A. (2006). Capturing complex 3D tissue physiology in vitro. *Nature Reviews Molecular Cell Biology, 7*(3), 211.

Griffitt, R. J., Weil, R., Hyndman, K. A., Denslow, N. D., Powers, K., Taylor, D., & Barber, D. S. (2007). Exposure to copper nanoparticles causes gill injury and acute lethality in zebrafish (*Danio rerio*). *Environmental Science & Technology, 41*(23), 8178–8186.

Gunalan, S., Sivaraj, R., & Rajendran, V. (2012). Green synthesized ZnO nanoparticles against bacterial and fungal pathogens. *Progress in Natural Science: Materials International, 22*(6), 693–700.

Hajipour, M. J., Fromm, K. M., Ashkarran, A. A., de Aberasturi, D. J., de Larramendi, I. R., Rojo, T., ... Mahmoudi, M. (2012). Antibacterial properties of nanoparticles. *Trends in Biotechnology, 30*(10), 499–511.

Han, G., Martinez, L. R., Mihu, M. R., Friedman, A. J., Friedman, J. M., & Nosanchuk, J. D. (2009). Nitric oxide releasing nanoparticles are therapeutic for *Staphylococcus aureus* abscesses in a murine model of infection. *PLoS One, 4*(11), e7804.

Hsueh, P.-R. (2010). New Delhi metallo-β-lactamase-1 (NDM-1): An emerging threat among Enterobacteriaceae. *Journal of the Formosan Medical Association, 109*(10), 685–687.

Huang, L., Dai, T., Xuan, Y., Tegos, G. P., & Hamblin, M. R. (2011). Synergistic combination of chitosan acetate with nanoparticle silver as a topical antimicrobial: Efficacy against bacterial burn infections. *Antimicrobial Agents and Chemotherapy, 55*(7), 3432–3438.

Huh, A. J., & Kwon, Y. J. (2011). "Nanoantibiotics": A new paradigm for treating infectious diseases using nanomaterials in the antibiotics resistant era. *Journal of Controlled Release, 156*(2), 128–145.

Jadhav, S., Gaikwad, S., Nimse, M., & Rajbhoj, A. (2011). Copper oxide nanoparticles: synthesis, characterization and their antibacterial activity. *Journal of Cluster Science, 22*(2), 121–129.

Jeong, M. S., Park, J. S., Song, S. H., & Jang, S. B. (2007). Characterization of antibacterial nanoparticles from the scallop, Ptinopecten yessoensis. *Bioscience, Biotechnology, and Biochemistry, 71*(9), 2242–2247.

Jiang, W., Mashayekhi, H., & Xing, B. (2009). Bacterial toxicity comparison between nano-and micro-scaled oxide particles. *Environmental pollution, 157*(5), 1619–1625.

Karlsson, H. L., Cronholm, P., Gustafsson, J., & Moller, L. (2008). Copper oxide nanoparticles are highly toxic: A comparison between metal oxide nanoparticles and carbon nanotubes. *Chemical Research in Toxicology, 21*(9), 1726–1732.

Kumar, B. N. P., Mahaboobi, S., & Satyam, S. (2017). Chitosan in Medicine–A Mini Review. *J Mol Pharm Org Process Res, 5*(134), 2.

Lakshminarayanan, R., Ye, E., Young, D. J., Li, Z., Loh, X. J. (2018). Recent advances in the development of antimicrobial nanoparticles for combating resistant pathogens. *Advanced Healthcare Materials,* 1701400.

Lam, S. J., Wong, E. H., Boyer, C., & Qiao, G. G. (2017). Antimicrobial polymeric nanoparticles. *Progress in Polymer Science, 63*, 561–570.

Lara, H. H., Ayala-Núñez, N. V., Turrent, L. D. C. I., & Padilla, C. R. (2010). Bactericidal effect of silver nanoparticles against multidrug-resistant bacteria. *World Journal of Microbiology and Biotechnology, 26*(4), 615–621.

Lee, J., Mahendra, S., & Alvarez, P. J. (2010). Nanomaterials in the construction industry: A review of their applications and environmental health and safety considerations. *ACS Nano, 4*(7), 3580–3590.

Lin, D., & Xing, B. (2008). Root uptake and phytotoxicity of ZnO nanoparticles. *Environmental Science & Technology, 42*(15), 5580–5585.

Liu, Z., Young, A. W., Hu, P., Rice, A. J., Zhou, C., Zhang, Y., & Kallenbach, N. R. (2007). Tuning the membrane selectivity of antimicrobial peptides by using multivalent design. *ChemBioChem, 8*(17), 2063–2065.

Liu, Y., He, L., Mustapha, A., Li, H., Hu, Z. Q., & Lin, M. (2009). Antibacterial activities of zinc oxide nanoparticles against Escherichia coli O157: H7. *Journal of applied microbiology, 107*(4), 1193–1201.

Liu, C., Xie, X., & Cui, Y. (2012). Antimicrobial nanomaterials for water disinfection. In *Nano-antimicrobials* (pp. 465–494). Heidelberg: Springer.

Luganini, A., Giuliani, A., Pirri, G., Pizzuto, L., Landolfo, S., & Gribaudo, G. (2010). Peptide-derivatized dendrimers inhibit human cytomegalovirus infection by blocking virus binding to cell surface heparan sulfate. *Antiviral Research, 85*(3), 532–540.

Luo, Y., Hossain, M., Wang, C., Qiao, Y., An, J., Ma, L., & Su, M. (2013). Targeted nanoparticles for enhanced X-ray radiation killing of multidrug-resistant bacteria. *Nanoscale, 5*(2), 687–694.

Lyon, D. Y., Adams, L. K., Falkner, J. C., & Alvarez, P. J. (2006). Antibacterial activity of fullerene water suspensions: Effects of preparation method and particle size. *Environmental Science & Technology, 40*(14), 4360–4366.

Magrez, A., Kasas, S., Salicio, V., Pasquier, N., Seo, J. W., Celio, M., … Forró, L. (2006). Cellular toxicity of carbon-based nanomaterials. *Nano Letters, 6*(6), 1121–1125.

Ocsoy, I., Temiz, M., Celik, C., Altinsoy, B., Yilmaz, V., & Duman, F. (2017). A green approach for formation of silver nanoparticles on magnetic graphene oxide and highly effective antimicrobial activity and reusability. *Journal of Molecular Liquids, 227*, 147–152.

Organization, W. H. (2016). *Consolidated guidelines on the use of antiretroviral drugs for treating and preventing HIV infection: Recommendations for a public health approach.* Geneva: World Health Organization.

Organization, W. H. (2018). High levels of antibiotic resistance found worldwide, new data shows. *Saudi Medical Journal, 39*(4), 430–431.

Patra, H. K., Banerjee, S., Chaudhuri, U., Lahiri, P., & Dasgupta, A. K. (2007). Cell selective response to gold nanoparticles. *Nanomedicine: Nanotechnology, Biology and Medicine, 3*(2), 111–119.

Pires, J., Siriwardena, T. N., Stach, M., Tinguely, R., Kasraian, S., Luzzaro, F., ... Endimiani, A. (2015). In vitro activity of the novel antimicrobial peptide dendrimer G3KL against multidrug-resistant Acinetobacter baumannii and Pseudomonas aeruginosa. *Antimicrobial Agents and Chemotherapy, 59*(12), 7915–7918.

Pushparaj Selvadoss, P., Nellore, J., Balaraman Ravindrran, M., Sekar, U., & Tippabathani, J. (2018). Enhancement of antimicrobial activity by liposomal oleic acid-loaded antibiotics for the treatment of multidrug-resistant Pseudomonas aeruginosa. *Artificial Cells, Nanomedicine, and Biotechnology, 46*(2), 268–273.

Qi, L., Xu, Z., Jiang, X., Hu, C., & Zou, X. (2004). Preparation and antibacterial activity of chitosan nanoparticles. *Carbohydrate research, 339*(16), 2693–2700.

Qiu, Z., Yu, Y., Chen, Z., Jin, M., Yang, D., Zhao, Z., ... Qian, D. (2012). Nanoalumina promotes the horizontal transfer of multiresistance genes mediated by plasmids across genera. *Proceedings of the National Academy of Sciences, 109*(13), 4944–4949.

Riley, R. S., & Day, E. S. (2017). Gold nanoparticle-mediated photothermal therapy: Applications and opportunities for multimodal cancer treatment. *Wiley Interdisciplinary Reviews: Nanomedicine and Nanobiotechnology, 9*(4), e1449.

Schwartz, B., Bell, D. M., & Hughes, J. M. (1997). Preventing the emergence of antimicrobial resistance: A call for action by clinicians, public health officials, and patients. *JAMA, 278*(11), 944–945.

Scorciapino, M. A., Serra, I., Manzo, G., & Rinaldi, A. C. (2017). Antimicrobial dendrimeric peptides: structure, activity and new therapeutic applications. *International journal of molecular sciences, 18*(3), 542.

Shiohara, A., Hoshino, A., Hanaki, K., Suzuki, K., & Yamamoto, K. (2004). On the cytotoxicity caused by quantum dots. *Microbiology and Immunology, 48*, 669–675.

Silva, L. F., Oliveira, M. L., Neace, E. R., O'Keefe, J. M., Henke, K. R., & Hower, J. C. (2011). Nanominerals and ultrafine particles in sublimates from the Ruth Mullins coal fire, Perry County, Eastern Kentucky, USA. *International Journal of Coal Geology, 85*(2), 237–245.

Simon-Deckers, A., Loo, S., Mayne-L'hermite, M., Herlin-Boime, N., Menguy, N., Reynaud, C., ...Carriere, M. (2009). Size-, composition-and shape-dependent toxicological impact of metal oxide nanoparticles and carbon nanotubes toward bacteria. *Environmental science & technology, 43*(21), 8423–8429.

Sperling, R. A., & Parak, W. J. (2010). Surface modification, functionalization and bioconjugation of colloidal inorganic nanoparticles. *Philosophical Transactions of the Royal Society of London A: Mathematical, Physical and Engineering Sciences, 368*(1915), 1333–1383.

Srisitthiratkul, C., Pongsorrarith, V., & Intasanta, N. (2011). The potential use of nanosilver-decorated titanium dioxide nanofibers for toxin decomposition with antimicrobial and self-cleaning properties. *Applied Surface Science, 257*(21), 8850–8856.

Wang, H., Kou, X., Pei, Z., Xiao, J. Q., Shan, X., & Xing, B. (2011). Physiological effects of magnetite (Fe3O4) nanoparticles on perennial ryegrass (*Lolium perenne* L.) and pumpkin (*Cucurbita mixta*) plants. *Nanotoxicology, 5*(1), 30–42.

Xia, T., Kovochich, M., Brant, J., Hotze, M., Sempf, J., Oberley, T., ... Nel, A. E. (2006). Comparison of the abilities of ambient and manufactured nanoparticles to induce cellular toxicity according to an oxidative stress paradigm. *Nano Letters, 6*(8), 1794–1807.

Zeyons, O., Thill, A., Chauvat, F., Menguy, N., Cassier-Chauvat, C., Oréar, C., ... Spalla, O. (2009). Direct and indirect CeO2 nanoparticles toxicity for *Escherichia coli* and Synechocystis. *Nanotoxicology, 3*(4), 284–295.

Zhang, L. W., William, W. Y., Colvin, V. L., & Monteiro-Riviere, N. A. (2008). Biological interactions of quantum dot nanoparticles in skin and in human epidermal keratinocytes. *Toxicology and Applied Pharmacology, 228*(2), 200–211.
Zhang, L., Pornpattananangkul, D., Hu, C.-M., & Huang, C.-M. (2010). Development of nanoparticles for antimicrobial drug delivery. *Current Medicinal Chemistry, 17*(6), 585–594.
Zhu, L., Chang, D. W., Dai, L., & Hong, Y. (2007). DNA damage induced by multiwalled carbon nanotubes in mouse embryonic stem cells. *Nano Letters, 7*(12), 3592–3597.
Zhu, X., Chang, Y., & Chen, Y. (2010). Toxicity and bioaccumulation of TiO2 nanoparticle aggregates in *Daphnia magna*. *Chemosphere, 78*(3), 209–215.

Therapeutic Leishmaniasis: Recent Advancement and Developments in Nanomedicines

Mallikarjuna Rao Gedda, Om Prakash Singh, Onkar Nath Srivastava, and Shyam Sundar

Abstract Leishmaniasis is a syndrome caused by the protozoan parasites which is transmitted by the bite of the female sand fly. The disease has three major forms: Cutaneous, Mucocutaneous and most fatal Visceral Leishmaniasis (VL) in affected individuals. Additionally, some of the VL patients (<1%) develop stigmatizing post-kala-azar dermal leishmaniasis (PKDL) characterized by skin lesions in which parasites can be identified, who is otherwise fully recovered from VL. On the Indian Sub-continent, a joint VL elimination initiative has been launched in 2005 by the Governments of India, Bangladesh and Nepal. The main strategy to achieve this is entrusted to the public sector primary health care (PHC) services that should ensure early diagnosis and treatment of the disease. However, the current therapeutic options for treatment remains limited and treatment of patients are complicated due to the paucity of effective drugs or due to the toxicity caused by the available anti-leishmanial drugs used. Other limitations are the duration of treatment and expenditure on hospitalisation. Besides this, drug resistance in clinical settings has further aggravated the problem to the next level. Therefore, there is a need for cost-effective therapeutic alternatives for an excellent leishmanicidal potential that can be conferred to the target cells with no side effects/toxicity to normal cells, higher efficacy and minimal cost. Nanoparticles, due to their outstanding physical and chemical properties, have shown effective environmental, biological and biomedical applications and could be helpful in the detection and elimination of vector-based infectious diseases. In this chapter, we have summarized the current challenges in diagnosis and treatment of leishmaniasis and discussed the wide range of nanomaterials showing promising applications in leishmaniasis.

Keywords Nanomedicine · Visceral leishmaniasis · Diagnosis · Drug resistance

---

M. R. Gedda · O. P. Singh (✉) · S. Sundar
Infectious Diseases Research Laboratory, Department of Medicine,
Institute of Medical Sciences, Banaras Hindu University, Varanasi, India

O. N. Srivastava
Department of Physics, Institute of Science, Banaras Hindu University, Varanasi, India

S. Singh, P. K. Maurya (eds.), *Nanotechnology in Modern Animal Biotechnology*, https://doi.org/10.1007/978-981-13-6004-6_6

Abbreviations

DDT	Dichlorodiphenyltrichloroethane
MDR1	Multidrug resistance protein 1
MRPA	Multi-drug resistance associated protein ABC transporter
P-gp	P-glycoprotein
PRP1	Pentamidine resistance protein 1
WHO	World Health Organization

1 Introduction

Leishmaniasis, a neglected tropical disease caused by the protozoan parasite of the genus *Leishmania*, which is transmitted by the bite of female sand fly of the genus *Phlebotomine* (old world) and *Lutzomyia* (new world) leading to cutaneous (CL), mucocutaneous (MCL) and visceral leishmaniasis (VL) in affected individuals with diverse morbidity and mortality having global foothold in 98 countries (Torres-Guerrero et al. 2017; Tiwari et al. 2018) (Fig. 1). *Leishmania* species (of 54 known, 21 are pathogenic) exhibit digenetic life cycle (Fig. 2) i.e. promastigote, a flagellated extracellular infective form, which reside in the gut of the sand-fly and a amastigote form, which holed up in the phagolysosomal compartment of macrophage, leading to prior said clinical manifestations (Akhoundi et al. 2017). VL, CL and MCL subjects show different immunopathologies with few peculiar clinical outcomes and nearly 0.2–0.4 and 0.7–1.2 million cases of new VL and CL in the world and ~310 million subjects are under the risk of infection (Alvar et al. 2012). The clinical manifestations of VL include hepatosplenomegaly, anemia, leukopenia, low-grade fever, muscle wasting, polyclonal hyper-gammaglobulinemia and weight loss (Pearson and Sousa 1996; Dedet and Pratlong 2008). Additionally, approx 1% of these VL patients develop skin form of disease known as post-kala-azar dermal

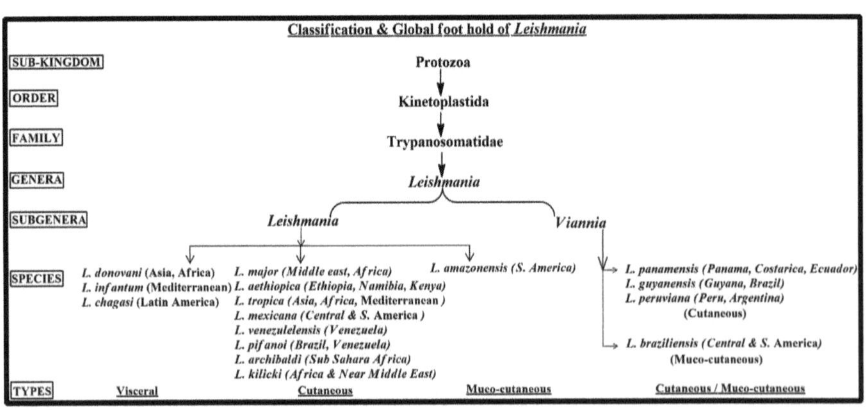

Fig. 1 Classification and global foot hold of different forms of Leishmaniasis

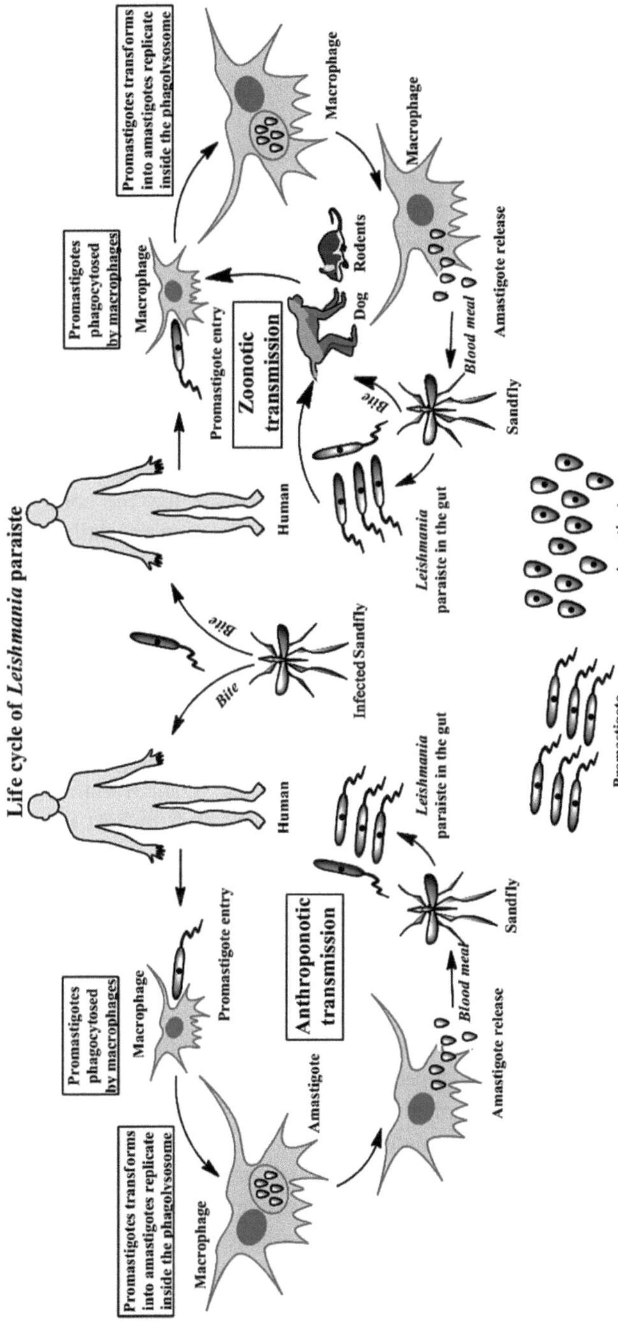

Fig. 2 Life cycle of *Leishmania* parasite in the vector i.e. Sand fly and the host i.e. Human

leishmaniasis (PKDL) after 6 months to 3 years of complete treatment (Mukhopadhyay et al. 2014). Importantly, infection with *L. donovani* does not always lead to clinical disease, for reasons that are not clearly understood.

Importantly, in endemic areas, a large number of exposed individuals are able to mount a protective cellular immune response against *Leishmania* and either eliminate infection or remain asymptomatic carriers (Ostyn et al. 2011; Hasker et al. 2013). The quantum of asymptomatic infection occurring in the endemic regions of the Indian subcontinent is not known, but available data from other regions show that the ratio is highly variable. These asymptomatic infected individuals can also jeopardize the VL elimination programme of 2020 due to their infectiousness to sand flies in non-endemic areas, besides their limited drug regimes. Moreover, the immune-compromised patients i.e. HIV subjects who are asymptomatic earlier may develop into symptomatic VL or CL (Alvar et al. 2008) and they are very difficult for treatment (Burza et al. 2014).

The incidence of the disease in subjects mainly depends on the *Leishmania* parasite ability to imbalance the host immunity for their luxuriant growth inside the macrophages by a survival strategy for evading the host innate and adaptive immune response (Gannavaram et al. 2016). Hence the parasite has the ability to switch between a pro-inflammatory Th1 type healing response to an anti-inflammatory Th2 type non-healing response, which prioritizes their survival and growth inside the macrophages. Additionally, the parasite has also the ability to inhibit the intracellular leishmanicidal activity by decreasing the production of ROS, nitric oxide and pro-inflammatory cytokines (Tiwari et al. 2018) leading for their better growth and survival by reduced proliferation of CD4+ and CD8+ T cells that eventually leads to enhanced Th2 response (Rodrigues et al. 2016). Furthermore, several co-inhibitory molecules; CTLA-4, PD-L1, CD200, and Tim-3, have shifted the balance of the immune system towards the non-healing Th2 response (Gannavaram et al. 2016).

The lack of knowledge regarding the Th1 to Th2 shift in host immunological response is due to the unidentified host or parasitic factors that contribute to the severe pathology of leishmaniasis. Due to the lack of demarcated entities for protective immunity of the host, generation of vaccines for the parasite has been a tough task for researchers. Several *Leishmania* vaccine candidates have been developed and evaluated in native and recombinant form like gp46, gp63, PSA2, LACK, TSA, LmsT1, Leish111f, m2, etc. and killed parasites. However, none of them have shown any outcomes towards prophylaxis (Nagill and Kaur 2011; Singh and Sundar 2012). Hence lack of prophylactic measures has been a concern in the elimination of this neglected tropical disease. Although the control measures for the elimination of *Leishmania* are limited, yet two strategies have been applied such as classical therapeutics interventions (use of miltefosine, amphotericin B, paromomycin, pentavalent antimonials and pentamidine) (Fig. 3a) and vector management through insecticides (DDT and synthetic pyrethroids) (Fig. 3b) for the control of the *Leishmania* parasite in disease-endemic regions (WHO 2018). The currently available therapeutic interventions are not effective antileishmanial drugs besides their enhanced number of cases with relapse and repercussions, have made the current situation critical for the elimination of leishmaniasis (Singh et al. 2014,

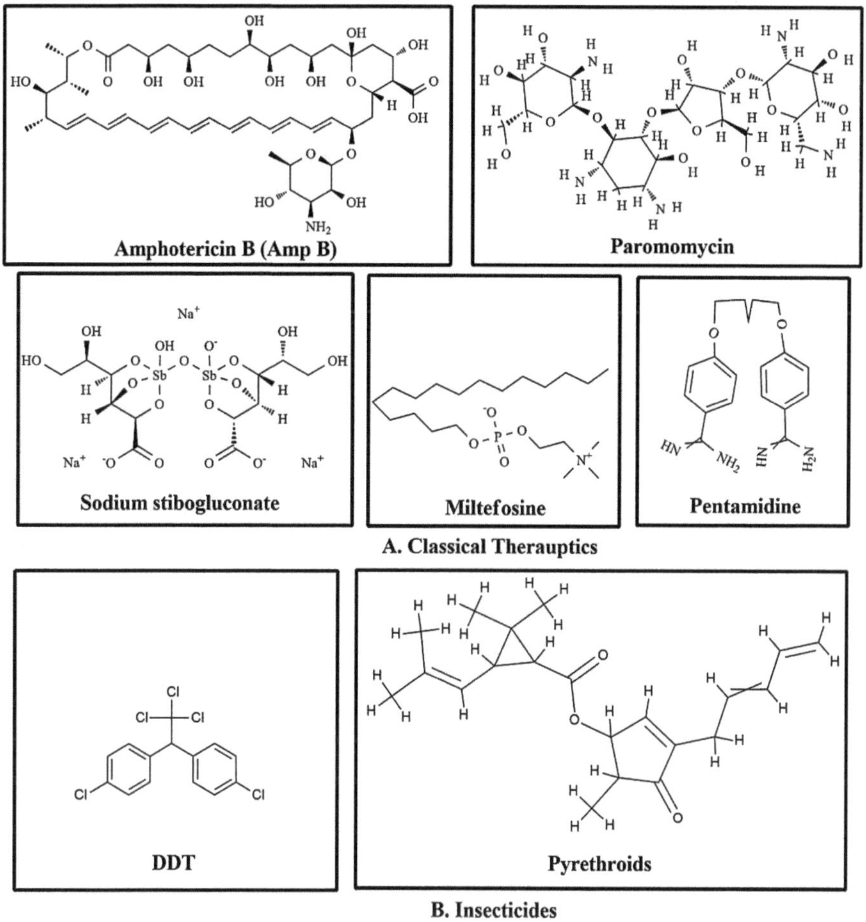

Fig. 3 Chemical structure of (**a**) Classical drugs used for the treatment of different forms of leishmaniasis and (**b**) Insecticides used for the vector management programme for mosquitoes

2016a). AmBisome, a liposomal formulation of amphotericin B has been in use through parenteral administration to VL patients in the endemic regions of Indian subcontinent (Singh et al. 2016b) with associated aftermath such as headache, nausea, dyspnea, liver dysfunctions and infusion reactions (Brand et al. 2017). Furthermore, PKDL cases have been a case of concern for both treatment and elimination strategies of the parasite, which further aggravate the issue to the next level. Besides, the DDT resistant in endemic areas, sand flies have even played a significant part in elevation of the problem. Hence, the altering epidemiology of the disease due to the migration of population i.e. asymptomatic subjects, PKDL and immune compromised infected individuals, unavailability of a prophylactic vaccine, limited drug regimens and their resistance, the emergence of sand flies resistant to insecticide, necessitate some other strategies to control the disease with the known nanomedicine.

2 *Leishmania* Diagnostics and Their Limitations

For the diagnosis and detection of *Leishmania* parasite from the samples, several clinical tests such as microscopic examinations, rK39 rapid diagnostic test (RDT), KATEX, Polymerase chain reaction (PCR) and Quantitative PCR (qPCR) are performed and they are still far from being ideal. Although, the approach like a microscopic observation of spleen aspirates of the subjects is considered the gold standard with high sensitivity, but their application is only limited to the lab settings and not feasible at primary health care centers (Srividya et al. 2011; Singh and Sundar 2015). rK39, a dipstick and rapid test which has been currently being used for the diagnosis of VL has the limitation of not being able to differentiate between active, cured VL and asymptomatic subjects (Boelaert et al. 2014). Therefore, in some cases, where the rK39 results were negative but showing VL manifestations or rK39 positive but not responding to the first line treatment then it has been prescribed for splenic/bone marrow aspirates examination using microscopy. Direct Agglutination Test (DAT), a non-invasive serological test for immunoglobulins have been extensively validated as per the recommendation by the World Health Organization (WHO) in endemic areas (Sundar and Rai 2002; Jacquet et al. 2006). The main issue associated with DAT is the requirement of expertise and facilities at the health centers for this diagnosis process. KATEX, a diagnostic test based on urinary leishmanial antigen have shown immune compromised co-infected patients of HIV-*Leishmania*. Importantly, KATEX results are not consistent and vary at different centers (Rijal et al. 2004; Sundar et al. 2005; Diro et al. 2007; Boelaert et al. 2008) representing its poor sensitivity.

Several molecular tools have been developed so far with good sensitivity and specificity. Polymerase Chain Reaction (PCR) is one of the molecular diagnosis tool that helps in amplification of *Leishmania* parasite DNA from blood and bone marrow using specific primers that amplifies a target sequence in the genome of the parasite, which can play a vital role in observing the drug efficacy and disease early reporting (Blackwell 1992; Brustoloni et al. 2007; Srivastava et al. 2011; Alam et al. 2012). The target sequences that has been amplified during several studies include repetitive kinetoplastid DNA (kDNA) (Salotra et al. 2001; Cortes et al. 2004; Maurya et al. 2005), ribosomal internal transcribed spacer (ITS) region (Schonian et al. 2003) and small subunit ribosomal rRNA gene (SSU rRNA) (Mathis and Deplazes 1995; Salotra et al. 2001; Reithinger and Dujardin 2007; Srivastava et al. 2010, 2011). Quantitative PCR/Real-time PCR diagnosis technique helps in detection of parasite load and even a single molecule from complex samples with high sensitivity and accuracy besides the treatment response of the drugs (Sudarshan et al. 2011) and asymptomatic parasitemia (Sudarshan et al. 2014). Although, these molecular-based methods have shown sensitivity and specificity the main issue remains the same due to their lack of adaptability in health centers.

3 Current Therapeutic Regimen and Their Limitations

The current limited drug regime for the control of leishmaniasis include only five drugs; pentavalent antimonials (true antileishmanial), miltefosine (anticancerous), amphotericin B (antifungal) and its liposomal formulations i.e. AmBisome, pentamidine (antimicrobial) and paromomycin (antiprotozoal) as shown in Fig. 3a (Markle and Makhoul 2004; Palumbo 2010; Freitas-Junior et al. 2012).

3.1 Pentavalent Antimony Compound

Pentavalent antimony has been in use for several decades as the first line of drug for the treatment of leishmaniasis (VL, CL, MCL) (Reithinger et al. 2007; Haldar et al. 2011) with commercial names such as meglumine antimoniate and sodium stibogluconate. Under low pH conditions, pentavalent antimony Sb(V), which is a pro-drug is converted into active trivalent Sb(III) with the ability to inhibit enzyme trypanothione reductase (Tiwari et al. 2018). In order to counter the host oxidants, the parasite produces trypanothione reductase, which converts the oxidized trypanothione to reduced trypanothione. Besides, this trypanothione reductase system, the parasite doesn't have any antioxidant machinery for its survival against host ROS and RNS. Hence, this enzyme would act as an excellent drug target for the clearance of the parasite in the treatment of leishmaniasis. Additionally, Sb(V) has the ability to inhibit the topoisomerase- I enzyme activity (No 2016). But in the past few years, the relapse cases of pentavalent antimony in endemic regions has compromised its use in the Indian subcontinent (Chakravarty and Sundar 2010).

3.2 Pentamidine

Although the precise mode of action of the drug pentamidine is not well known in the case of leishmaniasis, it is known to inhibit active transport system of *Leishmania* parasite (No 2016). It has been also reported through a study that pentamidine enters the parasite through the arginine and polyamine transporters and finally inhibits mitochondrial topoisomerase II enzyme activity leading to the death of the *Leishmania* parasite (Singh et al. 2016b). This drug has got serious aftermaths causing gastrointestinal and cardiac toxicity (Das et al. 2001). Though pentamidine resistance hasn't been reported in clinical settings yet but in experimental settings, the gradual increase in pentamidine concentration has evoked resistance in *L. donovani* (Mukherjee et al. 2006).

3.3 *Miltefosine*

Miltefosine, which is an anti-cancerous oral drug has been in use in the treatment of VL and PKDL (Sundar et al. 2015). This alkyl-phosphocholine drug affects the phosphoinositol 3-kinase-Akt/PKB signaling, which in turn affect the metastatic growth of tumor cells and in case of *L. donovani* promastigotes, miltefosine cause apoptosis-like condition with an unknown mechanism (Paris et al. 2004; Dorlo et al. 2012). Additionally, the death of the parasite also occurs due to the depolarization of mitochondrial membrane and inhibition of cytochrome-c oxidase leading to apoptosis of *L. amazonensis* promastigotes (Marinho et al. 2011). Due to severe toxicity, miltefosine hasn't been prescribed to children and pregnant women because it leads to gastrointestinal toxicity besides its teratogenic effect (Dorlo et al. 2012). About 2–3% of patients on miltefosine develop life-threatening side-effects (1% severe diarrhea/vomiting, 1.6% renal toxicity, 0.3–0.4% clinical hepatitis). Nepal proposes to hospitalize patients on miltefosine for 3 days. Only a few district-level health facilities are equipped to monitor renal and liver functions. Importantly, due to its 7 day half-life can shift the parasite towards resistance in clinical settings. But in lab settings, it has also been reported that enhanced expression of P-glycoprotein and ABC transporter besides down-regulation of proteins responsible for the miltefosine uptake (Mishra and Singh 2013).

3.4 *Amphotericin B (AmpB)*

In India, amphotericin B has replaced Sb in many districts. Amphotericin B is administered as intravenous infusions at 1 mg/kg on alternate days for 15 infusions, necessitating hospitalization for prolonged periods and thus limiting the number of patients who can be treated. In addition, infusion reactions are very common and occasional serious adverse events like nephrotoxicity, myocarditis, and deaths can occur with this amphotericin B.

Amphotericin B, an antifungal drug, has been used in the treatment including the endemic regions of the Indian subcontinent for leishmaniasis (Singh et al. 2016b; Sundar and Singh 2016) by targeting parasite cell membrane ergosterol leading to the death of the parasite due to altered membrane permeability and ion influx (Tiwari et al. 2018). However, in clinical settings AmpB importance has been threatened by the extensive and improper use that may further help in the development of resistance. The serious side effects such as nephrotoxicity and infusion reactions have been a concern for the clinical settings for the elimination of VL (Freitas-Junior et al. 2012), which has been successfully minimized by their liposomal formulation i.e. Ambisome in the endemic area (Singh et al. 2016b).

3.5 Liposomal Amphotericin-B

Experiences with lipid formulations of amphotericin B such as AmBisome or Albacet are excellent in treating VL cases. Now AmBisome is available at a WHO negotiated price of 10% of its original price and therefore has become affordable within the national elimination program. This is very significant because now AmBisome is being considered as a first line drug for VL and this could significantly change the way VL is managed in the near future. AmBisome has important advantages over the other VL drugs, miltefosine, paromomycin and antimonials as treatment do not need hospitalization of patients. However, most of side effects associated with AmBisome are infusion-induced reactions of a milder nature, which are seen in only one-third of the patients and these usually do not require any intervention. Moreover, a single dose of AmBisome enables treatment at the public health center level. This is also the most practical and less expensive drug for the patients and will be successful in reducing the burden on the larger district hospitals.

3.6 Paromomycin

Paromomycin, an antibiotic, which shows both antileishmanial and antibacterial activities has been known for leishmanicidal activity since 1960 and came into human trials in Kenya from the 1980s (NIu and Pershin 1966; Chunge et al. 1990). This drug shows poor absorption and the invasive intramuscular medication has been an issue that discouraged its use in the treatment of leishmanial patients. Although the mode of action of this drug is largely unknown, the parasitic death has been attributed due to the inhibition of protein synthesis and disruption of mitochondrial membrane potential (Chawla et al. 2011). Though resistance in clinical settings hasn't been reported yet but experimentally resistant *L. donovani* and *L. infantum* could be generated (Hendrickx et al. 2014), which may possess critical threat due to their extensive and improper use.

Besides these classical treatments for leishmaniasis, several natural compounds (sulphated polysaccharides, alkaloids, terpenoids, monoterpenes, sesquiterpenes, iridoids, diterpenes, triterpenes, sterols, phenolics and some metabolites from algae, fungi and other sources etc) have been evaluated at the in vitro and in vivo level to check their efficacy and the results have been encouraging for their future use as effective drug candidates against leishmaniasis (Iqbal et al. 2016; Tiwari et al. 2018). This vast reservoir of natural compounds can offer an excellent opportunity for the generation of new drug candidate for VL which can save humans from this vector born disease. However, lack of proper focused and prioritized research to identify the natural compounds has let us down in finding the one for VL thearuptics. Additionally, these natural compounds tagged with NPs (different) can serve this purpose for which every individual compound should be further tested exhaustively.

4 Nanomedicine

Innovations in interdisciplinary sciences have been moving the translational sciences to the next level for better control of infectious diseases. Nanomedicine is one of the promising fields in this area that has been continuously growing, keeping the hope for highly sensitive diagnostic tools and better drugs for VL in near future. In view of that, systematic evaluation of these nanomedicines based approaches are the current need to develop safe, effective and affordable VL drug delivery systems, that may also find applicability in other chronic infections. With a clear objective of developing high potential nanomedicinal approach for addressing the existing challenges of VL, several fine pieces of work have been published in recent years. NP-based diagnostics and drug delivery system both at the in vitro and in vivo levels using polymer NPs, polysaccharide polymers, polymeric micelles, quantum dots, nanoemulsions, niosomes, lipid cochleates, nano-discs, liposomes, solid lipid NPs and inorganic compounds (Fig. 4) have shown some encouraging results in case of experimental leishmaniasis (Yasinzai et al. 2013; Gutiérrez et al. 2016; Bruni et al. 2017). These studies have been covered extensively in detail in the coming sections.

4.1 *Nanoparticles Based Diagnosis of Leishmaniasis*

Improvement in current approaches and the discovery of appropriate biomarkers for the detection of *Leishmania* are very much crucial for the early diagnosis and therapy. Several crucial features such as sensitivity, specificity, ease of usage and cost-effectiveness are needed in the field of diagnostics for the identification of 21 pathogenic *Leishmania* species in humans (Akhoundi et al. 2017). Hence a non-gel and non-culture based technique with higher sensitivity have to be established in resource stringent endemic areas. NPs based diagnostics can fit into this bill having different physical and chemical characteristics with high sensitivity and specificity to decrease the cost burden for disease diagnosis in near future (Sharma et al. 2015). The NP-based biosensors are showing promising results in the detection of pathogenic infectious agents like *Salmonella*, *S. aureus*, *E. coli, Influenza*, etc. with great stability, sensitivity and selectivity (Torres-Sangiao et al. 2016). Though, the progress in this direction has been not up to the mark in case of leishmaniasis, few studies have generated a ray of hope for future diagnostic development for the disease (Mohan et al. 2011; Bose et al. 2015; Andreadou et al. 2016; Heli et al. 2016; Moradi et al. 2016; Sattarahmady et al. 2016; Toubanaki et al. 2016).

A recent technique based on Lateral Flow Immunoassay (LFIA), developed by Anfossi et al. for the detection of VL in canines is highly effective and convenient (Anfossi et al. 2018). It includes two components, gold NPs with protein A labelled as the reporter and a specific chimeric recombinant antigen which binds with anti-leishmanial antibodies with 98.4% sensitivity and 98.9% specificity. Souto et al. developed SPR immunosensor for *L. infantum* detection by using antibodies fixed

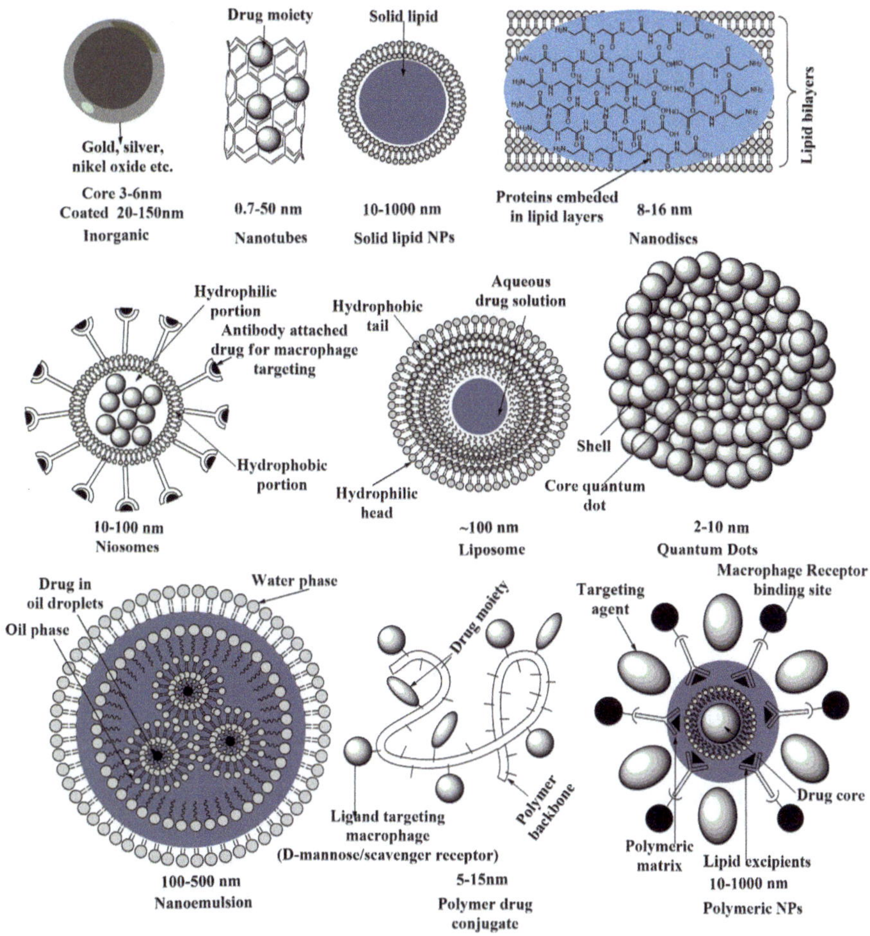

Fig. 4 Nanoparticles that has been already used for the drug delivery and diagnostics against leishmaniasis

on gold dendrimers against hypothetical C1 protein (Souto et al. 2015). Perinoto et al. developed an efficient diagnosis of leishmaniasis through nano structured films with specific antigen of *L. amazonensis* and *Trypanosoma cruzi* through impedance spectroscopy (Perinoto et al. 2010). In another study by Andreadou et al., when gold NPs tagged with four oligonucleotide probes that were able to detect *Leishmania* kinetoplastid minicircle DNA, which is a non-amplification detection assay (Andreadou et al. 2014). Toubanaki et al. used dog blood samples for detection of leishmaniasis through gold NP-based lateral flow biosensor that can detect the amplification of *Leishmania*-specific kinetoplast DNA (Toubanaki et al. 2016).

Sol-gel synthesized nickel oxide film coated onto glass plate with indium tin oxide has developed DNA biosensor for the diagnosis and detection of VL (Mohan et al. 2011). Gold NP-based detection has been successfully developed which can be helpful for SAG-susceptibility profiling in *Leishmania* endemic regions (Bose et al. 2015). In the next year, the same group worked on visual assessment of parasite burden (amastigotes) through *Leishmania*-specific small subunit rRNA using gold NPs by plasmonic detection (Bose and Kumar 2016). A combination of cadmium selenite QD probes has been employed for the *Leishmania*-specific surface antigens detection of proteins and DNA on cultured isolates of several microbial pathogens, which have shown 100% sensitivity and specificity that can be helpful in resource-poor settings (Andreadou et al. 2016). Magnetic cobalt-zinc ferriteQDhybridized with SS-DNA probe and methylene blue has shown high selectivity and sensitivity for the detection of *L. major* (Heli et al. 2016). Gold nano leaves synthesized by electrodeposition method tagged with DNA probe by immobilization has the ability to detect DNA with a detection limit of 1.8×10^{-20} $molL^{-1}$for *L. major* (Moradi et al. 2016). Gold NPs based biosensing of kDNA genome has been synthesized which can be helpful for diagnosis of *L. major* from other non-*Leishmania* species with a detection limit of 7.0 pg μL^{-1} (Sattarahmady et al. 2016). All these studies on nanodiagnostics have been summarized in Table 1, though a lot has to be explored for early diagnosis followed by early treatment, which ultimately serves towards the early elimination of this infectious disease.

4.2 Nanomedicine for Drug Delivery and Its Efficacy for Leishmaniasis

In order to reduce the side effects of AmpB and invasive mode of delivery and the high cost of Ambisome, several research groups have been in the process of developing NPs that enhances the efficacy and non-invasive (oral) mode based drug delivery for leishmaniasis. These NPs are targeted towards the *Leishmania* infected macrophages with specific properties and pharmacokinetics. Incidentally, to achieve this task researchers have been working rigorously at the in vitro and in vivo level for the treatment of leishmaniasis using non macrophagic and macrophage directed NPs drug delivery system (Akbari et al. 2017; Sarwar et al. 2017). Our lab has also worked at both in vitro and in vivo level (invasive and non-invasive mode) for the treatment of leishmaniasis using non-macrophagic and macrophage directed amine functionalized carbon nanotubes and graphene and chitosan NPs drug delivery system of AmpB (Prajapati et al. 2011a, b; Mudavath et al. 2014, 2016; Chaubey et al. 2018).

A nano-reservoir of lactoferrin appended poly (d, l-lactide- coglycolide) encapsulated with AmpB was fabricated with mean particle size ranging from 196.0 ± 5.28 nm. This formulation showed greater suppression of the parasite (88%) compared to conventional formulations in experimental VL model (Asthana et al. 2015). Nanocapsules loaded with doxorubicin (Fig. 5) have shown enhanced anti-antileishmanial activity with enhanced Th1 immune response (IL-12, IFN-γ and TNF-α) with parallel alleviation in Th2 immune response (IL-4, IL-10 and TGF-β)

Table 1 Sensitivity and specificity of nanoparticles in leishmaniasis

S.No	NPs	Leishmaniasis	Probe attached	Detection	Sensitivity & specificity (%)	Remarks
1	Quantum dots of magnetic cobalt-zinc ferrite	*L. major*	Specific single stranded DNA probe ((p-ssDNA) sequence: 5′-TGTTGGGTGACGCTTTAGTGGGTT-3′; t-ssDNA sequence: 5′-AACCCACTAAAG-CGTCACCCAACA-3′)	Electrochemical genosensor	NA	ssDNA target (10^{-11}–10^{-18} Mol/L with a limit in detection up to 2×10^{-19}) and genomic DNA (7.31×10^{-14}–7.31×10^{-6} ng/μL with a limit in detection up to 1.8×10^{-14})
2	Nickel oxide	*L. donovani*	23mer DNA sequence (5′GCCGAATAGAA-AAGATACGTAAG3′)	DNA biosensor	NA	Detection limit is 0.02 ± 0.002 ng/μl
3	Cadmium selenite quantum dot and magnetic beads	*L. infantum*	Oligonucleotide probes (LeishQD1 (5′-3′): biotin-AAGA-GGCGGTGTCACAGAGATGGG; LeishQD2 (5′-3′): biotin-ACAGCGACGT CCGTGGAAAG) anti-Leishmania LPG (IgM); antiLeishmania gp63 (IgG1a); biotin-labeled polyclonal anti-mouse antibody (IgM); biotinlabeled polyclonal anti-mouse IgG1	Leishmania DNA and its specific surface antigens	100 & 100	Lower limit of detection 3125 ng/μl
4	Nanostructured films	*L. amazonensis*	Purified anti-IgG	Specific *L. amazonensis* and *T. cruzi* antigens and employing impedance spectroscopy	NA	
5	Gold nanoleaves	*L. major*	Probe DNA sequence: 5′-SH-(CH2)6-TTTTTTTTTT-AACCCACTAAAGCGTCACCCAACA-3′	Electrochemical DNA biosensor	NA	Limit of detection of 1.8×10^{-20} mol/L

(continued)

Table 1 (continued)

S.No	NPs	Leishmaniasis	Probe attached	Detection	Sensitivity & specificity (%)	Remarks
6	Gold	*L. major*	Oligonucleotide (5′-SH-(CH2)6-T10-AACCC ACTAAAGCGTCACCCAACA-3′) of k DNA	Visual and spectrophotometric	NA	Detection limit of 7.0 pg/μL
7	Gold	Canine VL	Chimeric recombinant antigen and protein A	Lateral flow immunoassay	CVL endemic region (95.8 & 100)	
					CVL nonendemic region (100 & 98.8)	
					Both regions (98.4 & 98.9)	
8	Gold	*L. infantum*	Biotin 22-mer 5′-CTTTTCTGGTCCT CCGGGTAGG-3′ upstream primer for *Leishmania* kinetoplast DNA	Lateral flow biosensor	NA	
			Amplification, 24-mer 5′-CCACCCGGCCCTATTTTACACCAA-downstream primer. 24-mer 5′-TTTTCGCAGAACGCCCCTACCCGC-3′			
			Leishmania-specific probe			
9	Gold	*L. donovani*	5′GGACGCACTAAACCCCTCAA3′) of SSU rRNA gene and a-tubulin gene of LD (forward primer-5′TCTGCCTTGA GCACGGCATC3′, reverse primer 5′ACACCAGCTG CTCGGGGTTG3′)	Plasmonic detection of leishmania specific marker RNA	NA	
10	Gold and dendrimers	*L. infantum*	Antibodies against hypothetical C1 protein	Surface plasmon resonance immunosensor	NA	Lower limit of detection 7.37 nmol/L low limits of quantification 7.83 nmol/L

11	Gold	*L. donovani* SAG resistant	Chr13 Pos442924 (Sensitive-5′GCCGACCAAGCCTAATTT3′,	Plasmonic detection of PCR amplified genetic biomarker	NA	Lower limit of detection 212.5–6.64 ng
			Resistance-5′GCCGACCAAGCCTAATT**A**3′) Chr35Pos1192217(sensitive- 5′ GCGACGACTGCGGCGGTG3′,			
			Resistance-5′GCGACGACTGCGGCGGT**A**3′) Chr35Pos1619958 (sensitive- 5′ GGGCGTAGAGGAAGCTCG3′,			
			Resistance-5′GGGCGTAGAGGAAGCTC**A**3′) Chr35Pos1656169 (sensitive- 5′ CAGCGAACGTACTGCCTC			
			3′, Resistance-5′ CAGCGAACGTACTGCCT**T** 3′), kDNA (5′ GGCTTAGTGTTAGCCGTGTGT3′)			
12	Gold	Canine leishmaniasis	4 oligonucleotide sequence (5′–3′)	Targeting kinetoplastid minicircle DNA	92 & 100	Lower limit of detection 11.5 ng/μl
			LeishAu1			
			(GTTAGCCGATGGTGGTCTTG)			
			LeishAu2			
			(ACGGGTGTCTTTGATGATGC)			
			LeishAu3			
			(TAGTCTGGTGGGATGCTTCG)			
			LeishAu4			
			(GTGCCTTTGATGTGGGTGTT)			
13	Gold nanorods	*L. major*	NASBA/RT-PCR F primer 5′-CCAAAGTGTGGAGATCGAAG-3′	Colorimetric assay for 18S rRNA	100 & 80	
			RT-PCR R primer 5′-AGGGCCGGTAAAGGCCGAATAG-3′			
			NASBA R primer 5-AATTCTAATACGA CTCACTATAGGGA-GAAGGGCCGGTAA AGGCCGAATAG-3			
			Nucleic acid sequence-based amplification (NASBA)			

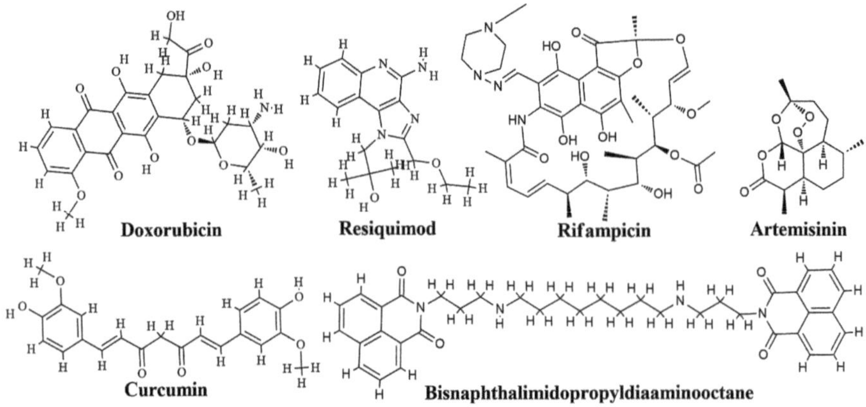

Fig. 5 Chemical structure of some drugs appended to nanoparticles used for the treatment of experimental leishmaniasis

(Chaurasia et al. 2015). In another study, highly pressurised AmpB with particle size ranging from 10 to 20 nm have shown significantly better results than AmpB with inhibition of the parasite replication (92.18% versus 74.57%, p = 0.005) and parasite suppression (99.18% versus 97.17%, p = 0.05) by several folds (Manandhar et al. 2008). Effective uptake of curcumin loaded mannosylated chitosan NPs of mean particle size 215 nm showed effective uptake by macrophages of reticulo-endothelial system by endocytosis (Chaubey et al. 2018). Stable emulsion based formulation of doxorubicin showed improved efficacy IC50 is almost 1.9-fold as compared to the plain drug, against *L. donovani* intracellular amastigotes (Kansal et al. 2012). Intra macrophagic amastigote directed delivery of doxorubicin was evaluated by using chitosan microparticles. The formulation can potentially resist lysosomal pH due to it's "acid-resistive" property and activation of macrophages resulting in high macrophage circulation and subsequent engulfment of microparticles (Kunjachan et al. 2011). Chitosan NPs of size 215 nm conjugated with rifampicin and D-mannose particles has been found to be effective for treatment of VL (Chaubey and Mishra 2014).

Gelatin NPs were modified by 1, 2-diacyl-sn-glycero- 3- phospho-l- serine and then complexed with AmpB showed targeted delivery with enhanced uptake in J774A.1 macrophage cell lines, higher extent of accumulation in macrophage-rich organs, particularly in liver and spleen and 85% inhibition of splenic parasitic burden in *L. donovani* infected hamsters when compared to the conventional AmpB (Khatik et al. 2014). Noncovalent complexation of AmpB with biodegradable, nontoxic and biocompatible polymer, poly (α-glutamic acid) (PGA) exhibited a potent and consistent antileishmanial activity equal to that of AmBisome, reduced toxicity compared to that of fungizone (AmpB) with increased selectivity for intracellular *L. donovani* amastigotes and is highly stable (Mohamed-Ahmed et al. 2013). The particle size (>100 nm), highly aggregated state, negative surface charge and colloidal nature of the complex enhanced the uptake by mononuclear phagocyte system (Sadat et al. 2016). Bis-naphthal-imido-propyldiaamino-octane (BNIPDaoct) loaded poly (d,l-lactide-co-glycolide) (PLGA) NPs of size ranging from 156.0 ± 2.8 nm with encapsulation efficiency of 85% by nano-precipitation method showed effective and selective antileishmanial activity against *L. infantum* with compelling evidence for complete internalization by the macrophages, the target cells for *Leishmania* infection (Costa Lima et al. 2012).

Nanospheres based drug delivery system for AmpB evidenced preferential accumulation in the visceral organs with striking immune-modulatory effect thereby significant reduction in parasite burden (Lima et al. 2014). This conjugate was shown to be stable over 60 days at 30 °C. A dual drug conjugate of AmpB and Miltefosine (HePC) loaded nanocochleates with a mean particle size of 250 ± 2 nm showed drug release preferentially in the intestinal medium containing bile salts hence showing a promise for oral applications (Pham et al. 2013). Electrospray was shown to greatly improve the encapsulation efficiency of resiquimod in polymer micro-particles of approximately 2 μm (Duong et al. 2013). Further when administered intravenously showed a significant reduction in the parasite load in mice infected with *L. donovani* parasites. Also, this technique appears to offer an elegant, scalable acid sensitive delivery vehicle in the treatment of fatal VL.

Liposomal resiquimod of 75.0 ± 30.7 nm in diameter has been prepared by lipid film hydration with extrusion method, didn't have any hepatic or renal toxicity and showed enhanced parasite clearance in experimental VL (Peine et al. 2013). AmpB loaded on functionalized poly (d,l-lactide-co-glycolide) nanospheres of 180.1 ± 5.1 nm size showed increased levels of IFN-γ, TNF-α and nitric oxide thereby substantial reduction in parasite burden both in vivo and in vitro (Shirkhani et al. 2015). Development of a formulation of polyester NPs for delivery of AmpB, which can be given by a non-invasive route was significantly more effective than the same dose of AmpB at suppressing parasite numbers compared to controls in bone marrow derived macrophages infected with *L. donovani* (Bruni et al. 2017). Further, nano mediated combinational drug delivery and the bi-functionalized nano-carrier system would overcome all the hurdles faced by the first line drugs. These receptor mediated targeted drug delivery using NPs may improve traditional chemotherapy efficacy by precisely targeting macrophage phagosomal *Leishmania* parasite.

5 Limitation and Future Prospects of Nanomedicine for Leishmaniasis

Nanoformulations with common limitation for drug delivery include compromised physicochemical characterization, without proper sterility and endotoxin, residual components, inactive pharmaceutical ingredients, improper batch to batch validations, NP instability in vivo and their improper drug release rate making them inefficient for invasive and non-invasive administration (Crist et al. 2013). Hence, the future of biomedical research has been directed towards NPs based drugs candidates that are safe, efficient and affordable with antileishmanial therapy having shorter duration. Most of the above studies discussed have been directed through the invasive route of NPs delivery. Hence much of the focus has to be shifted on non-invasive methods which are stable showing higher efficacy against leishmaniasis. Additionally, both in vivo and in vitro studies with limited pharmacokinetics, nanotoxicity, biodegradability and post exposure bio-persistence and NPbio-distribution haven't been explored yet in case of leishmaniasis in liver, spleen, kidney and other tissues. Several cancer studies have gone through extensively through then genotoxicity, biodegradability and post exposure biopersistence of NPs based drug delivery system for its future application in clinical settings (Poland et al. 2008; Singh et al. 2012; Ferreira et al. 2013; Hobson et al. 2016; Savaliya et al. 2016; Rahman et al. 2017), and it can be used as a lead for *Leishmania* infection and their NPs based drug delivery.

Although the nano diagnostic techniques in a clinical setting are dependable for their applications, yet there are limitations due to their false-negative results, inaccuracy and paucity of sensitivity in detection (Wang et al. 2017). Therefore, more and more novel strategies are being adopted for conquering the uncharted territories in VL diagnosis and the future objective of the biomedical research. The recent detection techniques utilize inorganic NPs (gold, nickel oxide, cobalt-zinc ferrite, cadmium selenide QD) focuses mainly on DNA/RNA based detections and lot has to be explored targeting glycomics and proteomics approaches in *Leishmania* parasite. Additionally, greater improvement can be also done if the upcoming techniques focus on disease detection alongside its drug resistance in clinical samples.

Several studies on leishmaniasis have found that the parasite is sensitive to the ROS/RNS and contains only the trypanothione/trypanothione reductase system to combat oxidative stress, which is not sufficient enough (Dumas et al. 1997; Mauël and Ransijn 1997; Tiwari et al. 2018). Hence targeting trypanothione reductase system with NPs may help in death of the parasite through oxidative/nitrosative burst and a recent study using meglumine antimoniate tagged to mannose loaded thiolated polymer-based NPs showed profound leishmanicidal activity by inhibiting the trypanothione reductase and P-gp (ABC transporter) efflux pump (Sarwar et al. 2018). This strategy enhances the efficacy of antimonial drug resistant leishmaniasis and it could also be feasible with other drug resistant leishmaniasis using NPs.

Leishmania affects the host's immunity by altering some major macrophage functions by adaptive differentiation and through exoproteins (Lamotte et al. 2017) which eventually affects the host immunity for the parasite survival. Studies have indicated that only systematic and focused approach targeting the sterol biosynthesis pathway, trypanothione metabolism, purine salvage pathway, glyoxalase system, protein kinases and proteases can curb the parasite growth because biochemical pathways of the *Leishmania* parasite and mammalian host are structurally and functionally different (Tiwari et al. 2018). The post-translation modification of histones by histone-modifying enzymes could be helpful in the study of VL pathogenesis as it enables the better understanding of epigenetic determinants in the parasite (Jha et al. 2017; Lamotte et al. 2017). This would also give a clue in designing biocompatible NPs for VL elimination.

A systematic study of different characteristics of NPs and their individual effect on the biological system would make them a safe, cost-effective, efficacious endocytic alternative drug delivery system with no cases of resistance. The enhanced expression of ABC transporters (MRPA, MDR1, PRP1) causes the drugs efflux out of the macrophages infected with the parasite (Sarwar et al. 2017; Tiwari et al. 2018). However, the use of thiolated surface modified NPs tagged with AmpB have shown many fold internalization of drugs in *L. donovani* due to the inhibition of AmpB efflux (Shahnaz et al. 2017). Similarly, an effective concentration of AmpB drug delivery through thiolated NPs carrier inhibits the *Leishmania* replication.

6 Conclusion

Highly sensitive diagnostics and less toxic and more efficacious therapies are required for treating the individuals suffering from chronic infectious diseases including VL. Due to the limited drug regimens, their resistance and side effects of the treatment for leishmaniasis is a matter of huge concern for the patients, practitioner and researchers. Besides this, no other drug candidates seem to be coming into the market for human use in near futures. The drug and diagnostic limitation in leishmaniasis have made the situation more favorable for *Leishmania* survival in the host and vectors jeopardizing the elimination programme of VL. Hence the need of the hour are NP tagged drugs, co-delivery of drug-drug and some other natural compounds, which have been found to be working well at the in vitro and animal level, although more studies have to be done in the field of nanotoxicity, biopersistence and biocompatibility in order to justify its use in human trails. Extensive translational researches with NPs based theranostics can make this unachievable goal possible by 2020 for leishmaniasis control. Additionally, cost-effective, sensitive and sensible use of nanodiagnostics may further help in the elimination of leishmaniasis in the endemic and non-endemic regions, which may improve the quality of life of millions of impoverished people.

Acknowledgement This work was supported by Department of Science & Technology (SR/NM/NS-57/2016), New Delhi (Under nano-mission), and in part by the Extramural Program of the National Institute of Allergy and Infectious Diseases, National Institutes of Health (TMRC grant number U19AI074321). The funders had no role in design, decision to publish, or preparation of the report.

Conflict of Interest The authors declare no conflict of interest.

References

Akbari, M., Oryan, A., & Hatam, G. (2017). Application of nanotechnology in treatment of leishmaniasis: A review. *Acta Tropica, 172*, 86–90. https://doi.org/10.1016/j.actatropica.2017.04.029.

Akhoundi, M., Downing, T., Votýpka, J., Kuhls, K., Lukeš, J., Cannet, A., et al. (2017). Leishmania infections: Molecular targets and diagnosis. *Molecular Aspects of Medicine, 57*, 1–29.

Alam, M. Z., Yasin, G., Kato, H., Sakurai, T., & Katakura, K. (2012). PCR-based detection of Leishmania donovani DNA in a Stray dog from a visceral Leishmaniasis endemic focus in Bangladesh. *The Journal of Veterinary Medical Science, 75*(1), 75–78. doi: DN/JST.JSTAGE/jvms/12-0134 [pii].

Alvar, J., Aparicio, P., Aseffa, A., Den Boer, M., Canavate, C., Dedet, J.-P., et al. (2008). The relationship between leishmaniasis and AIDS: The second 10 years. *Clinical Microbiology Reviews, 21*(2), 334–359.

Alvar, J., Velez, I. D., Bern, C., Herrero, M., Desjeux, P., Cano, J., et al. (2012). Leishmaniasis worldwide and global estimates of its incidence. *PLoS One, 7*(5), e35671. https://doi.org/10.1371/journal.pone.0035671.

Andreadou, M., Liandris, E., Gazouli, M., Taka, S., Antoniou, M., Theodoropoulos, G., et al. (2014). A novel non-amplification assay for the detection of Leishmania spp. in clinical samples using gold nanoparticles. *Journal of Microbiological Methods, 96*, 56–61.

Andreadou, M., Liandris, E., Gazouli, M., Mataragka, A., Tachtsidis, I., Goutas, N., et al. (2016). Detection of Leishmania-specific DNA and surface antigens using a combination of functionalized magnetic beads and cadmium selenite quantum dots. *Journal of Microbiological Methods, 123*, 62–67.

Anfossi, L., Di Nardo, F., Profiti, M., Nogarol, C., Cavalera, S., Baggiani, C., et al. (2018). A versatile and sensitive lateral flow immunoassay for the rapid diagnosis of visceral leishmaniasis. *Analytical and Bioanalytical Chemistry, 410*, 1–12.

Asthana, S., Gupta, P. K., Jaiswal, A. K., Dube, A., & Chourasia, M. K. (2015). Targeted chemotherapy of visceral leishmaniasis by lactoferrin-appended amphotericin B-loaded nanoreservoir: In vitro and in vivo studies. *Nanomedicine, 10*(7), 1093–1109.

Blackwell, J. M. (1992). Leishmaniasis epidemiology: All down to the DNA. *Parasitology, 104*(Suppl), S19–S34.

Boelaert, M., El-Safi, S., Hailu, A., Mukhtar, M., Rijal, S., Sundar, S., et al. (2008). Diagnostic tests for kala-azar: A multi-centre study of the freeze-dried DAT, rK39 strip test and KAtex in East Africa and the Indian subcontinent. *Transactions of the Royal Society of Tropical Medicine and Hygiene, 102*(1), 32–40. S0035-9203(07)00287-8 [pii]. https://doi.org/10.1016/j.trstmh.2007.09.003.

Boelaert, M., Verdonck, K., Menten, J., Sunyoto, T., Chappuis, F., Rijal, S. (2014). Rapid tests for the diagnosis of visceral leishmaniasis in patients with suspected disease. *Cochrane Database of Systematic Reviews,* (6):CD009135. https://doi.org/10.1002/14651858.CD009135.pub2.

Bose, P. P., & Kumar, P. (2016). Visual assessment of parasitic burden in infected macrophage by plasmonic detection of leishmania specific marker RNA. *Biochemical and Biophysical Research Communications, 480*(1), 81–86.

Bose, P. P., Kumar, P., & Munagala, N. (2015). Concurrent visual diagnosis and susceptibility profiling of the first line drug against visceral leishmaniasis by plasmonic detection of PCR amplified genetic biomarker. *Acta Tropica, 152*, 208–214.

Brand, W., Noorlander, C. W., Giannakou, C., De Jong, W. H., Kooi, M. W., Park, M. V., et al. (2017). Nanomedicinal products: A survey on specific toxicity and side effects. *International Journal of Nanomedicine, 12*, 6107–6129.

Bruni, N., Stella, B., Giraudo, L., Della Pepa, C., Gastaldi, D., & Dosio, F. (2017). Nanostructured delivery systems with improved leishmanicidal activity: A critical review. *International Journal of Nanomedicine, 12*, 5289–5311.

Brustoloni, Y. M., Lima, R. B., da Cunha, R. V., Dorval, M. E., Oshiro, E. T., de Oliveira, A. L., et al. (2007). Sensitivity and specificity of polymerase chain reaction in Giemsa-stained slides for diagnosis of visceral leishmaniasis in children. *Memórias do Instituto Oswaldo Cruz, 102*(4), 497–500. doi: S0074-02762007000400011 [pii].

Burza, S., Mahajan, R., Sinha, P. K., van Griensven, J., Pandey, K., Lima, M. A., et al. (2014). Visceral leishmaniasis and HIV co-infection in Bihar, India: Long-term effectiveness and treatment outcomes with liposomal amphotericin B (AmBisome). *PLoS Neglected Tropical Diseases, 8*(8), e3053.

Chakravarty, J., & Sundar, S. (2010). Drug resistance in leishmaniasis. *Journal of Global Infectious Diseases, 2*(2), 167–176.

Chaubey, P., & Mishra, B. (2014). Mannose-conjugated chitosan nanoparticles loaded with rifampicin for the treatment of visceral leishmaniasis. *Carbohydrate Polymers, 101*, 1101–1108.

Chaubey, P., Mishra, B., Mudavath, S. L., Patel, R. R., Chaurasia, S., Sundar, S., et al. (2018). Mannose-conjugated curcumin-chitosan nanoparticles: Efficacy and toxicity assessments against Leishmania donovani. *International Journal of Biological Macromolecules, 111*, 109–120.

Chaurasia, M., Pawar, V. K., Jaiswal, A. K., Dube, A., & Chourasia, M. K. (2015). Chondroitin nanocapsules enhanced doxorubicin induced apoptosis against leishmaniasis via Th1 immune response. *International Journal of Biological Macromolecules, 79*, 27–36.

Chawla, B., Jhingran, A., Panigrahi, A., Stuart, K. D., & Madhubala, R. (2011). Paromomycin affects translation and vesicle-mediated trafficking as revealed by proteomics of paromomycin -susceptible -resistant Leishmania donovani. *PLoS One, 6*(10), e26660. https://doi.org/10.1371/journal.pone.0026660.

Chunge, C., Owate, J., Pamba, H., & Donno, L. (1990). Treatment of visceral leishmaniasis in Kenya by aminosidine alone or combined with sodium stibogluconate. *Transactions of the Royal Society of Tropical Medicine and Hygiene, 84*(2), 221–225.

Cortes, S., Rolao, N., Ramada, J., & Campino, L. (2004). PCR as a rapid and sensitive tool in the diagnosis of human and canine leishmaniasis using Leishmania donovani s.l.-specific kinetoplastid primers. *Transactions of the Royal Society of Tropical Medicine and Hygiene, 98*(1), 12–17. doi: S0035920303000026 [pii].

Costa Lima, S. A., Resende, M., Silvestre, R., Tavares, J., Ouaissi, A., Lin, P. K. T., et al. (2012). Characterization and evaluation of BNIPDaoct-loaded PLGA nanoparticles for visceral leishmaniasis: In vitro and in vivo studies. *Nanomedicine, 7*(12), 1839–1849.

Crist, R. M., Grossman, J. H., Patri, A. K., Stern, S. T., Dobrovolskaia, M. A., Adiseshaiah, P. P., et al. (2013). Common pitfalls in nanotechnology: Lessons learned from NCI's Nanotechnology Characterization Laboratory. *Integrative Biology, 5*(1), 66–73.

Das, V., Ranjan, A., Sinha, A., Verma, N., Lal, C., Gupta, A., et al. (2001). A randomized clinical trial of low dosage combination of pentamidine and allopurinol in the treatment of antimony unresponsive cases of visceral leishmaniasis. *The Journal of the Association of Physicians of India, 49*, 609–613.

Dedet, J., & Pratlong, F. (2008). Leishmaniasis. In G. C. Cook & A. I. Zumla (Eds.), *Manson's tropical diseases* (22nd ed., pp. 1341–1365). London: Saunders.

Diro, E., Techane, Y., Tefera, T., Assefa, Y., Kebede, T., Genetu, A., et al. (2007). Field evaluation of FD-DAT, rK39 dipstick and KATEX (urine latex agglutination) for diagnosis of visceral leishmaniasis in northwest Ethiopia. *Transactions of the Royal Society of Tropical Medicine and Hygiene, 101*(9), 908–914. S0035-9203(07)00143-5 [pii]. https://doi.org/10.1016/j.trstmh.2007.05.002.

Dorlo, T. P., Balasegaram, M., Beijnen, J. H., & de Vries, P. J. (2012). Miltefosine: A review of its pharmacology and therapeutic efficacy in the treatment of leishmaniasis. *Journal of Antimicrobial Chemotherapy, 67*(11), 2576–2597.

Dumas, C., Ouellette, M., Tovar, J., Cunningham, M. L., Fairlamb, A. H., Tamar, S., et al. (1997). Disruption of the trypanothione reductase gene of Leishmania decreases its ability to survive oxidative stress in macrophages. *The EMBO Journal, 16*(10), 2590–2598.

Duong, A. D., Sharma, S., Peine, K. J., Gupta, G., Satoskar, A. R., Bachelder, E. M., et al. (2013). Electrospray encapsulation of toll-like receptor agonist resiquimod in polymer microparticles for the treatment of visceral leishmaniasis. *Molecular Pharmaceutics, 10*(3), 1045–1055.

Ferreira, A., Cemlyn-Jones, J., & Cordeiro, C. R. (2013). Nanoparticles, nanotechnology and pulmonary nanotoxicology. *Revista Portuguesa de Pneumologia (English Edition), 19*(1), 28–37.

Freitas-Junior, L. H., Chatelain, E., Kim, H. A., & Siqueira-Neto, J. L. (2012). Visceral leishmaniasis treatment: What do we have, what do we need and how to deliver it? *International Journal for Parasitology: Drugs and Drug Resistance, 2*, 11–19.

Gannavaram, S., Bhattacharya, P., Ismail, N., Kaul, A., Singh, R., & Nakhasi, H. L. (2016). Modulation of innate immune mechanisms to enhance leishmania vaccine-induced immunity: Role of coinhibitory molecules. *Frontiers in Immunology, 7*, 187. https://doi.org/10.3389/fimmu.2016.00187.

Gutiérrez, V., Seabra, A. B., Reguera, R. M., Khandare, J., & Calderón, M. (2016). New approaches from nanomedicine for treating leishmaniasis. *Chemical Society Reviews, 45*(1), 152–168.

Haldar, A. K., Sen, P., & Roy, S. (2011). Use of antimony in the treatment of leishmaniasis: Current status and future directions. *Molecular Biology International, 2011*, 1–23.

Hasker, E., Kansal, S., Malaviya, P., Gidwani, K., Picado, A., Singh, R. P., et al. (2013). Latent infection with Leishmania donovani in highly endemic villages in Bihar, India. *PLoS Neglected Tropical Diseases, 7*(2), e2053. https://doi.org/10.1371/journal.pntd.0002053.

Heli, H., Sattarahmady, N., Hatam, G., Reisi, F., & Vais, R. D. (2016). An electrochemical genosensor for Leishmania major detection based on dual effect of immobilization and electrocatalysis of cobalt-zinc ferrite quantum dots. *Talanta, 156*, 172–179.

Hendrickx, S., Boulet, G., Mondelaers, A., Dujardin, J., Rijal, S., Lachaud, L., et al. (2014). Experimental selection of paromomycin and miltefosine resistance in intracellular amastigotes of Leishmania donovani and L. infantum. *Parasitology Research, 113*(5), 1875–1881.

Hobson, D. W., Roberts, S. M., Shvedova, A. A., Warheit, D. B., Hinkley, G. K., & Guy, R. C. (2016). Applied nanotoxicology. *International Journal of Toxicology, 35*(1), 5–16.

Iqbal, H., Ishfaq, M., Wahab, A., Abbas, M. N., Ahmad, I., Rehman, A., et al. (2016). Therapeutic modalities to combat leishmaniasis, a review. *Asian Pacific Journal of Tropical Disease, 6*(1), 1–5.

Jacquet, D., Boelaert, M., Seaman, J., Rijal, S., Sundar, S., Menten, J., et al. (2006). Comparative evaluation of freeze-dried and liquid antigens in the direct agglutination test for serodiagnosis of visceral leishmaniasis (ITMA-DAT/VL). *Tropical Medicine & International Health, 11*(12), 1777–1784. TMI1743 [pii]. https://doi.org/10.1111/j.1365-3156.2006.01743.x.

Jha, P. K., Khan, M. I., Mishra, A., Das, P., & Sinha, K. K. (2017). HAT2 mediates histone H4K4 acetylation and affects micrococcal nuclease sensitivity of chromatin in Leishmania donovani. *PLoS One, 12*(5), e0177372.

Kansal, S., Tandon, R., Verma, P. R. P., Dube, A., & Mishra, P. R. (2012). Development of doxorubicin loaded novel core shell structured nanocapsules for the intervention of visceral leishmaniasis. *Journal of Microencapsulation, 30*(5), 441–450.

Khatik, R., Dwivedi, P., Khare, P., Kansal, S., Dube, A., Mishra, P. R., et al. (2014). Development of targeted 1, 2-diacyl-sn-glycero-3-phospho-l-serine-coated gelatin nanoparticles loaded with amphotericin B for improved in vitro and in vivo effect in leishmaniasis. *Expert Opinion on Drug Delivery, 11*(5), 633–646.

Kunjachan, S., Gupta, S., Dwivedi, A. K., Dube, A., & Chourasia, M. K. (2011). Chitosan-based macrophage-mediated drug targeting for the treatment of experimental visceral leishmaniasis. *Journal of Microencapsulation, 28*(4), 301–310.

Lamotte, S., Späth, G. F., Rachidi, N., & Prina, E. (2017). The enemy within: Targeting host–parasite interaction for antileishmanial drug discovery. *PLoS Neglected Tropical Diseases, 11*(6), e0005480.

Lima, S. A. C., Silvestre, R., Barros, D., Cunha, J., Baltazar, M. T., Dinis-Oliveira, R. J., et al. (2014). Crucial CD8+ T-lymphocyte cytotoxic role in amphotericin B nanospheres efficacy against experimental visceral leishmaniasis. *Nanomedicine: Nanotechnology, Biology and Medicine, 10*(5), e1021–e1030.

Manandhar, K. D., Yadav, T. P., Prajapati, V. K., Kumar, S., Rai, M., Dube, A., et al. (2008). Antileishmanial activity of nano-amphotericin B deoxycholate. *Journal of Antimicrobial Chemotherapy, 62*(2), 376–380.

Marinho, F. D. A., Gonçalves, K. C. D. S., Oliveira, S. S. D., Oliveira, A.-C. D. S. C., Bellio, M., d'Avila-Levy, C. M., et al. (2011). Miltefosine induces programmed cell death in Leishmania amazonensis promastigotes. *Memórias do Instituto Oswaldo Cruz, 106*(4), 507–509.

Markle, W., & Makhoul, K. (2004). Cutaneous leishmaniasis: Recognition and treatment. *American Family Physician, 69*(6), 1455–1460.

Mathis, A., & Deplazes, P. (1995). PCR and in vitro cultivation for detection of Leishmania spp. in diagnostic samples from humans and dogs. *Journal of Clinical Microbiology, 33*(5), 1145–1149.

Mauël, J., & Ransijn, A. (1997). Leishmaniaspp.: Mechanisms of toxicity of nitrogen oxidation products. *Experimental Parasitology, 87*(2), 98–111.

Maurya, R., Singh, R. K., Kumar, B., Salotra, P., Rai, M., & Sundar, S. (2005). Evaluation of PCR for diagnosis of Indian kala-azar and assessment of cure. *Journal of Clinical Microbiology, 43*(7), 3038–3041. 43/7/3038 [pii]. https://doi.org/10.1128/JCM.43.7.3038-3041.2005.

Mishra, J., & Singh, S. (2013). Miltefosine resistance in Leishmania donovani involves suppression of oxidative stress-induced programmed cell death. *Experimental Parasitology, 135*(2), 397–406.

Mohamed-Ahmed, A. H., Seifert, K., Yardley, V., Burrell-Saward, H., Brocchini, S., & Croft, S. L. (2013). Antileishmanial activity, uptake, and biodistribution of an amphotericin B and poly (α-glutamic acid) complex. *Antimicrobial Agents and Chemotherapy, 57*(10), 4608–4614.

Mohan, S., Srivastava, P., Maheshwari, S., Sundar, S., & Prakash, R. (2011). Nano-structured nickel oxide based DNA biosensor for detection of visceral leishmaniasis (Kala-azar). *Analyst, 136*(13), 2845–2851.

Moradi, M., Sattarahmady, N., Rahi, A., Hatam, G., Sorkhabadi, S. R., & Heli, H. (2016). A label-free, PCR-free and signal-on electrochemical DNA biosensor for Leishmania major based on gold nanoleaves. *Talanta, 161*, 48–53.

Mudavath, S. L., Talat, M., Rai, M., Srivastava, O. N., & Sundar, S. (2014). Characterization and evaluation of amine-modified graphene amphotericin B for the treatment of visceral leishmaniasis: In vivo and in vitro studies. *Drug Design, Development and Therapy, 8*, 1235–1247.

Mudavath, S. L., Talat, M., Rai, M., Srivastava, O. N., & Sundar, S. (2016). An oral formulation of amphotericin B for the treatment of visceral Leishmaniasis: f-Gr-AmB. *International Journal of Infectious Diseases, 45*, 367. https://doi.org/10.1016/j.ijid.2016.02.790.

Mukherjee, A., Padmanabhan, P. K., Sahani, M. H., Barrett, M. P., & Madhubala, R. (2006). Roles for mitochondria in pentamidine susceptibility and resistance in Leishmania donovani. *Molecular and Biochemical Parasitology, 145*(1), 1–10.

Mukhopadhyay, D., Dalton, J. E., Kaye, P. M., & Chatterjee, M. (2014). Post kala-azar dermal leishmaniasis: An unresolved mystery. *Trends in Parasitology, 30*(2), 65–74.

Nagill, R., & Kaur, S. (2011). Vaccine candidates for leishmaniasis: A review. *International Immunopharmacology, 11*(10), 1464–1488. https://doi.org/10.1016/j.intimp.2011.05.008.

NIu, M., & Pershin, G. (1966). Comparative study of the chemotherapeutic effect of paromomycin and monomycin in experimental cutaneous leischmaniasis in albino mice. *Farmakologiia i Toksikologiia, 29*(1), 90–94.

No, J. H. (2016). Visceral leishmaniasis: Revisiting current treatments and approaches for future discoveries. *Acta Tropica, 155*, 113–123.

Ostyn, B., Gidwani, K., Khanal, B., Picado, A., Chappuis, F., Singh, S. P., et al. (2011). Incidence of symptomatic and asymptomatic Leishmania donovani infections in high-endemic foci in India and Nepal: A prospective study. *PLoS Neglected Tropical Diseases, 5*(10), e1284. https://doi.org/10.1371/journal.pntd.0001284.

Palumbo, E. (2010). Treatment strategies for mucocutaneous leishmaniasis. *Journal of Global Infectious Diseases, 2*(2), 147–150.

Paris, C., Loiseau, P. M., Bories, C., & Bréard, J. (2004). Miltefosine induces apoptosis-like death in Leishmania donovani promastigotes. *Antimicrobial Agents and Chemotherapy, 48*(3), 852–859.

Pearson, R., & Sousa, A. (1996). Clinical spectrum of leishmaniasis. *Clinical Infectious Diseases, 22*(1), 1–11.

Peine, K. J., Gupta, G., Brackman, D. J., Papenfuss, T. L., Ainslie, K. M., Satoskar, A. R., et al. (2013). Liposomal resiquimod for the treatment of Leishmania donovani infection. *Journal of Antimicrobial Chemotherapy, 69*(1), 168–175.

Perinoto, A. C., Maki, R. M., Colhone, M. C., Santos, F. R., Migliaccio, V., Daghastanli, K. R., et al. (2010). Biosensors for efficient diagnosis of leishmaniasis: Innovations in bioanalytics for a neglected disease. *Analytical Chemistry, 82*(23), 9763–9768.

Pham, T., Barratt, G., Michel, J., Loiseau, P., & Saint-Pierre-Chazalet, M. (2013). Interactions of antileishmanial drugs with monolayers of lipids used in the development of amphotericin B–miltefosine-loaded nanocochleates. *Colloids and Surfaces B: Biointerfaces, 106*, 224–233.

Poland, C. A., Duffin, R., Kinloch, I., Maynard, A., Wallace, W. A., Seaton, A., et al. (2008). Carbon nanotubes introduced into the abdominal cavity of mice show asbestos-like pathogenicity in a pilot study. *Nature Nanotechnology, 3*(7), 423–428.

Prajapati, V. K., Awasthi, K., Gautam, S., Yadav, T. P., Rai, M., Srivastava, O. N., et al. (2011a). Targeted killing of Leishmania donovani in vivo and in vitro with amphotericin B attached to functionalized carbon nanotubes. *Journal of Antimicrobial Chemotherapy, 66*(4), 874–879.

Prajapati, V. K., Awasthi, K., Yadav, T. P., Rai, M., Srivastava, O. N., & Sundar, S. (2011b). An oral formulation of amphotericin B attached to functionalized carbon nanotubes is an effective treatment for experimental visceral leishmaniasis. *Journal of Infectious Diseases, 205*(2), 333–336.

Rahman, L., Jacobsen, N. R., Aziz, S. A., Wu, D., Williams, A., Yauk, C. L., et al. (2017). Multi-walled carbon nanotube-induced genotoxic, inflammatory and pro-fibrotic responses in mice: Investigating the mechanisms of pulmonary carcinogenesis. *Mutation Research, Genetic Toxicology and Environmental Mutagenesis, 823*, 28–44.

Reithinger, R., & Dujardin, J. C. (2007). Molecular diagnosis of leishmaniasis: Current status and future applications. *Journal of Clinical Microbiology, 45*(1), 21–25. JCM.02029-06 [pii]. https://doi.org/10.1128/JCM.02029-06.

Reithinger, R., Dujardin, J.-C., Louzir, H., Pirmez, C., Alexander, B., & Brooker, S. (2007). Cutaneous leishmaniasis. *The Lancet Infectious Diseases, 7*(9), 581–596.

Rijal, S., Boelaert, M., Regmi, S., Karki, B. M., Jacquet, D., Singh, R., et al. (2004). Evaluation of a urinary antigen-based latex agglutination test in the diagnosis of kala-azar in eastern Nepal. *Tropical Medicine & International Health, 9*(6), 724–729. https://doi.org/10.1111/j.1365-3156.2004.01251.xTMI1251[pii].

Rodrigues, V., Cordeiro-da-Silva, A., Laforge, M., Silvestre, R., & Estaquier, J. (2016). Regulation of immunity during visceral Leishmania infection. *Parasites & Vectors, 9*(1), 118. https://doi.org/10.1186/s13071-016-1412-x.

Sadat, S. M., Jahan, S. T., & Haddadi, A. (2016). Effects of size and surface charge of polymeric nanoparticles on in vitro and in vivo applications. *Journal of Biomaterials and Nanobiotechnology, 7*(02), 91–108.

Salotra, P., Sreenivas, G., Pogue, G. P., Lee, N., Nakhasi, H. L., Ramesh, V., et al. (2001). Development of a species-specific PCR assay for detection of Leishmania donovani in clinical samples from patients with kala-azar and post-kala-azar dermal leishmaniasis. *Journal of Clinical Microbiology, 39*(3), 849–854. https://doi.org/10.1128/JCM.39.3.849-854.2001.

Sarwar, H. S., Akhtar, S., Sohail, M. F., Naveed, Z., Rafay, M., Nadhman, A., et al. (2017). Redox biology of Leishmania and macrophage targeted nanoparticles for therapy. *Nanomedicine, 12*(14), 1713–1725.

Sarwar, H. S., Ashraf, S., Akhtar, S., Sohail, M. F., Hussain, S. Z., Rafay, M., et al. (2018). Mannosylated thiolated polyethylenimine nanoparticles for the enhanced efficacy of antimonial drug against Leishmaniasis. *Nanomedicine, 13*(1), 25–41.
Sattarahmady, N., Movahedpour, A., Heli, H., & Hatam, G. (2016). Gold nanoparticles-based biosensing of Leishmania major kDNA genome: Visual and spectrophotometric detections. *Sensors and Actuators B: Chemical, 235*, 723–731.
Savaliya, R., Singh, P., & Singh, S. (2016). Pharmacological drug delivery strategies for improved therapeutic effects: Recent advances. *Current Pharmaceutical Design, 22*(11), 1506–1520.
Schonian, G., Nasereddin, A., Dinse, N., Schweynoch, C., Schallig, H. D., Presber, W., et al. (2003). PCR diagnosis and characterization of Leishmania in local and imported clinical samples. *Diagnostic Microbiology and Infectious Disease, 47*(1), 349–358. doi: S0732889303000932 [pii].
Shahnaz, G., Edagwa, B. J., McMillan, J., Akhtar, S., Raza, A., Qureshi, N. A., et al. (2017). Development of mannose-anchored thiolated amphotericin B nanocarriers for treatment of visceral leishmaniasis. *Nanomedicine, 12*(2), 99–115.
Sharma, S., Zapatero-Rodríguez, J., Estrela, P., & O'Kennedy, R. (2015). Point-of-care diagnostics in low resource settings: Present status and future role of microfluidics. *Biosensors, 5*(3), 577–601.
Shirkhani, K., Teo, I., Armstrong-James, D., & Shaunak, S. (2015). Nebulised amphotericin B-polymethacrylic acid nanoparticle prophylaxis prevents invasive aspergillosis. *Nanomedicine: Nanotechnology, Biology and Medicine, 11*(5), 1217–1226.
Singh, B., & Sundar, S. (2012). Leishmaniasis: Vaccine candidates and perspectives. *Vaccine, 30*(26), 3834–3842.
Singh, O. P., & Sundar, S. (2015). Developments in diagnosis of visceral Leishmaniasis in the elimination era. *Journal of Parasitology Research, 2015*, 239469. https://doi.org/10.1155/2015/239469.
Singh, S., Sharma, A., & Robertson, G. P. (2012). Realizing the clinical potential of cancer nanotechnology by minimizing toxicologic and targeted delivery concerns. *Cancer Research, 72*, 5663–5668.
Singh, N., Mishra, B. B., Bajpai, S., Singh, R. K., & Tiwari, V. K. (2014). Natural product based leads to fight against leishmaniasis. *Bioorganic & Medicinal Chemistry, 22*(1), 18–45. https://doi.org/10.1016/j.bmc.2013.11.048.
Singh, O. P., Hasker, E., Boelaert, M., & Sundar, S. (2016a). Elimination of visceral leishmaniasis on the Indian subcontinent. *The Lancet Infectious Diseases, 16*(12), e304–e309.
Singh, O. P., Singh, B., Chakravarty, J., & Sundar, S. (2016b). Current challenges in treatment options for visceral leishmaniasis in India: A public health perspective. *Infectious Diseases of Poverty, 5*, 19. https://doi.org/10.1186/s40249-016-0112-2.
Souto, D. E., Fonseca, A. M., Barragan, J. T., de CS Luz, R., Andrade, H. M., Damos, F. S., et al. (2015). SPR analysis of the interaction between a recombinant protein of unknown function in Leishmania infantum immobilised on dendrimers and antibodies of the visceral leishmaniasis: A potential use in immunodiagnosis. *Biosensors and Bioelectronics, 70*, 275–281.
Srivastava, P., Dayama, A., Mehrotra, S., & Sundar, S. (2010). Diagnosis of visceral leishmaniasis. *Transactions of the Royal Society of Tropical Medicine and Hygiene, 105*(1), 1–6. S0035-9203(10)00219-1 [pii]. https://doi.org/10.1016/j.trstmh.2010.09.006.
Srivastava, P., Mehrotra, S., Tiwary, P., Chakravarty, J., & Sundar, S. (2011). Diagnosis of Indian visceral leishmaniasis by nucleic acid detection using PCR. *PLoS One, 6*(4), e19304. https://doi.org/10.1371/journal.pone.0019304.
Srividya, G., Kulshrestha, A., Singh, R., & Salotra, P. (2011). Diagnosis of visceral leishmaniasis: Developments over the last decade. *Parasitology Research, 110*(3), 1065–1078. https://doi.org/10.1007/s00436-011-2680-1.

Sudarshan, M., Weirather, J. L., Wilson, M. E., & Sundar, S. (2011). Study of parasite kinetics with antileishmanial drugs using real-time quantitative PCR in Indian visceral leishmaniasis. *The Journal of Antimicrobial Chemotherapy, 66*(8), 1751–1755. doi: dkr185 [pii]. https://doi.org/10.1093/jac/dkr185.

Sudarshan, M., Singh, T., Singh, A. K., Chourasia, A., Singh, B., Wilson, M. E., et al. (2014). Quantitative PCR in epidemiology for early detection of visceral leishmaniasis cases in India. *PLoS Neglected Tropical Diseases, 8*(12), e3366. https://doi.org/10.1371/journal.pntd.0003366.

Sundar, S., & Rai, M. (2002). Laboratory diagnosis of visceral leishmaniasis. *Clinical and Diagnostic Laboratory Immunology, 9*(5), 951–958.

Sundar, S., & Singh, A. (2016). Recent developments and future prospects in the treatment of visceral leishmaniasis. *Therapeutic Advances in Infectious Sdisease, 3*(3–4), 98–109.

Sundar, S., Agrawal, S., Pai, K., Chance, M., & Hommel, M. (2005). Detection of leishmanial antigen in the urine of patients with visceral leishmaniasis by a latex agglutination test. *The American Journal of Tropical Medicine and Hygiene, 73*(2), 269–271. doi: 73/2/269 [pii].

Sundar, S., Singh, A., Chakravarty, J., & Rai, M. (2015). Efficacy and safety of miltefosine in treatment of post-kala-azar dermal leishmaniasis. *The Scientific World Journal, 2015*, 414378. https://doi.org/10.1155/2015/414378. Epub 2015 Jan 1.

Tiwari, N., Gedda, M. R., Tiwari, V. K., Singh, S. P., & Singh, R. K. (2018). Limitations of current therapeutic options, possible drug targets and scope of natural products in control of leishmaniasis. *Mini Reviews in Medicinal Chemistry, 18*(1), 26–41.

Torres-Guerrero, E., Quintanilla-Cedillo, M. R., Ruiz-Esmenjaud, J., & Arenas, R. (2017). Leishmaniasis: A review. *F1000Research, 6*, 750. https://doi.org/10.12688/f1000research.11120.1. eCollection 2017.

Torres-Sangiao, E., Holban, A. M., & Gestal, M. C. (2016). Advanced nanobiomaterials: Vaccines, diagnosis and treatment of infectious diseases. *Molecules, 21*(7), 867. https://doi.org/10.3390/molecules21070867.

Toubanaki, D. K., Athanasiou, E., & Karagouni, E. (2016). Gold nanoparticle-based lateral flow biosensor for rapid visual detection of Leishmania-specific DNA amplification products. *Journal of Microbiological Methods, 127*, 51–58.

Wang, Y., Yu, L., Kong, X., & Sun, L. (2017). Application of nanodiagnostics in point-of-care tests for infectious diseases. *International Journal of Nanomedicine, 12*, 4789.

WHO (2018). *Leishmaniasis, background information*. [Online]. Available: http://www.who.int/leishmaniasis/en/. Accessed 19/02/2018.

Yasinzai, M., Khan, M., Nadhman, A., & Shahnaz, G. (2013). Drug resistance in leishmaniasis: Current drug-delivery systems and future perspectives. *Future Medicinal Chemistry, 5*(15), 1877–1888.

Nanomanipulation of Consumer Goods: Effects on Human Health and Environment

Ragini Singh and Sanjay Singh

Abstract Extensive use of nanotechnology in commercial products has lead to the generation of waste products consist of synthetic nanomaterials (NMs), also referred as "nanowaste". Disposal of nanowaste requires appropriate framework to ensure that it does not cause adverse effect on the human health and environment. Several NMs are used at industrial level such as TiO2 NPs provides white color to paints. The major amount of TiO2 NPs is released into water from building paint. NPs can also be used in textile industries and provide antimicrobial, flame resistance and self-cleaning properties to fabrics but can be released into the environment by general means like hard wash, and sweating etc. Cosmetic products are another sources of NPs such as ZnO and TiO2. In health sectors, NMs have found wide range of applications including the construction of theranostic systems and anti-infective agents. These NMs further ends up their life in water bodies, landfills, soil and air while contaminating surrounding environment. Thus, with the increasing commercial value of NMs, their release in environment and exposure to living system also increases which eventually lead to induce epigenetic toxicity. Therefore, this chapter has been designed to focus towards the possible ways of release of NMs from consumer nanoproducts to the environment and their subsequent harmful effects on human health and living ecosystems. Towards the end of the chapter, various strategies being developed for the removal of NMs from the environment has also been discussed.

Keywords Nanotechnology · Nanotoxicity · Paints · Textile · Cosmetics · Epigenome toxicity · e-waste

R. Singh · S. Singh (✉)
Division of Biological and Life Sciences, School of Arts and Sciences, Ahmedabad University Central Campus, Ahmedabad, Gujarat, India
e-mail: sanjay.singh@ahduni.edu.in

S. Singh, P. K. Maurya (eds.), *Nanotechnology in Modern Animal Biotechnology*, https://doi.org/10.1007/978-981-13-6004-6_7

1 Introduction

Due to the recent overwhelming research in the area of nanotechnology and realization of their applications in human health and environment, myriad of novel nanomaterials (NMs) and nanoproducts are being developed. Extensive use of NMs in the consumer and commercial products leads to its direct exposure to human and eventually to the environment, which has raised awareness about the occupational safety aspects of NMs in the workplace. Potential exposure to NMs can be reduced by applying certain suitable mitigation strategies. Nano-enabled products are utilized in various sectors like medical industry, 3D printing, water purification, electronics, coating and surface treatment processes (Vílchez et al. 2016). Data suggest that NMs such as silver nanoparticles (AgNPs), titanium dioxide (TiO_2) and carbon-based NMs are majorly used in the consumer products (Yang and Westerhoff 2014). However, uncontrolled and unethical use of NMs has created great debate in the scientific community.

The release of NMs from nano-enabled products depends on the several major factors like stress and environmental conditions to which nano-enabled products are exposed, their intended utilization, distribution of NMs in containing products. The release of NMs in the environment depends on the basis of these factors (a) chemical degradation and physical stresses that NMs undergo during their use, (b) unintended or intended release, and (c) NMs transformation during their use (Vílchez et al. 2016). Thus, the release study of NMs has been divided into three major domains such as consumers, occupational and environmental release. Several NMs are released into the environment by different means, for example, iron oxide NPs (IONPs) are utilized in groundwater remediation or in agrochemicals, similarly, TiO_2 NPs are incorporated in cosmetics and can enter the environment after showering (Vílchez et al. 2016). NMs can potentially have an impact on the environment by three different ways such as (a) direct interaction with microorganisms, invertebrates, fish and other species, (b) interaction with other pollutants that may influence its bioavailability, and (c) changes to non-living environmental structure (Dhawan et al. 2009). A list of different NMs used for the generation of consumer products with the potential applications is provided in Table 1. Figure 1 shows the different nano-enabled product and the release of NMs in the environment.

Growing application of NMs containing products leads to the generation of waste which contains synthetic NMs generally known as nanowaste. It is of great need to ensure that the disposal of nanowaste does not cause any adverse effect on the human health and the environment. Therefore, a clear policy and appropriate framework are needed in order to monitor the products containing NMs throughout their life cycle. Further, effective procedures and strategy are also needed for proper disposal and recycling of these NMs (Kolodziejczyk 2016). In this context, the practice of green nanotechnology has been suggested, which could to some extent, overcome the adverse effects induced by NMs especially to the environment. (Hutchison 2008; Viswanath and Kim 2017). Applications of green nanoscience are currently realized into several areas including nanoelectronics, thermoelectrics, and

Table 1 List of consumer products incorporating nanomaterials

Type of NMs	Consumer products	Applications	Refs.
Zinc oxide (ZnO)	Cosmetics, sunscreen, food packaging, paints, cleaning, personal care products	Antimicrobial activity, protection from sun	Schmid and Riediker (2008), Society (2004), and Zhang et al. (2015)
Titanium dioxide (TiO_2)	Sunscreen, water treatment process, electronic devices	Ultraviolet radiation absorption, acts as catalyst	Schmid and Riediker (2008), Society (2004), and Zhang et al. (2015)
Cerium oxide (CeO_2)	Automotive/fuel catalyst, biomedicine, coatings	Catalysts, optical property	Goharshadi et al. (2011), Society (2004), and Zhang et al. (2015)
Silver (Ag)	Surface coating, textiles, medical devices, transparent electrodes for flexible devices	Electrical property conductivity, antimicrobial action	Aitken et al. (2006) and Zhang et al. (2015)
Silicon dioxide (SiO_2)	Food packaging, coatings, paints	To extend life of coating and paints	Schmid and Riediker (2008), Society (2004), and Zhang et al. (2015)
Iron oxide (Fe_xO_y)	Environmental remediation, catalyst, food and beverages	Magnetic property, hardness and strength	Xu et al. (2012) and Zhang et al. (2015)
Aluminium oxide (Al_2O_3)	Cosmetics, food additives, cleaning, water treatment	Antimicrobial action, adsorbent, light absorption	Schmid and Riediker (2008) and Zhang et al. (2015)

nanocomposites etc. Thus, it is expected that green nanotechnology could help researchers to develop strategies for safe applications of NMs in human health and environment.

Numerous methods are available for the synthesis of NMs which includes chemical, physical and biological routes. Chemical methods are advantageous as they require a short time for synthesis, however, chemicals used in the synthesis and stabilization process are generally toxic and generate compound which can show the deleterious effect on the environment (Vithiya and Sen 2011). Various natural sources like microorganisms and plants are available which can be used in the biosynthesis of NMs.

Microorganisms hold immense potential for NMs synthesis as they can prove to be cost-effective, eco-friendly and also avoid harsh and toxic chemicals. Reductase enzymes present in microorganisms help in detoxification of heavy metals and also reduces metal salt to metal NPs having narrow size distribution (Singh et al. 2016c). Recently, various microorganisms including fungi, yeast, and bacteria are explored for intra and extracellular synthesis of NMs. Among various available methods, the extracellular method is proved to be more efficient as in this case downstream processing is not required for NMs recovery like sonication to break down the cells, and subsequent purification by centrifugation. Metal-resistant genes, proteins, peptides and reducing cofactors acts as a reducing agent and also very frequently offer to cap the synthesized NPs, thus prevent aggregation (Singh et al. 2016c). Recently, various bacteria such as *Pseudomonas deceptionensis, Listeria monocytogenes, Bacillus subtilis*, and *Streptomyces anulatus* have been used for the synthesis of

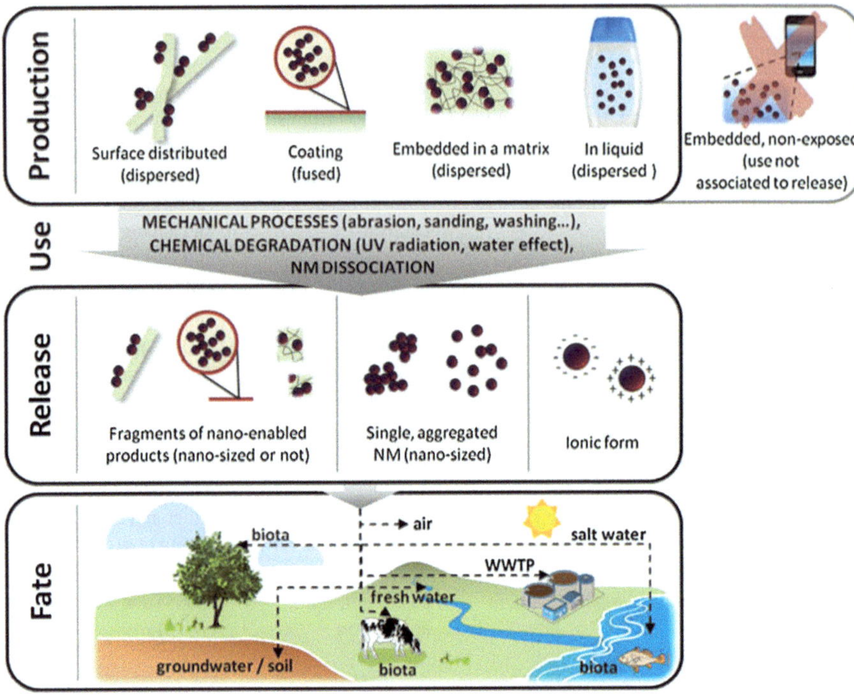

Fig. 1 Most common types of nano-enabled products considering NMs distribution and possible forms of NMs released during the use of these products that can end up into the different environmental compartments. (Reprinted with permission from Indoor and Outdoor Nanoparticles. Copyright, 2015 Springer nature)

NMs (Elbeshehy et al. 2015; Soni and Prakash 2015). Otari et al. demonstrated the intracellular synthesis of AgNPs by *Rhodococcus* species. Synthesis of AgNPs takes place in the cytoplasm, in presence of NADH-dependent nitrate reductase, which helps in reducing Ag^{+} ions to synthesize AgNPs (Otari et al. 2015).

Phytonanotechnology is an emerging field focusing on the synthesis of NPs. This technology is attractive because it offers a rapid, eco-friendly, biocompatible, and cost-effective process of NM synthesis. It uses universal solvent water as a reaction medium and naturally occurring plant products as reducing and capping agents (Noruzi 2015; Singh et al. 2016c). Leaf and root extract of medicinal plant *Panax ginseng* is used for the synthesis of AuNPs and AgNPs (Singh et al. 2016a, b). It has been suggested that various vitamins, amino acids, proteins and secondary metabolites such as flavonoids, terpenoids, polysaccharides play a major role in the reduction of salt precursors and act as capping and the stabilizing agent (Duan et al. 2015). El-kassas et al. have demonstrated that the hydroxyl functional group from polyphenols and the carbonyl group from proteins of *Corallina officinalis* extract could assist in the formation and stabilization of AuNPs (El-Kassas and El-Sheekh 2014). Further, different plant parts are reported to undergo different mechanisms for synthesizing NPs (Baker et al. 2013).

Although NMs can be synthesized by green route and thus capped with non-toxic molecules, however, their decomposition would lead to the generation of corresponding constituent ions, which may cause harm to human health as well as their release can contaminate the environment. Therefore, it is imperative to study the possible types of NMs toxicity as well as the potential nanoproducts which could release NMs in the environment. In last few decades, the area of nanotoxicology has been rapidly growing due to the ever-increasing use of NMs in myriads of applications (Dhawan and Sharma 2010). Although, several reports have been published explaining the methods of toxicity evaluation of NMs, however, reproducibility and reliability of these protocols remains a serious concern. Another major challenge with NMs is that they interfere with the toxicity assays, for e.g. NMs can react with one of the components of the assay/test (Bohnsack et al. 2012, Holder et al. 2012; Kroll et al. 2012; Guadagnini et al. 2015). Therefore, it is also equally important to comprehensively characterize the NMs and study the possible interference with the safety assessment methods (Di Bona et al. 2015). It is well known that there are no set guidelines so far, for safety assessment of NMs and therefore nanotoxicologists generally use the guidelines of classical toxicology (Oberdorster 2010). It is further important to consider that since NMs are significantly different from their bulk counterparts, the guidelines for classical toxicology do not fit well with the accurate analysis of NMs toxicity data. This issue could be ascribed due to the high number of surface atoms present on the NM surface, thus the reactivity (Elsaesser et al. 2010). Since there are several reports published on this topic, therefore, we will not discuss it here rather this chapter will focus more and comprehensively cover the area of release of NMs from the nanoproducts.

2 Release of Nanomaterials from Paints and Inks

Paints, polishes, and coating materials are frequently used to impart protection to machinery, vehicles, and buildings from normal wear and tear and other environmental factors. NMs have been found applications as additives to such polishing materials to make the surfaces scratch resistant, impact surface hardness to coating, make surfaces UV-radiation resistant, and protection from corrosion, microbial, chemical and physical deterioration (Banerjee et al. 2014; Munafò et al. 2015; Wei et al. 2015). Although NPs of barium sulfate ($BaSO_4$) and iron oxides are used as coloring pigment in the automotive sector, whereas TiO_2 is found to provide bright white color to paints and thus to the coated surfaces. To make antifouling paints, NMs conjugated with biocides are also synthesized which lead to the controlled release of biocides from the paints (Shtykova et al. 2009). Additionally, certain metal NPs (such as AgNPs) are known for their antimicrobial activity and high electron conductivity are being used in printing electronics like inks leading to the highly conductive patterns using low-cost and roll-to-roll processes. Other extremely tiny particles, such as quantum dots (such as CdSe, CdTe), are also used in covert security printing. Such applications are due to the unique optical properties of quantum dots at a specified wavelength.

Generally, TiO_2 (rutile) in a size range of 200–300 nm is responsible for the opaque and white bright color of paints, which is not claimed by many manufacturers. It is also reported by several researchers that a major amount of TiO_2 NPs released into the water from the building paints is of 100 nm size (Weir et al. 2012). Kaegi et al. (2008) first reported the release of TiO_2 in environment by performing transmission electron microscopy imaging equipped with energy dispersive X-ray spectroscopy (TEM-EDX) from the collected run-off water from aged façade and gave real estimation of synthetic NPs mass flux in urban environment (as high as 600 μg TiO_2/L). They also found that these TiO_2 NPs were found to be mainly embedded in an organic binder. Later, the same research group investigated the release of AgNPs (>10 nm size) from paints, which are exposed to the environmental condition for 1 year (Kaegi et al. 2010). They found that after 8 run-off event, the release of AgNPs was drastically decreased to 1 μg/L than the release during the first run-off (10,000 μg/L). TEM-EDX data suggest that silver was released in the form of Ag_2S and not in pure AgNPs form.

Pigment volume content (PVC) is the representation of volumetric ratio between solid binder and pigment, which determines various properties of paints like the gloss, permeability, thermal, and mechanical properties etc. (Gaylarde et al. 2011; Rodrı et al. 2004). PVC has been found as a crucial factor in the release of silica (SiO_2) NPs from paints (Zuin et al. 2014). Authors showed that the paint formulated with high PVC value (~63%) exhibit a high release of SiO_2 NPs (~1.7%). However, the paint sample consisting of a high amount of binder and less content of calcite filler with low PVC values (~35%) exhibit a small amount of SiO_2 NPs release. This observation could be explained on the basis that large content of binder in the paint formulation forms an appropriate matrix to hold SiO_2 NPs within (Zuin et al. 2014). Thus, by adjusting the common pigment in combination with properties of binder, it becomes very much possible to reduce the SiO_2 NPs release in the environment.

Smuders and co-workers performed a comparative study to evaluate the in vivo effect of three pristine NPs (TiO_2, Ag and SiO_2) and also these NPs contained in paints (Smulders et al. 2014). They found that the exposure of these NPs, contained in the paint, caused major pulmonary or systemic toxicity in BALB/C mice, however, unformulated pristine NPs exposure induces significant pulmonary toxicity. A detailed mechanistic study revealed that AgNPs exposure causes an increase in neutrophil count and pro-inflammatory cytokine secretion whereas, AgNPs contained in paints did not show these signatures of toxicity. Thus, direct NPs exposure induced toxic effect but when incorporated in a complex matrix of paints, no appreciable toxicity was observed (Smulders et al. 2014). As discussed above, the release of NPs can also be controlled by tuning the paint formulations, which could be considered as one of the important aspects of the formation of "safe-by-design" products (Fiorentino et al. 2015). The two very common components of the paints containing acrylic copolymer and a styrene-acrylic copolymer as a binder are also studied. From scanning electron microscopy and EDX (SEM-EDX) analysis, it was found that paints containing acrylic copolymer may undergo more chemical degradation than styrene-acrylic copolymer. Therefore, the components of paint formulation also play important role in the stability of the paint, which could affect the stability and release of incorporated NMs.

Airborne NMs can also pose threat to the human health and environment. Studies in this area have provided ample evidence of emission of airborne particles from common printers which need public awareness (Bello et al. 2013; He et al. 2007; Stephens et al. 2013). Several NPs like alumina, SiO_2, zinc oxide (ZnO), copper oxide (CuO) and TiO_2 are found to be incorporated in toner formulations and are released into the air during the printing process. Toner contains high amount of organic carbon (OC, 42–89%), elemental carbon (EC, 0.33–12% and metal/metal oxides (1–33%) (Pirela et al. 2015). Printers emitted NPs (PEPs) are also found in similar composition and thus contain 50–90% OC, 0.001–0.5% EC and 1–3% metals/metal oxides. Pioneer work done by He et al. summarizes the particle emission characteristic of 60 different commercial laser printers (He et al. 2007). They found that ~60% printers did not emit sub-micrometre particles, ~40% emitted particles, and ~27% were recorded to be the high particle emitters with maximum value reaching up to 38,000 particles/cm^3. Size of the airborne particles was found to be in the range of 40–60 nm, which indicate the release of a large number of ultrafine particles. Microscopy studies also revealed that NPs of SiO_2, TiO_2, IONPs and other metals which are released into the air were more from the original toner, suggesting the importance of the toner formulation (Bello et al. 2013). Studies have also performed to investigate the toxicological impact of PEPs released in the environment. In this context, Sisler et al. have shown that PEPs exhibit toxicological potential on animal models and induce chronic inflammation and fibrosis (Sisler et al. 2015). They also found that the direct exposure of Human Small Airway Epithelial Cells (SAEC) to a low concentration of PEPs induce a morphological change of actin remodeling and gap formation within the endothelial monolayer. Furthermore, Human Microvascular Endothelial Cells (HMVEC) exposed to PEPs showed increased production of reactive oxygen species (ROS) and angiogenesis (Sisler et al. 2015). Additionally, Khatri et al. have also investigated the early human responses by considering the exposure of photocopiers in busy photocopy center for 6 h. It was concluded that NPs emitted from photocopiers induce oxidative stress and upper airway inflammation in exposed humans (Khatri et al. 2013).

3 Use of Nanomaterials in Textile Industry

Owing to the unique properties of NMs, it is expected that their use in textile industry would provide desirable properties to fabrics such as antimicrobial, strain, dirt and flame resistance, radiation protection and possibly self-cleaning. Such fabrics would definitely be of great value for consumers. Among these fabrics, major emphasis has been given to AgNPs-based fabrics, due to the antimicrobial functionality. Fabrics in which nanoscale entity are incorporated after base textile production are known as "nano finished textile". These nano finished textiles can either be manufactured by adding the NMs during the fabric preparation or post-manufacturing treatment with NMs where the latter is most common practice in the textile industry (Vílchez et al. 2016). At the industrial level, most frequently used process for deposition of NMs over fabric surface is dip-pad-dry cure process, sol-gel process and physical-vapor deposition (Radetić 2013). Fibers with embedded NMs can be obtained by methods such as (i) melt compounding, where weaving or knitting are used to obtain textile from fibers; and (ii) electrospinning, in which the synthesized nanofibers can be used for the production of nonwoven fabrics (like filtration membranes) (Haydon and Eng 2013).

Due to the major concern of the release of NMs from fabrics to the environment, several reports are available focusing on the release of NMs (mainly Ag and TiO_2) from available textile products in the market (Nowack et al. 2011). The common methods of NMs release from textiles could be abrasion and mechanical rub, sweat, harsh wash and changes in temperature etc. Additionally, the major cause of release of NMs from textile industry can be classified broadly in three categories; during finishing process of manufacturing phase, laundry process while in use phase and finally disposal phase by customers. Ag ions released from silver NPs are reported to re-form AgNPs under certain environmental conditions (Glover et al. 2011). It is reported that one of the major causes of NMs release from fabrics is perspiration (in case of sportswear), which remains in the direct and constant contact. During this process, the dissociated species diffuse out of the fabric matrix, however, the leached and lightly adsorbed particles and their aggregates can be washes off from clothes due to several reasons including mechanical abrasion forces and washing (Vílchez et al. 2016).

The safe use of NMs in the textile industry requires careful investigation of the release of incorporated NMs. The rational design and the method of functionalization of NMs within the textile could also be considered as one of the most important parameters governing the release of NPs from textiles during the above-mentioned events. To prove this, Wagner et al. have demonstrated the release of total Ag ions and AgNPs from textile into the artificial sweat from two types of textiles, in one case, AgNPs were embedded within the textile fibers (composites) and another, AgNPs were studded on the surface of fibers. Result revealed that with respect to the textiles with surface coated AgNPs, composite textiles exhibited less release of total Ag ions. Interestingly, the quantity of particulate fraction in artificial sweat was found to be negligible, which clearly suggest that the released silver was mainly in form of Ag ions in dissolved form (Wagener et al. 2016). Further, the release of silver from textiles was also monitored during sequential washing (Limpiteeprakan et al. 2016). In this study, authors used AgNPs (~4–5 nm) coated over different types of fabrics such as cotton, polyester (PEC) and cotton blended with polyester (TC) and performed sequential washing experiment using water at neutral pH and also with detergents at alkaline pH. Similar experiments were also performed on the fabrics available for consumer use. It was found that most of the released silver was in the non-dissolved form and in the size range of >3–5 nm (Limpiteeprakan et al. 2016). After over 20 wash cycles, it was observed that 48–72% of silver was lost during washing with water whereas, the percentage was increased to 84–94% in case of washing with detergent. Increase in the release of silver after washing with detergent could be due to the presence of peroxide bleach activator [tetraacethylethylenediamine (4% w/w)] present in the detergent. TC fabric was found to retain a high percentage (~52%) of the silver content, which suggest that fabric blending can avoid significant silver release.

Due to the intrinsic UV-light absorbance and antimicrobial properties, TiO_2 NPs are an attractive material for incorporation in textiles because such materials may also offer these properties to the developed fabrics as well (Dastjerdi and Montazer 2010; Paul et al. 2010). TiO_2 NPs are also used in the textile industry as a delustrant (which remove lustre) to synthetic fibers (Barker 1975). Among several other NMs, TiO_2 NPs was found to be present at the highest concentration in soil, water, and sediments (Gottschalk et al. 2009), thus creating an issue with the accurate detection of released TiO_2 NPs in the environment. Therefore, in order to investigate the release of TiO_2 NPs from textile, Windler et al. have evaluated the six different functionalized textiles and quantified the release of TiO_2 NPs during washing (Windler et al. 2012). Result revealed that the five of the tested textiles with sun-protection property had released a very low amount of TiO_2 NPs (0.01–0.06 wt % of total Ti) during single wash because particles were located in the polymer matrix and thus strongly attached (Fig. 2). Whereas, the textile (textile 6) with antimicrobial properties was found to release a much higher amount of TiO_2 NPs (5 mg/L, corresponding to 3.4 wt % of total Ti in one wash). It was realized that in order to impart antimicrobial property in textile 6, TiO_2 NPs were attached to the surface of the fabrics (Gao and Cranston 2008), which resulted in weak binding to the surface and thus facilitate the particle release. Therefore, it can be concluded that these textiles are just a small source of TiO_2 NPs release into the environment. Since the major amount of TiO_2 NPs are present in textile and may be associated with it till the end of the life of the textile, therefore, the possible release of TiO_2 NPs depends on the end-of-life treatment of the textile. To prove this, Mackevica et al. quantitatively estimated the release of TiO_2 NPs from different fibers using single particle inductively coupled plasma mass spectrometry (ICP-MS) (Mackevica et al. 2017). Result showed that in the case of synthetic fibre sample ~80% of the total released titania was found to be in the size range of ~27–200 nm, whereas, from the natural fibre sample (wool, cotton, and blends), negligible amount of NPs release was found (Mackevica et al. 2017). The release of TiO_2 NPs from textile was also reported to be negligible when compared to other sources, such as food product containing TiO_2 NPs as an additive.

Further, AgNPs and TiO_2 NPs release into an artificial sweat were investigated from different textiles and also in various types of perspiration solution (von Goetz et al. 2013). Fabrics with antimicrobial property were shown to release more silver (<450 nm), whereas, fabrics containing TiO_2 NPs, imparting UV protection, did not show any significant release of particles. Additionally, fabrics with antimicrobial property were found releasing both TiO_2 and AgNPs (<450 nm). AgNPs were found to be released both in dissolved and particulate form, however, TiO_2 NPs only released in particulate form. These fabrics integrating the particles were used for the direct dermal exposure to the human population (von Goetz et al. 2013). Cotton

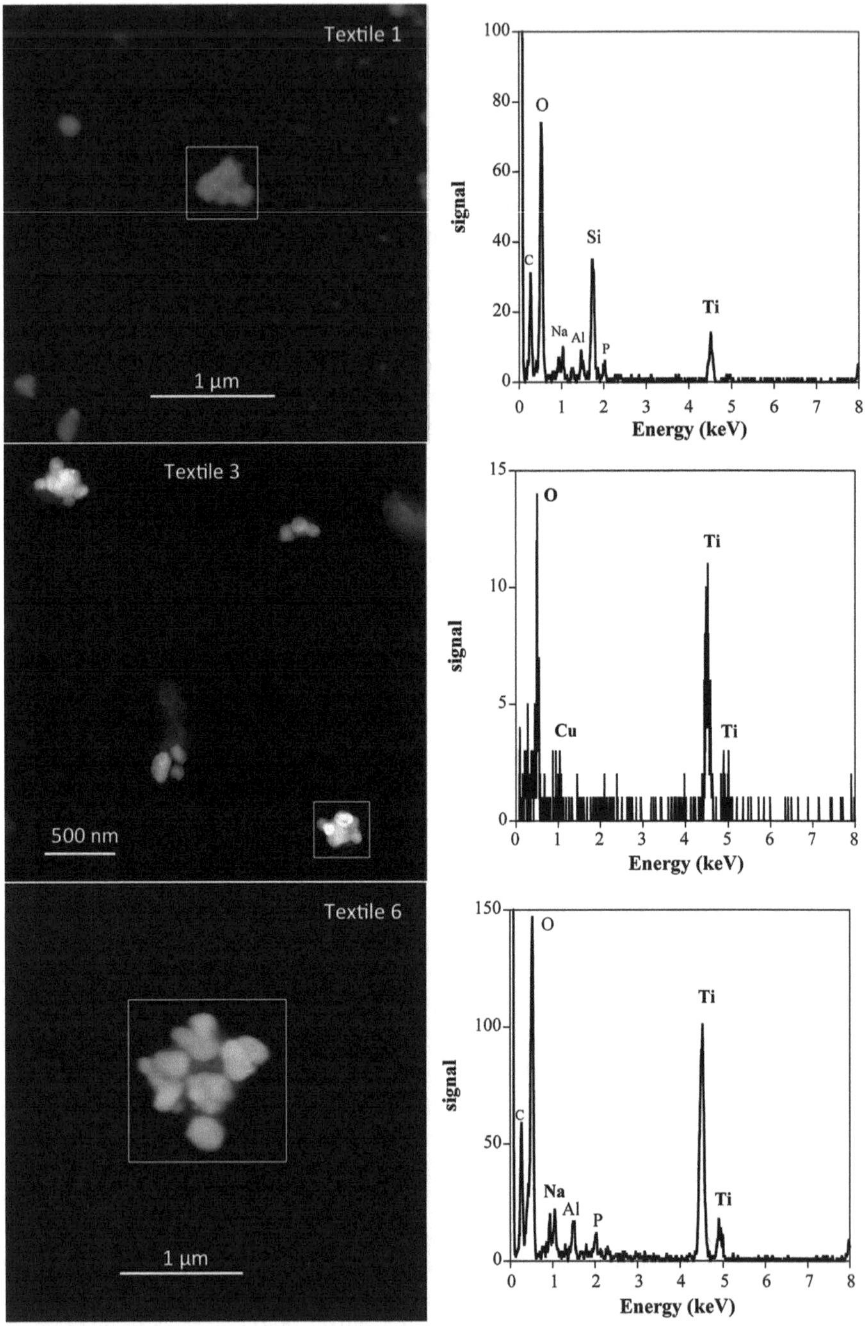

Fig. 2 Transmission electron microscopy of TiO2 particles in the washing solution of textiles 1, 3, and 6. The EDX spectra of the particles inside the square are shown at the right

yarns containing ZnO NPs were also reported to tolerate the knitting procedure and offer moderate to high UV protection factor (UPF) values (Paul et al. 2010).

4 Nanomaterials Release from Cosmetics

Currently, NMs are extensively being used in cosmetic products and incorporated in almost 81% of sunscreens, 7.5% in facial moisturizers, 5.7% in foundations and 3.1% in hair coloring products (Keller et al. 2014). In cosmetic products, NMs are generally incorporated in suspension form such as TiO_2 in sunscreens, nanoliposomes in anti-aging cream, carbon black as hair colorants, however, occasionally they can be found attached on the surfaces such as AgNPs in hair straightener (Keller et al. 2013). According to EU cosmetic regulation, authorities must be informed about the NMs properties such as toxicological profile, concentration limits before placing the nano-enabled consumer products (EP 2009).

Most of the sunscreens are reported to contain inorganic NMs like TiO_2, SiO_2, ZnO, IONPs as UV absorber. Since, these NMs are usually applied by spraying, which may lead to the exposure to lungs through direct inhalation (Vílchez et al. 2016). About 90% of NMs are expected to get released from the products and thus can affect the skin and lungs of the users and contaminate the environment by directly releasing in the water bodies. According to a report, ~40% NMs incorporated in personal care products finally end up in landfills, ~30% are released into the water bodies, and rest ~30% into the soil and ~0.8% into the air (Keller et al. 2013). Gondikas et al. estimated the release of TiO_2 NPs from sunscreen into surface water by collecting the water sample from Old Danube Lake (Vienna, Austria) and analyzed for the presence of TiO_2 by electron microscopy and ICPMS (Gondikas et al. 2014). The lake was extensively used in the recreational activities including bathing and water sports in the summer season. Analysis of suspended particulate matter (SPM) clearly revealed higher concentrations of Ti-containing materials during the summer season. Health risk of TiO_2 and ZnO NPs released in a swimming pool from a commercial sunscreen product was also investigated by Jeon et al. (2016). They found that a significantly high amount of NPs was released into the water which could have come from sunscreens applied over the skin during the swimming and other water sports activity. Further, these NPs were also found to produce hydrogen peroxide (H_2O_2) as a consequence of UV radiation and sunlight, however, the produced concentration (0.011–0.139 μM) of H_2O_2 was much lowered than the concentration required to cause an adverse effect on the human health.

Data demonstrate that only in 1/5th of the cases NPs are reported to penetrate into the healthy skin (Labouta and Schneider 2013). Some studies have demonstrated the penetration of TiO_2 and CdTe QDs under in vivo experimental models (Labouta and Schneider 2013) as well as percutaneous penetration of SiO_2 (<74 nm) (Nafisi et al. 2015). Chang et al. studied the TiO_2 NPs coated with fatty acids. A fatty acid coating was applied to minimize the toxicity and intracellular penetration of NPs (Chang et al. 2016). They performed the experiments in two different cell lines (human fibroblast skin cells and adenocarcinoma lung cells). Result demonstrated that TiO_2 NPs with fatty acids exhibited significantly reduced toxicity (~80–88% less) as well as less intracellular penetration than corresponding bare (uncoated) TiO_2 NPs. Thus, a coating of fatty acid was suggested to be an important component to avoid the adverse effect of NPs on human health (Chang et al. 2016). Antiaging products available in the market are also reported to incorporate NPs. For example, anti-wrinkle creams are reported to utilize the nanosomes of Pro-Retinol-A, which is expected to reduce the wrinkles of the skin. Similarly, use of Retinol is suggested to enhance the epidermal water content, cell renewal, epidermal hyperplasia along with enhanced collagen synthesis (Draelos 2005).

Additionally, hair, lip, and nail care, anti-aging creams are some of another cosmetic product in which NPs are extensively utilized (Lohani et al. 2014). Specifically, in hair care, it is reported that unlike normal hair straightening products, use of nanoemulsions does not destroy the outer structures of hair fibers (Ereno). Further, sericin NPs are reported to be used in hair products, which can facilitate the repair of the damaged hair cuticles by adhering to the surface (Carmen et al. 2012). Currently, NPs are also extensively used in hair coloring products to impart wash-resistant and long-lasting color to hair without causing much damage to keratin fibers. For e.g. carbon nanotubes (CNTs) are shown to enhance the affinity of carbon black to hair fibers and thus impart natural black look to hairs (Huang et al. 2007; Rosen et al. 2015). It was found that if the principal component of permanent hair dye i.e. p-phenylenediamine (PDA) was incorporated with hyaluronic NPs, the damaging effect was significantly reduced with the significant increase in cell viability with respect to PDA alone (Lee et al. 2013). Considering the biocompatibility of AuNPs and its precursors, an alkaline solution of chloroauric acid ($HAuCl_4$) was utilized for the synthesis of AuNPs inside human hairs (Haveli et al. 2012). When white hair fibers were treated with this solution, they showed different shades (from pale yellow to deep brown) of hairs, which could be controlled by varying the time (Fig. 3). This coloring effect is long-lasting and reported to be intact for at least several washes. Owing to the penetration of NPs into the pores of hairs, it is also expected that they can be used in the treatment of androgenic alopecia, which is male pattern hair loss, affecting ~50% of the male population. In this context, incorporation of fullerenes in hair products is reported to stimulate the growth of new hairs along with new hair follicle formation within the dermis of the skin (Zhou et al. 2009). This effect was suggested due to the free radical scavenging property of fullerenes, which might have protected hairs follicles from apoptosis and aging.

Additionally, the use of NPs in cosmetics for nail polish has shown several advantages over the traditional products. A study revealed that the nail paints containing NPs imparted improved toughness and impact resistance to nails (Amato et al. 2010). Another major advantage of this field is the antimicrobial activity of incorporated NPs (AgNPs and other metal oxides NPs) in nail polish to avoid or treat fungal infections in nails (Lohani et al. 2014). NPs can also impart color to the

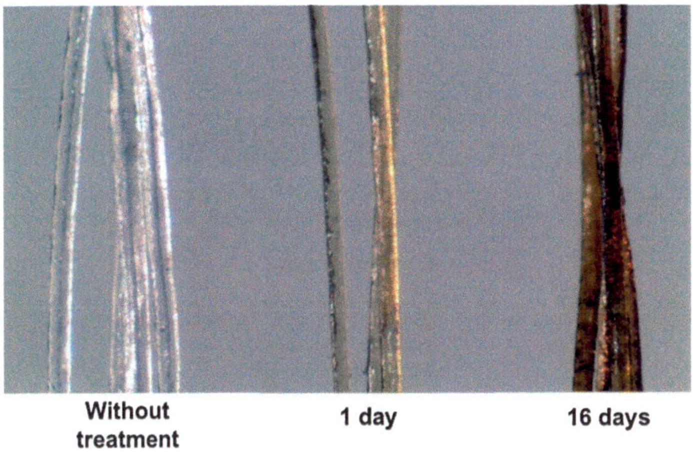

Fig. 3 Hairs after different time intervals of dyeing. (Reprinted with permission from Nano Letters. Copyright 2012, American Chemical Society)

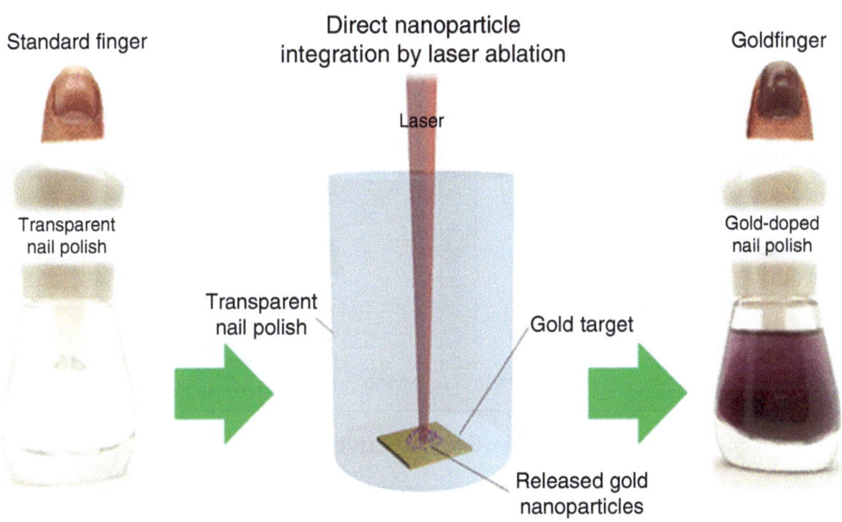

Fig. 4 Integration of AuNPs in nail polish. (Reprinted with permission from Industrial & Engineering Chemistry Research. Copyright 2017, American Chemical Society)

transparent nail paint where the NPs are synthesized with laser exposure (Lau et al. 2017) (Figs. 4 and 5). Such applications are very simple and give easy access to NPs containing polishes with optical properties, which otherwise is difficult to achieve by dispersing powders into highly viscous liquids. Additionally, NPs incorporated in varnishes provide coloring for cosmetic purposes as well as offer its usability for painting on any other solid surfaces. Plasmonic and antibacterial properties of NPs

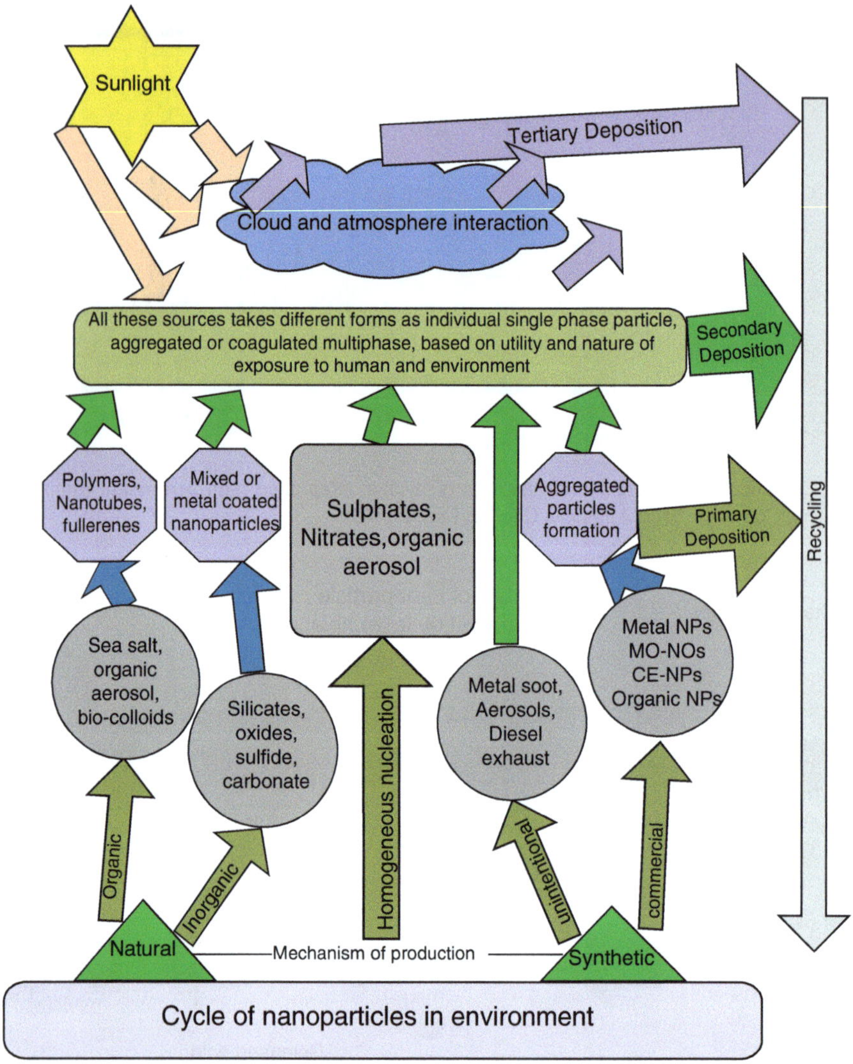

Fig. 5 Cycle of nanoparticles in environment. (Reprinted with permission from Environmental Science and Pollution Research. Copyright 2014, Springer Nature)

can easily be imparted to the surfaces by coating, which can be used for medical and optical applications. This approach minimizes the use of chemicals which affect the nail polish functionalization. Thus, the authors demonstrated the real "(nano-) Goldfinger" (Lau et al. 2017). In another study, ZnO NPs synthesized from the extracts of rose petals were incorporated into nail paint (Tiwari et al. 2017). These ZnO NPs exhibited strong antifungal activity against *Trichophyton mentagrophytes and Microsporum canis* which cause onychomycosis, which is a type of fungal infection in nails. Thus, incorporation of antimicrobial NMs in nail pains could offer multiple advantages to consumer goods.

5 Nanomaterial Release from the Products Used in the Health Sector

The health sector has been gaining tremendous attention for the use of NMs in a wide range of applications including the development of theranostics and as anti-infective material during operations in hospitals. Considering all the consumer products available in the market, health and fitness products contain the highest fraction of NMs (Ng et al. 2017; Yang and Westerhoff 2014). Among them, metal oxide NPs are widely used in the construction of medical devices. The magnetic property of IONPs is employed in magnetic resonance imaging (MRI), photoacoustic imaging, hyperthermia and ultrasonic techniques (Liu et al. 2016; Ramos et al. 2017). The unique electronic structure of ZnO NPs is used in several biomedical applications including the intrinsic fluorescence properties of ZnO nanowires for cancer cell imaging (Hong et al. 2011). ZnO NPs produce visible light upon X-ray irradiation, which has been shown to be used for photodynamic therapy (PDT) (Youssef et al. 2017). It can also undergo self-lighting PDT due to the UV-light emission upon X-ray excitation, which matches with the UV absorbance of most of the photosensitizer. Thus, it serves as a good candidate for PDT radiation source on internal tumor (Sadjadpour et al. 2016). Other than ZnO, TiO_2 NPs, fullerenes and graphene oxide (GO) are also much investigated for their PDT activity (Youssef et al. 2017). Among them, TiO_2 NPs are generally considered because of wide band gap semiconducting material, which is also photoactive upon UV light exposure. During PDT construction, the photosensitizer can be grafted over TiO_2 NPs surface serving as a support system. Due to grafting, it is possible to use visible light instead of UV light during the activation of TiO_2 NPs conjugated with a photosensitizer (Jia and Jia 2012). TiO_2 nanotubes are also investigated for their use in bone regeneration because the hierarchical structure of bones falls in the nanometer regime (Brammer et al. 2012). Utilizing the electrochemical anodization, TiO_2 nanotubes can result in unique 3D nanostructure shape which could be used to stimulate the growth of osteogenic cells and thus create surface designs for novel orthopedic materials. Such applications require optimal dimensions of TiO_2 nanotubes, which has been reported by several research groups, however, the dimension of nanotubes vary significantly between the groups. In this context, Wang et al. performed the in vivo study on minipigs and followed the osteogenesis response by exposing them by 30, 70 and 100 nm TiO_2 nanotube (Wang et al. 2011). Among these, 70 nm TiO_2 nanotube exposure resulted in a significant increase in bone-implant contact (BIC) and related gene expression in bone attached to implant with TiO_2 nanotube. Conversely, 30 nm TiO_2 nanotubes exposure resulted in optimal growth in vitro but did not impact the bone growth under in vivo condition (von Wilmowsky et al. 2009).

Zirconium oxide (ZrO_2) NPs were used to create bone-like nanocomposites and therefore found suitable for bone tissue engineering applications (Bhowmick et al. 2017). The synthesized nanocomposite consisted of chitosan, polyethylene glycol (PEG) and nano-hydroxyapatite. Incorporation of ZrO_2 NPs in above-made nanocomposites offers better water absorption capacity and mechanical strength, which is much similar to that of the spongy bone found in humans. Further, this nanocomposite did not induce any hemolysis and was found cytocompatible to osteoblastic cells with antimicrobial property (Bhowmick et al. 2017).

6 Release of Nanomaterials from Consumer Products and Their Impact on Environment and Human Health

Owing to the commercial values of NMs their release in the environment and thus exposure to humans and other organisms has increased. This raises concerns of researchers to evaluate the toxic effects of NMs on human health and environment. It is also suggested that the toxicological profile of NMs must be mapped before they are employed for any potential application. Despite this concern, there has been a lack of information regarding the effect of NMs on human health and environmental implications (Sajid et al. 2015). It is well established that the toxicity of NMs is controlled by size, shape, composition, and agglomeration state (Sajid et al. 2015). Due to the tiny size, high surface area to volume ratio of NMs, the reactivity of the particles increases exponentially, which makes particles more reactive and thus toxic. Additionally, due to the small size, NMs are easily internalized in cells through the cell membrane. According to one report, NMs with a diameter of ~35 nm can cross the blood-brain barrier, however, particles with <35 nm and >100 nm can enter into the cell nuclei and accumulate in organs, respectively (Oberdorster et al. 2004). Shapes of NPs also contribute to the toxicity and several reports have been published. A comparative study focussing on internalization and thus cytotoxicity of spherical and needle-shaped PLGA-PEG NPs (Zhang et al. 2017) showed that both types of NPs were internalized via endocytosis, however, the needle-shaped NPs induce significantly more toxicity to tested mammalian cells by lysosome disruption. It is well known that NMs are very reactive and this feature has been linked with their toxicity profile through the generation of ROS in the cell cytoplasm. The so generated intracellular ROS can induce oxidative stress, which results in oxidative damage to proteins and DNA (Abudayyak et al. 2017). Stability of NMs is another major toxicity controlling factor. It has been reported that in biological system NMs undergo oxidation and subsequent release of constituent ions. These ions then react with the intracellular biomolecules, which leads to the alteration in their normal biochemical reactions. Although solubility of NMs strongly depends on the pH of the environment, however, some NMs are easily ionizable than others. One such study compared the toxicity of ZnO NPs (slightly soluble) and TiO_2 NPs (insoluble). It was found that ZnO NPs induced more cell death when exposed to 15 ppm or higher concentration, whereas in case of TiO_2 NPs the toxicity was seen at very high concentrations (Roy et al. 2003). Surface modification of NMs can also modulate the surface charge and therefore the toxicity. In this context, it was demonstrated that AgNPs exhibited surface charge dependent toxicity to bacterial species (El Badawy et al. 2011). Among them, negatively charged AgNPs were found to be less toxic to the bacterial cells than the positive charged. NMs are prone to aggregation, which can also impart significant alteration to internalization and thus toxicity to mammalian cells (Sajid et al. 2015). The process of NMs synthesis and medium of storage are also some of the important factors dictating the toxicity. Toxicity of NPs can also be influenced by the length of storage time of NMs in medium with varying ionic strengths. Ag ions released in solution after long-term storage of AgNPs are shown to be toxic to cells (Kittler et al. 2010).

6.1 Nanomaterial Exposure to Human Organs

NMs can be exposed to the human population and environment by different mechanisms. At first, exposure occurs to workers (scientists, technicians, and engineers) employed in the field of NMs-based research involving the scale-up synthesis of NPs as well as working towards commercial applications of NMs. At the second level, consumers get exposed to NPs during its use and other applications, which may lead to harmful and toxic effect (Tsuji et al. 2006). NPs can interact with the components of the human body through different routes such as penetration through skin nodes, intake in the respiratory and digestive system by inhalation and ingestion etc. Further, the internalized particles can accumulate in different organs by respiratory and blood circulation system (Miller et al. 2017). NPs can be exposed to the skin by different means such as by applying NPs containing cosmetic products (lotions, sunscreens, exposed to vehicle tailpipe emissions and welding fumes, and burning of natural gas and coal) (Sajid et al. 2015).

Skin proves to be an effective barrier against NPs penetration, however, due to the presence of hair follicles and sweat glands this barrier facilitate the penetration of small-sized NPs. However, there are some ambiguous reports about the skin penetration of TiO_2 NPs. A study indicates that surface passivated TiO_2 NPs undergo indirect skin damage which further leads to penetration into the skin (Teow et al. 2011). Conversely, another study reports that TiO_2 NPs cannot penetrate the intact or even damaged skin and therefore are limited to epidermis and stratum corneum levels (Crosera et al. 2015). The quantitative data analysis revealed that TiO_2 was found in the epidermal layer of skin even after 24 h of exposure, whereas, the dermal layer did not show any detectable limit of Ti. This concentration of TiO_2 has also supported their lower toxicity. Studies from the use of NMs for skin diseases and wounds have shown that under these conditions the penetration of NMs was enhanced (Tak et al. 2015).

Recent studies revealed that NPs can penetrate deeper into lungs and interact with epithelium, which causes inflammation and chronic disease after penetration into interstitium and transfer of particles to lymph nodes (Donaldson et al. 1998; Sajid et al. 2015). In this context, after inhalation and intratracheal instillation, the pulmonary toxicity study of ZnO NPs was conducted under in vivo experimental condition (Morimoto et al. 2016). Result revealed that well dispersed ZnO NPs showed low toxicity and persistent inflammation. It was also shown that IONPs exposure induces the inflammatory response, respiratory toxicity, and cytotoxic damage to Wistar rats. IONPs were found cleared from the respiratory tract but after 28 days of installation (Hurbankova et al. 2017). NPs exposure causes damage in nasal epithelium and mucous membrane which further leads to a decrease in smell sensation and nasal humidification. In case of small size NPs, the effect is more pronounced as it is difficult to eliminate these particles, which damages the nasal air passages defense system and finally NPs can enter to the brain (Oberdorster et al. 2004). The adverse effects of NPs inhalation in lungs depend on several factors such as NPs exposure dose, amount of NMs deposition in lungs, as well as on the clearance mechanism (Borm and Kreyling 2004).

Ingestion of NPs through food and food products and drugs are consumed orally followed by absorption in the gastrointestinal tract. Absorbed NMs can subsequently enter in the lymphatic draining of the system (Sajid et al. 2015). In this context, a high concentration of ingested TiO_2 NPs was found to induce oxidative stress and lipid peroxidation in glands of the digestive system (Valant et al. 2012). TiO_2 NPs are repeatedly ingested through toothpaste has supported its role in pathogenesis in inflammatory bowel disease (IBD) and related disorder (Bouwmeester et al. 2018). TiO_2 NPs administration to rodents was found to induce the inflammation in the small intestine (Nogueira et al. 2012), promote colitis-associated tumor formation (Urrutia-Ortega et al. 2016), exacerbate colitis (Ruiz et al. 2017), and colonic inflammation and preneoplastic lesions (Bettini et al. 2017).

6.2 *Environmental Factors and Physicochemical Properties of Nanomaterials*

Several environmental factors such as temperature, geographical latitude, wind flow rate, nature of light and humidity also govern the toxicity of NPs. Dispersion of NPs at higher temperatures is well-known than at lower temperatures. NPs can also behave differently under the visible and UV light. Wind speed also helps in the penetration of NMs to plants and animal tissues (Sajid et al. 2015). Key issues regarding the behavior and transport of NMs are studied by several groups (Hartmann et al. 2014; Lowry et al. 2012; Viswanath and Kim 2017). NPs suspended in the air have a high chance of UV light exposure, which may lead to alteration in their photochemical transformation (Mitrano et al. 2015). It has been suggested that larger NPs are found to have high deposition rate than of smaller particles. This could be attributed due to the fact that the diameter of particles is directly proportional to the gravitational settling velocity which controls the deposition of NPs in the air. Agglomeration tendency of NMs also found to increase the chances of their deposition. However, when compared with photochemical reactions other processes such as agglomeration and deposition in the air are found to be less significant (Soni et al. 2015). The fate of NMs in the environment mainly depends on the potential source, chemical properties and their interaction with environmental pollutants (Viswanath and Kim 2017). Three major sources are reported to be responsible for atmospheric release of NPs, (i) emission from road traffic exhaust and industrial combustion; (ii) compression of low volatility vapors due to oxidation of atmospheric gases; and (iii) diesel exhaust dilution (Baalousha and Lead 2009).

6.3 Environmental Issues Caused by Nanomaterials

Due to the frequent incorporation of NMs in consumer products, aquatic environment is prone to get contaminated by the degradation and subsequent release of particles. Contamination of aquatic environment with these NMs can pose toxicity to aquatic animals such as fishes and daphnia, however, the extent varies from species to species (Sajid et al. 2015). It is reported that smaller NMs are involved with more production of ROS, which ultimately causes damage to DNA, proteins and cell membrane. Further, a report published by Mueller et al. revealed that the release of NMs resulted in the increase of AgNPs concentration (0.03 μg/L) in natural water (Mueller and Nowack 2008). It is also reported that Ag ions released from AgNPs induce toxicity to aquatic creatures at the concentrations as low as 0.1–5 μg/L (Bondarenko et al. 2013). The uptake of these Ag ions in organisms of different trophic levels may pose threat to the ecological systems (Fabrega et al. 2011). Toxicity of AgNPs has been shown to be influenced by the shape and composition. Triangular Ag nanoplates are shown to cause more damage to the cell membrane and thus cell death (Limpiteeprakan et al. 2016). Additionally, in the sewage treatment process, Ag ions accumulate in sewage sludge in form of AgCl and Ag_2S. Presence of sulfhydryl group containing molecules in water leads to the formation of Ag_2S, which has lower solubility in water and thus impose lesser toxicity during short-term exposure (Levard et al. 2012). Sewage sludge containing AgNPs usually get dumped in landfills and also its use in agriculture field as fertilizers lead to the contamination of groundwater and soil (Reidy et al. 2013). In this context, Li et al. have demonstrated that Ag_2S can undergo morphological transformation and dissolution in the aqueous system within a short duration of 96 h (Li et al. 2016). Photoinduced transformation of Ag_2S can be greatly influenced by pH, inorganic salts, dissolved organic matter (DOM) and dissolved oxygen (DO). Phyto-toxicity of Ag_2S NPs has been demonstrated by Wang et al., where they report that Ag_2S NPs get internalized into cowpea and wheat plant leading to the reduction in the growth ~52% in 2 weeks period (Wang et al. 2015). They also concluded that the toxicity of Ag_2S NPs was not due to the dissolution and release of Ag^+ ions because the soluble Ag concentration was extremely low. Conversely, the toxicity was thought to arise due to the accumulation of Ag_2S NPs in the root and shoot system of the plant.

6.4 Soil as a Sink for Nanomaterials

Soil contamination by NMs is another important area of study. In this context, the interaction of NMs with soil components is imperative to investigate the fate and future effects of particles over agricultural products. Such studies become further essential considering that the soil as a major sink of NMs than aqueous and atmospheric systems. NMs discharged into the soil can be sorbed into soil particles, and components may undergo degradation by abiotic and biotic processes, and also get transported to groundwater, and finally run off and drain flow (Rajput et al. 2018). Recently, it has been reported that soil contamination by IONPs significantly reduces the microbial biomass, and microbial colony counts (Rashid et al. 2017). Further investigation revealed that IONPs reduces the microbial activity, which further lowers the nitrogen decomposition and mineralization in applied litter in the soil. Thus, it is necessary to rethink before using NPs in agriculture as fertilizer or for bioremediation applications. Since the fertility of soil is mainly dependent on the soil microbial activity (Rajput et al. 2018) therefore, there have been several reports suggesting that metal and metal oxide NPs induce the modification in soil microbes enzyme activity (Simonin et al. 2015; Xu et al. 2015). Additionally, Shen et al. have demonstrated the ecotoxicity of ZnO NPs on soil microorganisms by investigating dehydrogenase enzyme activity, microbial respiration, fluorescent diacetate hydrolase activity and ammonification in soil (Shen et al. 2015). They observed that ammonification and respiration were significantly inhibited within 3 and 1 month, respectively. Enzymes activity was inhibited in a dose-dependent exposure of ZnO NPs but the trend of the curve was varied over time due to the difference in ZnO NPs toxicity. It was also found that soil type can moderate the toxicity of ZnO NPs as more toxicity was observed in acidic soil than neutral and alkaline soil (Shen et al. 2015).

7 Toxic Effect of Nanomaterial on Human Health

It is well documented that toxicity of NMs is much higher than its counter bulk parts, for example, nano-sized TiO_2 may induce pulmonary inflammation when compared to the larger TiO_2 particles. NMs can readily interact with immune cells and induce immunotoxicity (Hussain et al. 2012). Direct damage to immune cells leads to apoptosis, which may result in the activation of specific immune signalling pathway. It has also been demonstrated that inflammasome activation can occur via different mechanisms such as lysosome rupture or direct recognition of NMs by toll-like receptors (Dusinska et al. 2017). NMs, which come in direct contact with the genetic materials, cause chemical modification as well as physical damage. ROS induction by NMs also causes significant damage to DNA, protein, lipid and other cellular componets. Excessive ROS generation can also damage the mitochondrial DNA (mt DNA) (Esposito et al. 1999). A Secondary mechanism of NMs induced toxicity is also reported due to the extracellular ROS generation by inflammatory responses produced by macrophages and neutrophils (Magdolenova et al. 2014). Carbon-based NMs are reported to enhance oxidative stress in fish brain cells as well as pulmonary inflammation in rats (Sharifi et al. 2012). NPs accumulation in the reticuloendothelial system (RES) with various phagocytic cells causes an imbalance in ROS homeostasis and antioxidant defence, which further facilitates oxidative stress in liver and spleen.

Toxicity of metal-based NMs is caused mainly due to the DNA damage. Recent reports suggest that metal NMs induce the persistent epigenomic changes (Smolkova et al. 2017). Although, AuNPs have shown promise involvement in several biomedical applications but some of the recent studies have reported the epigenetic toxicity. Both in vitro and in vivo studies have shown that AuNPs exposure leads to the DNA methylation. Further, AuNPs exposure to BALB/c mice was found to induce the hypermethylation of some genes, however, hypomethylation of several genes, including *GPX, CDK, ATM, GSR* in lung tissues of mouse (Tabish et al. 2017). These genes are thought to be involved in the double strand DNA damage sensing, and transcription modulation in response to various intra and extracellular factors (Smolkova et al. 2017). Further, Mazumder et al. demonstrated that AuNPs is also involved in the reorganization of chromatin (Mazumder and Shivashankar 2007). Authors showed the modulation of heterochromatin connections with lamin protein and histones after exposure of AuNPs to HeLa cells. Later, it was reported that AuNPs exposure leads to the change in miRNA expression and transplacental epigenetic and clastogenic effect in mouse foetus (Balansky et al. 2013). Further, AgNPs are also shown to be utilized in several antimicrobial applications such as infected wound treatment, and dental plaque biofilms catheters dressings. It has been reported that augmentation of 5-methylcytosine, DNA methyltransferase (DNMT1, DNMT2, DNMT3a and DNMT3b) level occurs in hippocampal HT22 mouse neuronal cells (Mytych et al. 2017). Sub-lethal concentrations of AgNPs were found to decrease H3K4me3 and H3K79me1 methylation, which leads to the change in haemoglobin level in mouse erythroleukemia cells. (Qian et al. 2015; Eom et al. 2014). Recently, Blanco et al. evaluated the effect of AgNPs over tumor suppressor protein, p53 (Blanco et al. 2017). Exposure of high AgNPs concentration to A549 cells for 72 h, induces high DNA methylation and histone H3 deacetylation. Cadmium (Cd) containing compounds are classified as carcinogenic and reported to induce their harmful effects by epigenetic mechanism (Cheng et al. 2012). However, cadmium telluride quantum dots (CdTe QDs) have been utilized in several biomedical application such as molecular imaging and fluorescent tags for therapeutic targeting (Smolkova et al. 2017). Lovric et al. demonstrated that the toxicity of QDs is mainly due to the membrane blebbing and chromatin condensation (Lovric et al. 2005). Later, Choi et al. also demonstrated that exposure of cells to CdTe QDs for 4 and 24 h leads to the global hypo acetylation. Its treatment also leads to an increase in the expression of apoptotic genes via p53 activation (Choi et al. 2008).

The adverse effects of NMs can be efficiently eliminated by surface coating, which stabilizes the particles and avoid agglomeration. Dense coating has been shown to prevent the dissolution and subsequently release of the toxic ions (Kirchner et al. 2005). Surface modification of CNTs is reported to reduce its in vivo toxicity. In this context, Lacerda et al. have reported that functionalized MWCNTs form a stable dispersion with high excretion rate when injected intravenously (Lacerda et al. 2008). Additionally, Oleszczuk et al. have also demonstrated that the coating of NPs with surfactant leads to the decrease in its toxicity of ZnO, TiO_2 and nickel (Ni) NPs to *Daphnia magna*. Result revealed that coating of ZnO with cetyltrimethylammonium bromide (CTAB) reduces its toxicity by >60%, however, TiO_2 NP, showed reduction in toxicity when coated with Triton-100. In case of Ni NPs, surfactant effect was found depending on the concentration of NPs (Oleszczuk et al. 2015). Therefore, it may be be concluded that the use of biocompatible surfactants for NMs coating could impart biocompatibility as well as stabilization of NPs in eco-toxicological studies.

8 Removal of NMs from Environment

Due to the extremely small size, settling dynamics of NMs cannot be predicted easily and also such particles resist aggregation because of their surface charge (Kunhikrishnan et al. 2015). NMs, which tends to aggregate, can be collected in a wastewater treatment plant (WTP), whereas, NMs resisting aggregation can end up in WTP effluent (Westerhoff et al. 2013). For removal of NPs, the focus has been shifted towards the agents capable of inducing aggregation. Well documented reports suggest that NMs do not undergo aggregation in wastewater effluents (Zhou et al. 2015), however, with dissolved organic matter exhibit rapid aggregation. A detailed investigation in this context revealed that change in surface chemistry may be induced by organic matter, which affects the NMs stability (Solomon et al. 2017). Most of the NMs captured in WTP process which uses organic processes such as activated sludge or trickling filter. Sorption of NMs to activated sludge is an important removal mechanism for pollutants including synthesized NPs in conventional WTP (Kiser et al. 2010). Different types of NPs were removed from water but to different extent by biosorption like 97% AgNPs, 88% aqueous fullerenes, 23% TiO_2 NPs, 39% of functionalized AgNPs and 13% fullerol NPs. This study also reported that the different surface functionalities of carbonaceous NPs affect the biosorption process differently. The earlier report suggested that N-bromosuccinimide (NBS) can be used for recycling of AuNPs (Singh and Prasad 2007). Authors showed that NBS can oxidize octadecylamine (ODA)-capped AuNPs to Au^{3+} ions. In this process, it was suggested that the released molecular bromine species can strip off the ODA molecules from the AuNPs surface and thus react with surface Au^0 atoms to form Au (III) bromide which releases into solution and allow fresh bromine molecules to react with new surface of Au^0 which leads to complete oxidation of Au^0 to Au^{3+}. It was also reported that these Au^{3+} ions can again be re-reduced to AuNPs, thus are capable of completely recycle the AuNPs (Singh and Prasad 2007).

8.1 Electronic Waste

Currently, electronic waste (e-waste) is a major concerned pollution source as it contains the toxic substance which can affect human health and the environment (Kiddee et al. 2013). E-waste mainly comprises of waste electric and electronic equipment (EEE) or goods which are not appropriate for their original intended use. EEEs includes the phones, refrigerators, TV, AC, computers, printers, DVDs, CDs etc. Government office machines, industrial machines, medical equipment are some of the major sources of e-waste (Kaya 2016). Due to the frequent replacement of EEEs, e-waste quantity increases rapidly and are thought to consist of 8% of municipal waste (Widmer et al. 2005; Zhou and Qiu 2010). E-waste consists lots of heavy metals which can act as significant polymetallic secondary source and also an environmental contaminant. Due to rapidly growing disposal problems recycling of e-waste is compulsory in many developing/developed countries. United Nations (UN) recorded the constant growth in e-waste and is estimated to be more than 50 million tons (Sakunda 2013). E-waste contains the various hazardous substance which is harmful to human health i.e. computers and rechargeable battery contains lead (Pb), transformers and condenser contain polychlorinated biphenyls (PBBs), chlorofluorocarbon (CFC) are found in insulation foams and cooling units, Polyvinyl chloride (PVC) is a major component of plastics. Pb is known to cause damage in kidney, central and peripheral nervous system, blood system, PVC produces dioxin after burning which causes reproductive and developmental complications as well as damage to the immune system. E-waste pollutes the groundwater, produces toxic fumes while burning, cause soil acidification and releases the carcinogenic substance in the environment. Thus, it is essential to conduct proper e-waste management in all countries (Kaya 2016). Recycling of valuable metallic and non-metallic components from e-waste is one step towards its management.

8.2 Process of Materials Recycling and Recovery

Recycling of materials can be done both by physical/mechanical and chemical techniques. The physical processes are generally employed during upgrading stage where metal and non-metals can be separated by crushing and shredding processes. Physical recycling techniques consist of a crusher, pulverizer, classifier and separator. Recycling process result can be evaluated by two aspects: efficiency of material (metal and non-metal) recovery and environmental impact of the process (Kaya

2016). Chemical recycling is decomposing of waste polymers into their monomeric unit or into useful chemical by means of chemical reactions. Chemical recycling techniques consist of pyrolysis, gasification, depolymerisation with supercritical fluids and hydrogenic degradation (Guo et al. 2009). It separates metallic and organic materials. Recently, Jeon et al. demonstrated the recovery of Au from shredder light fraction (SLF) of recycling plant by flotation and leaching (Jeon et al. 2018).

Recovery of the precious components of materials, including NMs, is also considered as next important step. NMs are adsorbed to the biomass, but they maintain their characteristic as several particles will remain in the nano dimension or may be in an agglomerated form (Solomon et al. 2017). After the capture of NPs in the solid system, its disposal process is complete, but they maintain their unique NPs properties. Final disposal for sludge management is to spread in the field which allows the NPs to return back to the environment. In order to reuse the NPs for commercial purposes, they need to be separated from bulk mass (Solomon et al. 2017). Thus, further research is needed to gain more knowledge about the recycle/reuse of NPs.

Thus, there is an urgent need to develop techniques which are involved in capturing and removing NPs. Removal and capture of NPs occur in a series of steps and consist of two main steps: surface chemistry of NPs can be changed which lead to aggregation followed by settling of the NPs. In the next step, the so obtained solid mass is dried and the intermediate material is removed (used to enhance the aggregation), which was used in the first step. Several trials are required in order to completely establish a practical and working process to recycle the NMs from consumer goods.

9 Conclusion and Future Prospects

Considering the potential of NMs to be actively incorporated in consumer products, it is almost essential to quantify the material incorporated. NMs incorporated in a variety of consumer goods such as sporting accessories, medical devices, cosmetic products, sunscreen lotions and anti-aging creams, therefore, prone to be released in the environment due to the weathering of these products. This raises the concerns about environmental pollution and potential hazard for humans. It is also assumed that the released particles would be of different shape, size and composition, therefore, they may exhibit some unusual properties and ultimately damage unexpectedly. Although several mechanisms for the removal of NMs from the environment has been devised, it is still not enough and more such strategies are needed to effectively avoid the toxicity of NMs. One of the methods could be the use of green chemistry to synthesize NMs following "safe-by-design" approach. However, it has been observed that the NMs developed with green chemistry methods are deficient in some of the key properties and also require sophisticated instruments for complete characterization. Further, size and shape control synthesis of NMs remains a challenge with green chemistry methods, however, through chemical synthesis, there are several reports for an extremely controlled synthesis of nanoparticles. Safety evaluation method is another challenge for effectively quantifying the toxicity of NMs to the human health and environment. As discussed in the chapter, different methods are needed to establish the toxicity of NMs incorporated in different consumer goods. A single safety assessment method may not be enough to measure the toxic potential of NMs used in different consumer products. Considering the above issues, it is expected that experts from several fields such as materials synthesis, biologists, engineers, immunologists and toxicologists need to brainstorm and develop methods for characterizing the materials in biologically relevant buffers and to measure the impact on human health and environment. This team also need to formulate policy for safe handling of NMs either in free-standing form or incorporated in consumer goods. Although we have made significant progress in the development of materials with desired properties until their future impact on human health and the environment is addressed the real potential of nanotechnology will not be realized.

Acknowledgements R. Singh thanks to the Department of Science and Technology, New Delhi, for providing INSPIRE Senior Research Fellowship (SRF). The financial assistance for the Centre for Nanotechnology Research and Applications (CENTRA) by The Gujarat Institute for Chemical Technology (GICT) is acknowledged. The funding from the Department of Science and Technology - Science and Engineering Research Board (SERB) (Grant No.: ILS/SERB/2015-16/01) to Dr. Sanjay Singh under the scheme of a Start-Up Research Grant (Young Scientists) in Life Sciences is also gratefully acknowledged.

Conflict of Interest The authors declare no competing financial interest.

References

Abudayyak, M., Guzel, E., & Ozhan, G. (2017). Nickel oxide nanoparticles induce oxidative DNA damage and apoptosis in kidney cell line (NRK-52E). *Biological Trace Element Research, 178*, 98–104. https://doi.org/10.1007/s12011-016-0892-z.

Aitken, R. J., Chaudhry, M. Q., Boxall, A. B., & Hull, M. (2006). Manufacture and use of nanomaterials: Current status in the UK and global trends. *Occupational Medicine, 56*, 300–306. https://doi.org/10.1093/occmed/kql051.

Amato, S. W., Farer, A., Hoyte, W. M., Pavlovsky, M., & Smith, R., Valdiviezo, G. (2010). *Coatings for mammalian nails that include nanosized particles*. Google Patents

Baalousha, M., & Lead, J. R. (2009). *Overview of nanoscience in the environment environmental and human health impacts of nanotechnology* (pp. 1–25). Hoboken: Wiley-Blackwell.

Baker, S., Rakshith, D., Kavitha, K. S., Santosh, P., Kavitha, H. U., Rao, Y., & Satish, S. (2013). Plants: Emerging as nanofactories towards facile route in synthesis of nanoparticles. *BioImpacts: BI, 3*, 111–117. https://doi.org/10.5681/bi.2013.012.

Balansky, R., et al. (2013). Transplacental clastogenic and epigenetic effects of gold nanoparticles in mice. *Mutation Research, 751–752*, 42–48. https://doi.org/10.1016/j.mrfmmm.2013.08.006.

Banerjee, D. A., Kessman, A. J., Cairns, D. R., & Sierros, K. A. (2014). Tribology of silica nanoparticle-reinforced, hydrophobic sol–gel composite coatings. *Surface and Coatings Technology, 260*, 214–219. https://doi.org/10.1016/j.surfcoat.2014.07.091.

Barker, R. H. (1975). Additives in fibers and fabrics. *Environmental Health Perspectives, 11*, 41–45.
Bello, D., Martin, J., Santeufemio, C., Sun, Q., Lee Bunker, K., Shafer, M., & Demokritou, P. (2013). Physicochemical and morphological characterisation of nanoparticles from photocopiers: Implications for environmental health. *Nanotoxicology, 7*, 989–1003. https://doi.org/10.3109/17435390.2012.689883.
Bettini, S., et al. (2017). Food-grade TiO2 impairs intestinal and systemic immune homeostasis, initiates preneoplastic lesions and promotes aberrant crypt development in the rat colon. *Scientific Reports, 7*, 40373. https://doi.org/10.1038/srep40373.
Bhowmick, A., Pramanik, N., Jana, P., Mitra, T., Gnanamani, A., Das, M., & Kundu, P. P. (2017). Development of bone-like zirconium oxide nanoceramic modified chitosan based porous nanocomposites for biomedical application. *International Journal of Biological Macromolecules, 95*, 348–356. https://doi.org/10.1016/j.ijbiomac.2016.11.052.
Blanco, J., Lafuente, D., Gomez, M., Garcia, T., Domingo, J. L., & Sanchez, D. J. (2017). Polyvinyl pyrrolidone-coated silver nanoparticles in a human lung cancer cells: Time- and dose-dependent influence over p53 and caspase-3 protein expression and epigenetic effects. *Archives of Toxicology, 91*, 651–666. https://doi.org/10.1007/s00204-016-1773-0.
Bohnsack, J. P., Assemi, S., Miller, J. D., & Furgeson, D. Y. (2012). The primacy of physicochemical characterization of nanomaterials for reliable toxicity assessment: A review of the zebrafish nanotoxicology model. *Methods in molecular biology, 926*, 261–316.
Bondarenko, O., Juganson, K., Ivask, A., Kasemets, K., Mortimer, M., & Kahru, A. (2013). Toxicity of Ag, CuO and ZnO nanoparticles to selected environmentally relevant test organisms and mammalian cells in vitro: A critical review. *Archives of Toxicology, 87*, 1181–1200. https://doi.org/10.1007/s00204-013-1079-4.
Borm, P. J., & Kreyling, W. (2004). Toxicological hazards of inhaled nanoparticles – potential implications for drug delivery. *Journal of Nanoscience and Nanotechnology, 4*, 521–531.
Bouwmeester, H., van der Zande, M., & Jepson, M. A. (2018). Effects of food-borne nanomaterials on gastrointestinal tissues and microbiota. *Wiley Interdisciplinary Reviews: Nanomedicine and Nanobiotechnology, 10*.
Brammer, K. S., Frandsen, C. J., & Jin, S. (2012). TiO2 nanotubes for bone regeneration. *Trends in Biotechnology, 30*, 315–322. https://doi.org/10.1016/j.tibtech.2012.02.005.
Carmen, M., Pereda, V., & Polezel, A. (2012). *Sericin cationic nanoparticles for application in products for hair and dyed hair*. US Patent 20120164196, June.
Chang, J., et al. (2016). Role of fatty acid composites in the toxicity of titanium dioxide nanoparticles used in cosmetic products. *The Journal of Toxicological Sciences, 41*, 533–542. https://doi.org/10.2131/jts.41.533.
Cheng, T. F., Choudhuri, S., & Muldoon-Jacobs, K. (2012). Epigenetic targets of some toxicologically relevant metals: A review of the literature. *Journal of Applied Toxicology: JAT, 32*, 643–653. https://doi.org/10.1002/jat.2717.
Choi, A. O., Brown, S. E., Szyf, M., & Maysinger, D. (2008). Quantum dot-induced epigenetic and genotoxic changes in human breast cancer cells. *Journal of Molecular Medicine, 86*, 291–302. https://doi.org/10.1007/s00109-007-0274-2.
Crosera, M., et al. (2015). Titanium dioxide nanoparticle penetration into the skin and effects on HaCaT cells. *International Journal of Environmental Research and Public Health, 12*, 9282–9297. https://doi.org/10.3390/ijerph120809282.
Dastjerdi, R., & Montazer, M. (2010). A review on the application of inorganic nano-structured materials in the modification of textiles: Focus on anti-microbial properties. *Colloids and Surfaces B, Biointerfaces, 79*, 5–18. https://doi.org/10.1016/j.colsurfb.2010.03.029.
Dhawan, A., & Sharma, V. (2010). Toxicity assessment of nanomaterials: Methods and challenges. *Analytical and Bioanalytical Chemistry, 398*, 589–605. https://doi.org/10.1007/s00216-010-3996-x.
Dhawan, A., Sharma, V., & Parmar, D. (2009). Nanomaterials: A challenge for toxicologists. *Nanotoxicology, 3*, 1–9. https://doi.org/10.1080/17435390802578595.

Di Bona, K. R., et al. (2015). Short- and long-term effects of prenatal exposure to iron oxide nanoparticles: Influence of surface charge and dose on developmental and reproductive toxicity. *International Journal of Molecular Sciences, 16*, 30251–30268. https://doi.org/10.3390/ijms161226231.

Donaldson, K., Li, X. Y., & MacNee, W. (1998). Ultrafine (nanometre) particle mediated lung injury. *Journal of Aerosol Science, 29*, 553–560. https://doi.org/10.1016/S0021-8502(97)00464-3.

Draelos, Z. (2005). Retinoids in cosmetics. *Cosmetic Dermatology, 18*, 3–5.

Duan, H., Wang, D., & Li, Y. (2015). Green chemistry for nanoparticle synthesis. *Chemical Society Reviews, 44*, 5778–5792. https://doi.org/10.1039/c4cs00363b.

Dusinska, M., et al. (2017). Immunotoxicity, genotoxicity and epigenetic toxicity of nanomaterials: New strategies for toxicity testing? *Food and Chemical Toxicology, 109*, 797–811.

El Badawy, A. M., Silva, R. G., Morris, B., Scheckel, K. G., Suidan, M. T., & Tolaymat, T. M. (2011). Surface charge-dependent toxicity of silver nanoparticles. *Environmental Science & Technology, 45*, 283–287. https://doi.org/10.1021/es1034188.

Elbeshehy, E. K., Elazzazy, A. M., & Aggelis, G. (2015). Silver nanoparticles synthesis mediated by new isolates of Bacillus spp., nanoparticle characterization and their activity against bean yellow mosaic virus and human pathogens. *Frontiers in Microbiology, 6*, 453.

El-Kassas, H. Y., & El-Sheekh, M. M. (2014). Cytotoxic activity of biosynthesized gold nanoparticles with an extract of the red seaweed *Corallina officinalis* on the MCF-7 human breast cancer cell line. *Asian Pacific Journal of Cancer Prevention: APJCP, 15*, 4311–4317.

Elsaesser, A., Taylor, A., de Yanes G. S., McKerr, G., Kim E. M., O'Hare E., et al. (2010). Quantification of nanoparticle uptake by cells using microscopical and analytical techniques. *Nanomedicine. 5*:1447–1457.

Eom, H. J., Chatterjee, N., Lee, J., & Choi, J. (2014). Integrated mRNA and micro RNA profiling reveals epigenetic mechanism of differential sensitivity of Jurkat T cells to AgNPs and Ag ions. *Toxicology Letters, 229*, 311–318. https://doi.org/10.1016/j.toxlet.2014.05.019.

EP. (2009). Regulation (EC) No 1223/2009 of the European Parliament and of the Council of 35 November 2009 on cosmetic products. Europe Ereno D Well-grounded Beauty.

Esposito, L. A., Melov, S., Panov, A., Cottrell, B. A., & Wallace, D. C. (1999). Mitochondrial disease in mouse results in increased oxidative stress. *Proceedings of the National Academy of Sciences of the United States of America, 96*, 4820–4825.

Fabrega, J., Luoma, S. N., Tyler, C. R., Galloway, T. S., & Lead, J. R. (2011). Silver nanoparticles: Behaviour and effects in the aquatic environment. *Environment International, 37*, 517–531. https://doi.org/10.1016/j.envint.2010.10.012.

Fiorentino, B., Golanski, L., Guiot, A., Damlencourt, J.-F., & Boutry, D. (2015). Influence of paints formulations on nanoparticles release during their life cycle. *Journal of Nanoparticle Research, 17*, 149. https://doi.org/10.1007/s11051-015-2962-0.

Gao, Y., & Cranston, R. (2008). Recent advances in antimicrobial treatments of textiles. *Textile Research Journal, 78*, 60–72. https://doi.org/10.1177/0040517507082332.

Gaylarde, C. C., Morton, L. H. G., Loh, K., & Shirakawa, M. A. (2011). Biodeterioration of external architectural paint films – A review. *International Biodeterioration & Biodegradation, 65*, 1189–1198. https://doi.org/10.1016/j.ibiod.2011.09.005.

Glover, R. D., Miller, J. M., & Hutchison, J. E. (2011). Generation of metal nanoparticles from silver and copper objects: Nanoparticle dynamics on surfaces and potential sources of nanoparticles in the environment. *ACS Nano, 5*, 8950–8957. https://doi.org/10.1021/nn2031319.

Goharshadi, E. K., Samiee, S., & Nancarrow, P. (2011). Fabrication of cerium oxide nanoparticles: Characterization and optical properties. *Journal of Colloid and Interface Science, 356*, 473–480. https://doi.org/10.1016/j.jcis.2011.01.063.

Gondikas, A. P., von der Kammer, F., Reed, R. B., Wagner, S., Ranville, J. F., & Hofmann, T. (2014). Release of TiO2 nanoparticles from sunscreens into surface waters: A one-year survey at the old Danube recreational Lake. *Environmental Science & Technology, 48*, 5415–5422. https://doi.org/10.1021/es405596y.

Gottschalk, F., Sonderer, T., Scholz, R. W., & Nowack, B. (2009). Modeled environmental concentrations of engineered nanomaterials (TiO2, ZnO, Ag, CNT, fullerenes) for different regions. *Environmental Science & Technology, 43*, 9216–9222. https://doi.org/10.1021/es9015553.

Guadagnini, R., Halamoda Kenzaoui, B., Walker, L., Pojana, G., Magdolenova, Z., Bilanicova, D., et al. (2015). Toxicity screenings of nanomaterials: challenges due to interference with assay processes and components of classic in vitro tests. *Nanotoxicology. 9*(1):13–24.

Guo, J., Guo, J., & Xu, Z. (2009). Recycling of non-metallic fractions from waste printed circuit boards: A review. *Journal of Hazardous Materials, 168*, 567–590. https://doi.org/10.1016/j.jhazmat.2009.02.104.

Hartmann, N. I. B., Skjolding, L. M., Hansen, S. F., Baun, A., Kjølholt, J., & Gottschalk, F. (2014). *Environmental fate and behaviour of nanomaterials: New knowledge on important transfomation processes*. Copenhagen: Danish Environmental Protection Agency.

Haveli, S. D., et al. (2012). Hair fiber as a nanoreactor in controlled synthesis of fluorescent gold nanoparticles. *Nano Letters, 12*, 6212–6217. https://doi.org/10.1021/nl303107w.

Haydon, B., & Eng, P. (2013). *Nanomaterials and their applications in textiles, standards: domestic standardization for Canadian Manufacturers and Importers and International Standardization Developments*. Industry Canada.

He, C., Morawska, L., & Taplin, L. (2007). Particle emission characteristics of office printers. *Environmental Science & Technology, 41*, 6039–6045.

Holder, A. L., Goth-Goldstein, R., Lucas, D., & Koshland, C. P. (2012). Particle-induced artifacts in the MTT and LDH viability assays. *Chemical research in toxicology, 25*, 1885–1892.

Hong, H., et al. (2011). Cancer-targeted optical imaging with fluorescent zinc oxide nanowires. *Nano Letters, 11*, 3744–3750. https://doi.org/10.1021/nl201782m.

Huang, X., Kobos, R. K., & Xu, G. (2007). *Hair coloring and cosmetic compositions comprising carbon nanotubes*. Google Patents

Hurbankova, M., Volkovova, K., Hraskova, D., Wimmerova, S., & Moricova, S. (2017). Respiratory toxicity of Fe3O4 nanoparticles: Experimental study. *Reviews on Environmental Health, 32*, 207–210. https://doi.org/10.1515/reveh-2016-0022.

Hussain, S., Vanoirbeek, J. A., & Hoet, P. H. (2012). Interactions of nanomaterials with the immune system. *Wiley Interdisciplinary Reviews: Nanomedicine and Nanobiotechnology, 4*, 169–183.

Hutchison, J. E. (2008). Greener nanoscience: A proactive approach to advancing applications and reducing implications of nanotechnology. *ACS Nano, 2*, 395–402. https://doi.org/10.1021/nn800131j.

Jeon, S. K., Kim, E. J., Lee, J., & Lee, S. (2016). Potential risks of TiO2 and ZnO nanoparticles released from sunscreens into outdoor swimming pools. *Journal of Hazardous Materials, 317*, 312–318. https://doi.org/10.1016/j.jhazmat.2016.05.099.

Jeon, S., Ito, M., Tabelin, C. B., Pongsumrankul, R., Kitajima, N., Park, I., & Hiroyoshi, N. (2018). Gold recovery from shredder light fraction of E-waste recycling plant by flotation-ammonium thiosulfate leaching. *Waste Management, 77*, 195–202. https://doi.org/10.1016/j.wasman.2018.04.039.

Jia, X., & Jia, L. (2012). Nanoparticles improve biological functions of phthalocyanine photosensitizers used for photodynamic therapy. *Current Drug Metabolism, 13*, 1119–1122.

Kaegi, R., et al. (2008). Synthetic TiO2 nanoparticle emission from exterior facades into the aquatic environment. *Environmental Pollution, 156*, 233–239. https://doi.org/10.1016/j.envpol.2008.08.004.

Kaegi, R., et al. (2010). Release of silver nanoparticles from outdoor facades. *Environmental Pollution, 158*, 2900–2905. https://doi.org/10.1016/j.envpol.2010.06.009.

Kaya, M. (2016). Recovery of metals and nonmetals from electronic waste by physical and chemical recycling processes. *Waste Management, 57*, 64–90. https://doi.org/10.1016/j.wasman.2016.08.004.

Keller, A. A., McFerran, S., Lazareva, A., & Suh, S. (2013). Global life cycle releases of engineered nanomaterials. *Journal of Nanoparticle Research, 15*, 1692. https://doi.org/10.1007/s11051-013-1692-4.

Keller, A. A., Vosti, W., Wang, H., & Lazareva, A. (2014). Release of engineered nanomaterials from personal care products throughout their life cycle. *Journal of Nanoparticle Research, 16*, 2489. https://doi.org/10.1007/s11051-014-2489-9.

Khatri, M., Bello, D., Gaines, P., Martin, J., Pal, A. K., Gore, R., & Woskie, S. (2013). Nanoparticles from photocopiers induce oxidative stress and upper respiratory tract inflammation in healthy volunteers. *Nanotoxicology, 7*, 1014–1027. https://doi.org/10.3109/17435390.2012.691998.

Kiddee, P., Naidu, R., & Wong, M. H. (2013). Electronic waste management approaches: An overview. *Waste Management, 33*, 1237–1250. https://doi.org/10.1016/j.wasman.2013.01.006.

Kirchner, C., et al. (2005). Cytotoxicity of colloidal CdSe and CdSe/ZnS nanoparticles. *Nano Letters, 5*, 331–338. https://doi.org/10.1021/nl047996m.

Kiser, M. A., Ryu, H., Jang, H., Hristovski, K., & Westerhoff, P. (2010). Biosorption of nanoparticles to heterotrophic wastewater biomass. *Water Research, 44*, 4105–4114. https://doi.org/10.1016/j.watres.2010.05.036.

Kittler, S., Greulich, C., Diendorf, J., Köller, M., & Epple, M. (2010). Toxicity of silver nanoparticles increases during storage because of slow dissolution under release of silver ions. *Chemistry of Materials, 22*, 4548–4554. https://doi.org/10.1021/cm100023p.

Kolodziejczyk, B. (2016). Nanotechnology, nanowaste and their effects on ecosystems: A need for efficient monitoring, disposal and recycling brief for GSDR – 2016 update.

Kroll, A., Pillukat, M. H., Hahn, D., & Schnekenburger, J. (2012). Interference of engineered nanoparticles with in vitro toxicity assays. *Archives of toxicology, 86*, 1123–1136.

Kunhikrishnan, A., Shon, H. K., Bolan, N. S., El Saliby, I., & Vigneswaran, S. (2015). Sources, distribution, environmental fate, and ecological effects of nanomaterials in wastewater streams. *Critical Reviews in Environmental Science and Technology, 45*, 277–318. https://doi.org/10.1080/10643389.2013.852407.

Labouta, H. I., & Schneider, M. (2013). Interaction of inorganic nanoparticles with the skin barrier: Current status and critical review. *Nanomedicine: Nanotechnology, Biology, and Medicine, 9*, 39–54. https://doi.org/10.1016/j.nano.2012.04.004.

Lacerda, L., et al. (2008). Dynamic imaging of functionalized multi-walled carbon nanotube systemic circulation and urinary excretion. *Advanced Materials, 20*, 225–230.

Lau, M., Waag, F., & Barcikowski, S. (2017). Direct integration of laser-generated nanoparticles into transparent nail polish: The plasmonic "Goldfinger". *Industrial & Engineering Chemistry Research, 56*, 3291–3296. https://doi.org/10.1021/acs.iecr.7b00039.

Lee, H. Y., Jeong, Y. I., Kim, D. H., & Choi, K. C. (2013). Permanent hair dye-incorporated hyaluronic acid nanoparticles. *Journal of Microencapsulation, 30*, 189–197. https://doi.org/10.3109/02652048.2012.714412.

Levard, C., Hotze, E. M., Lowry, G. V., & Brown, G. E. (2012). Environmental transformations of silver nanoparticles: Impact on stability and toxicity. *Environmental Science & Technology, 46*, 6900–6914. https://doi.org/10.1021/es2037405.

Li, L., Wang, Y., Liu, Q., & Jiang, G. (2016). Rethinking stability of silver sulfide nanoparticles (Ag2S-NPs) in the aquatic environment: Photoinduced transformation of Ag2S-NPs in the presence of Fe(III). *Environmental Science & Technology, 50*, 188–196. https://doi.org/10.1021/acs.est.5b03982.

Limpiteeprakan, P., Babel, S., Lohwacharin, J., & Takizawa, S. (2016). Release of silver nanoparticles from fabrics during the course of sequential washing. *Environmental Science and Pollution Research International, 23*, 22810–22818. https://doi.org/10.1007/s11356-016-7486-3.

Liu, H., Zhang, J., Chen, X., Du, X. S., Zhang, J. L., Liu, G., & Zhang, W. G. (2016). Application of iron oxide nanoparticles in glioma imaging and therapy: From bench to bedside. *Nanoscale, 8*, 7808–7826. https://doi.org/10.1039/c6nr00147e.

Lohani, A., Verma, A., Joshi, H., Yadav, N., & Karki, N. (2014). Nanotechnology-based cosmeceuticals. *ISRN Dermatology, 2014*, 843687. https://doi.org/10.1155/2014/843687.

Lovric, J., Bazzi, H. S., Cuie, Y., Fortin, G. R., Winnik, F. M., & Maysinger, D. (2005). Differences in subcellular distribution and toxicity of green and red emitting CdTe quantum dots. *Journal of Molecular Medicine, 83*, 377–385. https://doi.org/10.1007/s00109-004-0629-x.

Lowry, G. V., Gregory, K. B., Apte, S. C., & Lead, J. R. (2012). Transformations of nanomaterials in the environment. *Environmental Science & Technology, 46*(13), 6893–6899. ACS Publications.
Mackevica, A., Olsson, M. E., & Hansen, S. F. (2017). Quantitative characterization of TiO2 nanoparticle release from textiles by conventional and single particle ICP-MS. *Journal of Nanoparticle Research, 20*, 6. https://doi.org/10.1007/s11051-017-4113-2.
Magdolenova, Z., Collins, A., Kumar, A., Dhawan, A., Stone, V., & Dusinska, M. (2014). Mechanisms of genotoxicity. A review of in vitro and in vivo studies with engineered nanoparticles. *Nanotoxicology, 8*, 233–278.
Mazumder, A., & Shivashankar, G. V. (2007). Gold-nanoparticle-assisted laser perturbation of chromatin assembly reveals unusual aspects of nuclear architecture within living cells. *Biophysical Journal, 93*, 2209–2216. https://doi.org/10.1529/biophysj.106.102202.
Miller, M. R., et al. (2017). Inhaled nanoparticles accumulate at sites of vascular disease. *ACS Nano, 11*, 4542–4552. https://doi.org/10.1021/acsnano.6b08551.
Mitrano, D. M., Motellier, S., Clavaguera, S., & Nowack, B. (2015). Review of nanomaterial aging and transformations through the life cycle of nano-enhanced products. *Environment International, 77*, 132–147.
Morimoto, Y., et al. (2016). Evaluation of pulmonary toxicity of zinc oxide nanoparticles following inhalation and intratracheal instillation. *International Journal of Molecular Sciences, 17*, 1241. https://doi.org/10.3390/ijms17081241.
Mueller, N. C., & Nowack, B. (2008). Exposure modeling of engineered nanoparticles in the environment. *Environmental Science & Technology, 42*, 4447–4453. https://doi.org/10.1021/es7029637.
Munafò, P., Goffredo, G. B., & Quagliarini, E. (2015). TiO2-based nanocoatings for preserving architectural stone surfaces: An overview. *Construction and Building Materials, 84*, 201–218. https://doi.org/10.1016/j.conbuildmat.2015.02.083.
Mytych, J., Zebrowski, J., Lewinska, A., & Wnuk, M. (2017). Prolonged effects of silver nanoparticles on p53/p21 pathway-mediated proliferation, DNA damage response, and methylation parameters in HT22 hippocampal neuronal cells. *Molecular Neurobiology, 54*, 1285–1300. https://doi.org/10.1007/s12035-016-9688-6.
Nafisi, S., Schafer-Korting, M., & Maibach, H. I. (2015). Perspectives on percutaneous penetration: Silica nanoparticles. *Nanotoxicology, 9*, 643–657. https://doi.org/10.3109/17435390.2014.958115.
Ng, C. T., Baeg, G. H., Yu, L. C., Ong, C. N., & Bay, B. H. (2017). Biomedical applications of nanomaterials as therapeutics. *Current Medicinal Chemistry*. https://doi.org/10.2174/0929867324666170331120328.
Nogueira, C. M., et al. (2012). Titanium dioxide induced inflammation in the small intestine. *World Journal of Gastroenterology, 18*, 4729–4735. https://doi.org/10.3748/wjg.v18.i34.4729.
Noruzi, M. (2015). Biosynthesis of gold nanoparticles using plant extracts. *Bioprocess and Biosystems Engineering, 38*, 1–14. https://doi.org/10.1007/s00449-014-1251-0.
Nowack, B., Krug, H. F., & Height, M. (2011). 120 years of nanosilver history: Implications for policy makers. *Environmental Science & Technology, 45*, 1177–1183. https://doi.org/10.1021/es103316q.
Oberdorster, G. (2010). Safety assessment for nanotechnology and nanomedicine: Concepts of nanotoxicology. *Journal of Internal Medicine, 267*, 89–105. https://doi.org/10.1111/j.1365-2796.2009.02187.x.
Oberdorster, G., Sharp, Z., Atudorei, V., Elder, A., Gelein, R., Kreyling, W., & Cox, C. (2004). Translocation of inhaled ultrafine particles to the brain. *Inhalation Toxicology, 16*, 437–445. https://doi.org/10.1080/08958370490439597.
Oleszczuk, P., Jośko, I., & Skwarek, E. (2015). Surfactants decrease the toxicity of ZnO, TiO2 and Ni nanoparticles to *Daphnia magna*. *Ecotoxicology, 24*, 1923–1932. https://doi.org/10.1007/s10646-015-1529-2.

Otari, S. V., Patil, R. M., Ghosh, S. J., Thorat, N. D., & Pawar, S. H. (2015). Intracellular synthesis of silver nanoparticle by actinobacteria and its antimicrobial activity. *Spectrochimica Acta Part A, Molecular and Biomolecular Spectroscopy, 136*(Pt B), 1175–1180. https://doi.org/10.1016/j.saa.2014.10.003.

Paul, R., Bautista, L., Varga, M. D., Botet, J. M., Casals, E., Puntes, V., & Marsal, F. (2010). Nano-cotton fabrics with high ultraviolet protection. *Textile Research Journal, 80*, 454–462. https://doi.org/10.1177/0040517509342316.

Pirela, S. V., et al. (2015). Consumer exposures to laser printer-emitted engineered nanoparticles: A case study of life-cycle implications from nano-enabled products. *Nanotoxicology, 9*, 760–768. https://doi.org/10.3109/17435390.2014.976602.

Qian, Y., et al. (2015). Silver nanoparticle-induced hemoglobin decrease involves alteration of histone 3 methylation status. *Biomaterials, 70*, 12–22. https://doi.org/10.1016/j.biomaterials.2015.08.015.

Radetić, M. (2013). Functionalization of textile materials with silver nanoparticles. *Journal of Materials Science, 48*, 95–107. https://doi.org/10.1007/s10853-012-6677-7.

Rajput, V. D., et al. (2018). Effects of zinc-oxide nanoparticles on soil, plants, animals and soil organisms: A review. *Environmental Nanotechnology, Monitoring & Management, 9*, 76–84. https://doi.org/10.1016/j.enmm.2017.12.006.

Ramos, A. P., Cruz, M. A. E., Tovani, C. B., & Ciancaglini, P. (2017). Biomedical applications of nanotechnology. *Biophysical Reviews, 9*, 79–89. https://doi.org/10.1007/s12551-016-0246-2.

Rashid, M. I., et al. (2017). Toxicity of iron oxide nanoparticles to grass litter decomposition in a sandy soil. *Scientific Reports, 7*, 41965. https://doi.org/10.1038/srep41965.

Reidy, B., Haase, A., Luch, A., Dawson, K. A., & Lynch, I. (2013). Mechanisms of silver nanoparticle release, transformation and toxicity: A critical review of current knowledge and recommendations for future studies and applications. *Materials, 6*, 2295–2350. https://doi.org/10.3390/ma6062295.

Rodrı, X., Guez, M. T., Gracenea, J. J., Kudama, A. H., & Suay, J. J. (2004). The influence of pigment volume concentration (PVC) on the properties of an epoxy coating: Part I. Thermal and mechanical properties. *Progress in Organic Coatings, 50*, 62–67. https://doi.org/10.1016/j.porgcoat.2003.10.013.

Rosen, J., Landriscina, A., & Friedman, A. J. (2015). Nanotechnology-based cosmetics for hair care. *Cosmetics, 2*, 211–224.

Roy, I., et al. (2003). Ceramic-based nanoparticles entrapping water-insoluble photosensitizing anticancer drugs: A novel drug-carrier system for photodynamic therapy. *Journal of the American Chemical Society, 125*, 7860–7865. https://doi.org/10.1021/ja0343095.

Ruiz, P. A., et al. (2017). Titanium dioxide nanoparticles exacerbate DSS-induced colitis: Role of the NLRP3 inflammasome. *Gut, 66*, 1216–1224. https://doi.org/10.1136/gutjnl-2015-310297.

Sadjadpour, S., Safarian, S., Zargar, S. J., & Sheibani, N. (2016). Antiproliferative effects of ZnO, ZnO-MTCP, and ZnO-CuMTCP nanoparticles with safe intensity UV and X-ray irradiation. *Biotechnology and Applied Biochemistry, 63*, 113–124. https://doi.org/10.1002/bab.1344.

Sajid, M., Ilyas, M., Basheer, C., Tariq, M., Daud, M., Baig, N., & Shehzad, F. (2015). Impact of nanoparticles on human and environment: Review of toxicity factors, exposures, control strategies, and future prospects. *Environmental Science and Pollution Research, 22*, 4122–4143. https://doi.org/10.1007/s11356-014-3994-1.

Sakunda, P. (2013). Strategy of e-waste management.

Schmid, K., & Riediker, M. (2008). Use of nanoparticles in Swiss industry: A targeted survey. *Environmental Science & Technology, 42*, 2253–2260.

Sharifi, S., Behzadi, S., Laurent, S., Laird Forrest, M., Stroeve, P., & Mahmoudi, M. (2012). Toxicity of nanomaterials. *Chemical Society Reviews, 41*, 2323–2343. https://doi.org/10.1039/C1CS15188F.

Shen, Z., Chen, Z., Hou, Z., Li, T., & Lu, X. (2015). Ecotoxicological effect of zinc oxide nanoparticles on soil microorganisms. *Frontiers of Environmental Science & Engineering, 9*, 912–918. https://doi.org/10.1007/s11783-015-0789-7.

Shtykova, L., et al. (2009). Adsorption of antifouling booster biocides on metal oxide nanoparticles: Effect of different metal oxides and solvents. *Progress in Organic Coatings, 64*, 20–26.
Simonin, M., Guyonnet, J. P., Martins, J. M., Ginot, M., & Richaume, A. (2015). Influence of soil properties on the toxicity of TiO(2) nanoparticles on carbon mineralization and bacterial abundance. *Journal of Hazardous Materials, 283*, 529–535. https://doi.org/10.1016/j.jhazmat.2014.10.004.
Singh, S., & Prasad, B. L. V. (2007). Nearly complete oxidation of Au° in Hydrophobized nanoparticles to Au3+ ions by N-Bromosuccinimide. *The Journal of Physical Chemistry C, 111*, 14348–14352. https://doi.org/10.1021/jp073717i.
Singh, P., Kim, Y. J., Wang, C., Mathiyalagan, R., & Yang, D. C. (2016a). The development of a green approach for the biosynthesis of silver and gold nanoparticles by using *Panax ginseng* root extract, and their biological applications. *Artificial Cells, Nanomedicine, and Biotechnology, 44*, 1150–1157. https://doi.org/10.3109/21691401.2015.1011809.
Singh, P., Kim, Y. J., & Yang, D. C. (2016b). A strategic approach for rapid synthesis of gold and silver nanoparticles by *Panax ginseng* leaves. *Artificial Cells, Nanomedicine, and Biotechnology, 44*, 1949–1957. https://doi.org/10.3109/21691401.2015.1115410.
Singh, P., Kim, Y. J., Zhang, D., & Yang, D. C. (2016c). Biological synthesis of nanoparticles from plants and microorganisms. *Trends in Biotechnology, 34*, 588–599. https://doi.org/10.1016/j.tibtech.2016.02.006.
Sisler, J. D., et al. (2015). Small airway epithelial cells exposure to printer-emitted engineered nanoparticles induces cellular effects on human microvascular endothelial cells in an alveolar-capillary co-culture model. *Nanotoxicology, 9*, 769–779. https://doi.org/10.3109/17435390.2014.976603.
Smolkova, B., Dusinska, M., & Gabelova, A. (2017). Nanomedicine and epigenome. Possible health risks. *Food and Chemical Toxicology: An International Journal Published for the British Industrial Biological Research Association, 109*, 780–796. https://doi.org/10.1016/j.fct.2017.07.020.
Smulders, S., et al. (2014). Toxicity of nanoparticles embedded in paints compared with pristine nanoparticles in mice. *Toxicological Sciences: An Official Journal of the Society of Toxicology, 141*, 132–140. https://doi.org/10.1093/toxsci/kfu112.
Society R. (2004). *Nanoscience and nanotechnologies: opportunities and uncertainties*. Royal Society.
Solomon, W. L., Bradley, W., Karl De, J., & James, C. K. L. (2017). Critical review of removal of nano materials in waste streams. *IOP Conference Series: Earth and Environmental Science, 68*, 012019.
Soni, N., & Prakash, S. (2015). Antimicrobial and mosquitocidal activity of microbial synthesized silver nanoparticles. *Parasitology Research, 114*, 1023–1030. https://doi.org/10.1007/s00436-014-4268-z.
Soni, D., Naoghare, P. K., Saravanadevi, S., & Pandey, R. A. (2015). Release, transport and toxicity of engineered nanoparticles. In *Reviews of environmental contamination and toxicology* (pp. 1–47). Cham: Springer.
Stephens, B., Azimi, P., El Orch, Z., & Ramos, T. (2013). Ultrafine particle emissions from desktop 3D printers. *Atmospheric Environment, 79*, 334–339. https://doi.org/10.1016/j.atmosenv.2013.06.050.
Tabish, A. M., et al. (2017). Changes in DNA methylation in mouse lungs after a single intratracheal administration of nanomaterials. *PLoS One, 12*, e0169886. https://doi.org/10.1371/journal.pone.0169886.
Tak, Y. K., Pal, S., Naoghare, P. K., Rangasamy, S., & Song, J. M. (2015). Shape-dependent skin penetration of silver nanoparticles: Does it really matter? *Scientific Reports, 5*, 16908. https://doi.org/10.1038/srep16908.
Teow, Y., Asharani, P. V., Hande, M. P., & Valiyaveettil, S. (2011). Health impact and safety of engineered nanomaterials. *Chemical Communications, 47*, 7025–7038. https://doi.org/10.1039/C0CC05271J.

Tiwari, N., Pandit, R., Gaikwad, S., Gade, A., & Rai, M. (2017). Biosynthesis of zinc oxide nanoparticles by petals extract of *Rosa indica* L., its formulation as nail paint and evaluation of antifungal activity against fungi causing onychomycosis. *IET Nanobiotechnology, 11*, 205–211. https://doi.org/10.1049/iet-nbt.2016.0003.

Tsuji, J. S., Maynard, A. D., Howard, P. C., James, J. T., Lam, C. W., Warheit, D. B., & Santamaria, A. B. (2006). Research strategies for safety evaluation of nanomaterials, part IV: Risk assessment of nanoparticles. *Toxicological Sciences: An Official Journal of the Society of Toxicology, 89*, 42–50. https://doi.org/10.1093/toxsci/kfi339.

Urrutia-Ortega, I. M., et al. (2016). Food-grade titanium dioxide exposure exacerbates tumor formation in colitis associated cancer model. *Food and Chemical Toxicology: An International Journal Published for the British Industrial Biological Research Association, 93*, 20–31. https://doi.org/10.1016/j.fct.2016.04.014.

Valant, J., Drobne, D., & Novak, S. (2012). Effect of ingested titanium dioxide nanoparticles on the digestive gland cell membrane of terrestrial isopods. *Chemosphere, 87*, 19–25. https://doi.org/10.1016/j.chemosphere.2011.11.047.

Vílchez, A., Fernández-Rosas, E., González-Gálvez, D., & Vázquez-Campos, S. (2016). Nanomaterials release from nano-enabled products. In M. Viana (Ed.), *Indoor and outdoor nanoparticles: Determinants of release and exposure scenarios* (pp. 127–158). Cham: Springer International Publishing. https://doi.org/10.1007/698_2015_409.

Viswanath, B., & Kim, S. (2017). Influence of nanotoxicity on human health and environment: The alternative strategies. In P. de Voogt (Ed.), *Reviews of environmental contamination and toxicology* (Vol. 242, pp. 61–104). Cham: Springer International Publishing. https://doi.org/10.1007/398_2016_12.

Vithiya, K., & Sen, S. (2011). Biosynthesis of nanoparticles. *International Journal of Pharmaceutical Sciences and Research, 2*, 2781.

von Goetz, N., Lorenz, C., Windler, L., Nowack, B., Heuberger, M., & Hungerbuhler, K. (2013). Migration of Ag- and TiO2-(Nano)particles from textiles into artificial sweat under physical stress: Experiments and exposure modeling. *Environmental Science & Technology, 47*, 9979–9987. https://doi.org/10.1021/es304329w.

von Wilmowsky, C., et al. (2009). In vivo evaluation of anodic TiO2 nanotubes: An experimental study in the pig. *Journal of Biomedical Materials Research Part B, Applied Biomaterials, 89*, 165–171. https://doi.org/10.1002/jbm.b.31201.

Wagener, S., et al. (2016). Textile functionalization and its effects on the release of silver nanoparticles into artificial sweat. *Environmental Science & Technology, 50*, 5927–5934. https://doi.org/10.1021/acs.est.5b06137.

Wang, N., Li, H., Lu, W., Li, J., Wang, J., Zhang, Z., & Liu, Y. (2011). Effects of TiO2 nanotubes with different diameters on gene expression and osseointegration of implants in minipigs. *Biomaterials, 32*, 6900–6911. https://doi.org/10.1016/j.biomaterials.2011.06.023.

Wang, P., et al. (2015). Silver sulfide nanoparticles (Ag2S-NPs) are taken up by plants and are phytotoxic. *Nanotoxicology, 9*, 1041–1049. https://doi.org/10.3109/17435390.2014.999139.

Wei, H., et al. (2015). Advanced micro/nanocapsules for self-healing smart anticorrosion coatings. *Journal of Materials Chemistry A, 3*, 469–480. https://doi.org/10.1039/C4TA04791E.

Weir, A., Westerhoff, P., Fabricius, L., Hristovski, K., & von Goetz, N. (2012). Titanium dioxide nanoparticles in food and personal care products. *Environmental Science & Technology, 46*, 2242–2250. https://doi.org/10.1021/es204168d.

Westerhoff, P. K., Kiser, M. A., & Hristovski, K. (2013). Nanomaterial removal and transformation during biological wastewater treatment. *Environmental Engineering Science, 30*, 109–117.

Widmer, R., Oswald-Krapf, H., Sinha-Khetriwal, D., Schnellmann, M., & Böni, H. (2005). Global perspectives on e-waste. *Environmental Impact Assessment Review, 25*, 436–458.

Windler, L., Lorenz, C., von Goetz, N., Hungerbühler, K., Amberg, M., Heuberger, M., & Nowack, B. (2012). Release of titanium dioxide from textiles during washing. *Environmental Science & Technology, 46*, 8181–8188. https://doi.org/10.1021/es301633b.

Xu, P., et al. (2012). Use of iron oxide nanomaterials in wastewater treatment: A review. *The Science of the Total Environment, 424*, 1–10. https://doi.org/10.1016/j.scitotenv.2012.02.023.
Xu, C., Peng, C., Sun, L., Zhang, S., Huang, H., Chen, Y., & Shi, J. (2015). Distinctive effects of TiO2 and CuO nanoparticles on soil microbes and their community structures in flooded paddy soil. *Soil Biology and Biochemistry, 86*, 24–33. https://doi.org/10.1016/j.soilbio.2015.03.011.
Yang, Y., & Westerhoff, P. (2014). Presence in, and release of, nanomaterials from consumer products. *Advances in Experimental Medicine and Biology, 811*, 1–17. https://doi.org/10.1007/978-94-017-8739-0_1.
Youssef, Z., et al. (2017). The application of titanium dioxide, zinc oxide, fullerene, and graphene nanoparticles in photodynamic therapy. *Cancer Nanotechnology, 8*, 6. https://doi.org/10.1186/s12645-017-0032-2.
Zhang, Y., Leu, Y. R., Aitken, R. J., & Riediker, M. (2015). Inventory of engineered nanoparticle-containing consumer products available in the Singapore retail market and likelihood of release into the aquatic environment. *International Journal of Environmental Research and Public Health, 12*, 8717–8743. https://doi.org/10.3390/ijerph120808717.
Zhang, B., Sai Lung, P., Zhao, S., Chu, Z., Chrzanowski, W., & Li, Q. (2017). Shape dependent cytotoxicity of PLGA-PEG nanoparticles on human cells. *Scientific Reports, 7*, 7315. https://doi.org/10.1038/s41598-017-07588-9.
Zhou, Y., & Qiu, K. (2010). A new technology for recycling materials from waste printed circuit boards. *Journal of Hazardous Materials, 175*, 823–828. https://doi.org/10.1016/j.jhazmat.2009.10.083.
Zhou, Z., Lenk, R., Dellinger, A., MacFarland, D., Kumar, K., Wilson, S. R., & Kepley, C. L. (2009). Fullerene nanomaterials potentiate hair growth. *Nanomedicine: Nanotechnology, Biology, and Medicine, 5*, 202–207. https://doi.org/10.1016/j.nano.2008.09.005.
Zhou, X. H., Huang, B. C., Zhou, T., Liu, Y. C., & Shi, H. C. (2015). Aggregation behavior of engineered nanoparticles and their impact on activated sludge in wastewater treatment. *Chemosphere, 119*, 568–576. https://doi.org/10.1016/j.chemosphere.2014.07.037.
Zuin, S., Massari, A., Ferrari, A., & Golanski, L. (2014). Formulation effects on the release of silica dioxide nanoparticles from paint debris to water. *The Science of the Total Environment, 476-477*, 298–307. https://doi.org/10.1016/j.scitotenv.2014.01.029.

MIX
Papier aus verantwortungsvollen Quellen
Paper from responsible sources
FSC® C105338

If you have any concerns about our products,
you can contact us on
ProductSafety@springernature.com

In case Publisher is established outside the EU,
the EU authorized representative is:
Springer Nature Customer Service Center GmbH
Europaplatz 3, 69115 Heidelberg, Germany

Printed by Libri Plureos GmbH
in Hamburg, Germany